Sediment Transport

FLUID MECHANICS AND ITS APPLICATIONS
Volume 82

Aims and Scope of the Series

The purpose of this series is to focus on subjects in which fluid mechanics plays a fundamental role.

As well as the more traditional applications of aeronautics, hydraulics, heat and mass transfer etc., books will be published dealing with topics which are currently in a state of rapid development, such as turbulence, suspensions and multiphase fluids, super and hypersonic flows and numerical modelling techniques.

It is a widely held view that it is the interdisciplinary subjects that will receive intense scientific attention, bringing them to the forefront of technological advancement. Fluids have the ability to transport matter and its properties as well as transmit force, therefore fluid mechanics is a subject that is particulary open to cross fertilisation with other sciences and disciplines of engineering. The subject of fluid mechanics will be highly relevant in domains such as chemical, metallurgical, biological and ecological engineering. This series is particularly open to such new multidisciplinary domains.

The median level of presentation is the first year graduate student. Some texts are monographs defining the current state of a field; others are accessible to final year undergraduates; but essentially the emphasis is on readability and clarity.

For a list of related mechanics titles, see final pages.

Sediment Transport

A Geophysical Phenomenon

by

ALBERT GYR

and

KLAUS HOYER

Institute of Environmental Engineering
Swiss Federal Institute of Technology
Zürich, Switzerland

Springer

A C.I.P. Catalogue record for this book is available from the Library of Congress.

ISBN-13 978-90-481-7256-6
ISBN-10 1-4020-5016-X (e-book)
ISBN-13 978-1-4020-5016-9 (e-book)

Published by Springer,
P.O. Box 17, 3300 AA Dordrecht, The Netherlands.

www.springer.com

Printed on acid-free paper

Table of Contents

Preface

A major part of this sediment transport representation closely follows the progress made in understanding the interactions between a turbulent flow and transportable solid particles. Introducing new aspects found in the research of turbulent flows, this book updates the theory of sediment transport, e.g., using new representations for flow separations, and coherent structures thought to be relevant and confronts the problem that existent theories do not relate directly the relevant quantities involved in the physical processes.

A review of the complex matter suggests that a closer cooperation between engineers and physicists would benefit the problem and our concept tries to acknowledge this fact. Having this in mind the book was organized in four parts. The engineer who is interested in predicting sediment transport will find the classical as well as statistical approaches in the first part (Chaps. 1–4). The second part (Chaps. 5–8) critically reviews the most problematic issues like rheology, turbulence, topological aspects of flow separations, vortical dynamics, and scaling parameters. This part is mainly addressed to the physicist interested in the geophysical aspects of river dynamics but will also support the engineers in their decision making process, when constructing a simulation scheme. The third part (Chaps. 9–11) presents the sediment transport using micromechanical principles, using new results from turbulence research and introducing flow separation as the main self-organization mechanism observed in the formation of bedforms. This part separately treats three sediment classes: fine sand, for which coherent structure dynamics are relevant; grain mixtures, which interact also with secondary flows which cannot be neglected; and gravel beds, for which flow separations have to be introduced into the theory. The fourth part compiles practical calculation advice supported by Tables and Graphs.

Sediment transport has so many aspects, making it impossible to treat everything in the framework of a book. Therefore, we have added an extended list for the readers confronted with complex problems needing information on specially covered detail. This list comprises a personal selection exceeding the direct citations covered in this book. We recommend it as an interdisciplinary textbook treating the engineering as well as the scientific problem aspects equally important.

Our main acknowledgement goes to Prof. Wolfgang Kinzelbach, who has contributed several very valuable ideas and helped a lot as a lector. We also acknowledge very much the help of our colleague Dr. Hannes Bühler and Prof. Arkady Tsinober for the several fruitful discussions.

Zurich, February 2006

1 Introduction

1.1 Why a New Book on Sediment Transport

Ever since its historical appearance, mankind had to be concerned with sediment and sediment transport. Preferred settlement areas were in the vicinity of rivers, with which they were naturally interacting, and the protection of riverbanks from erosion and the farming on alluvial land were dominant. The deposition of sediment produced damages to the farming land. However, as the Nile valley showed, a well-managed deposition could also give rise to high productivity. Today, the riparians rather fear the results of a high sediment transport and try to get it under control. On the small scale, the erosion of sediment around pillars and other structures built in the river bed or the deposition of sediment in shipping lanes are problems. On the medium scale, the silting up of lakes and reservoirs with sediment is the main concern. Extreme deposition rates cause riverbeds to rise and make expensive dike buildings indispensable. The result of sediment deposition on an even larger scale can be seen, for example, along the lower reaches of the Yellow River in China, whose decrease in slope has become existential for millions of people. While in the past the river looked for a new bed this option is not feasible today due to the high population density.

Compared to those problems, the ones occurring in channels especially built for transport are rather small. The plugging of a channel flow by sediment is in its worst case a local catastrophe and in any event connected to costs for the population concerned. Sediment transport is a subject, which concerns geologists, hydraulic engineers as well as chemical engineers; all of whom must have a deeper knowledge of its physics to fulfill their specific task. Therefore, there is a need for textbooks addressed to application engineers, researchers as well as students. A series of good books on the whole theme as well as on particular aspects exist such as books on the bed-load transport or the transport of suspensions.

To be useful a new book therefore has to include the new research results on sediment transport achieved over the last years. This is the more necessary, as over decades there was stagnation in the incorporation of new ideas. These new ideas are helping to predict transport more accurately and are also helpful in situations not treated in the classical textbooks. To underscore the situation, let us quote Müller (1996), as he made a reference to what Kennedy (1971) said in his general report at the 14th IAHR congress in Paris. Kennedy gave his report the provocative title: "Too many doctors and too many remedies," meaning that we are flooded by an incredible amount of papers without new information- and moreover irrelevant for the essential problems. Müller, 24 years later comes to the conclusion that the situation remained the same. He formulates as follows:

- Sediment transport involves accelerated motions. Therefore the instantaneous forces on the particles, given by the integrated fluid pressure and the impacts of other particles, are the relevant physical quantities. The concepts we use, however, typically relate a bed shear stress to the particle transport rate. Moreover, the flow is normally assumed to be steady and uniform and not affected by the sediment transport itself. The effect of turbulence on particle transport is considered only in terms of the averaged Reynolds stresses or, more recently, in terms of large eddy simulation.

- Our theories do not relate directly the relevant quantities involved in the physical processes. We model those quantities indirectly by using statistical quantities derived from them. Only under well-defined equilibrium conditions are the assumed relations between the relevant quantities and the derived ones valid.

However, when we only look at the instantaneous and local 3Dflow, we do not solve the problem. We must recognize that the complexity of the micromechanics can be understood only for some special simplified situations. For those situations, we can study the topology of the instantaneous flow and can describe turbulence by coherent structures. Further, we can observe and understand the stabilizing and destabilizing feedback loops between flow, bed topography and sediment transport.

- We are not successful in linking this understanding at small scales to large engineering scales, and in developing the predictive tools the river engineer is looking for (Müller).

From this analysis he draws the following conclusions for further research on sediment transport:

- Honestly admit the intrinsic difficulties and recognize the deficiencies of our theories and concepts.
- Acknowledge that the experience and engineering judgment of river engineers must compensate for these deficiencies; and,
- Continue to study the micromechanics on a scientific level (Müller).

A new book therefore makes sense, if it shows new aspects helping to formulate a paradigmatic change in theory taking into consideration what Müller said. This is by far not an easy task, and it makes sense to remember what happened the last time a paradigmatic change has been undertaken by H.A. Einstein.

One of the authors (A. Gyr) had the opportunity to discuss this matter at length with Einstein, and this discussion will be reported here as a fictitious dialogue between the two. The contents are based on notes made at the time, only half a year before Einstein's death.

It is an often-heard anecdote that when H.A. Einstein told his famous father that he would like to investigate into sediment transport, Albert Einstein tore his hair and asked his son ironically: can it not be something more complicated? This conversation never happened, and it is a construction of frustrated researchers. The contrary was the case. Albert Einstein wrote a very modest letter to Meyer-Peter asking him for the position of a doctoral student for his son, where he would have to investigate just these phenomena. See a copy of this letter in Chap. 14 (Sect. 14.1).

This digression, reported for historical reasons, is relevant because H.A. Einstein quits the old thinking by replacing deterministic functions by statistical distribution. The statistical categories, which entered into the subject, were strongly influenced by father

Einstein, with whom H.A. Einstein discussed the matter. In fact, this was also one of the reasons why Einstein became curious about sediment transport, which resulted in a paper on meandering. He postulated the Coriolis forces originating from the rotation of the earth as the main mechanism causing the meanders.

Now to the fictitious dialogue:

It was in 1972 just after Hans-Albert Einstein retired from Berkeley that he visited us at the Institute of Hydromechanics and Water Resources Management (IHW) at the Swiss Federal Institute of Technology (ETH) in Zurich. As he told us, it was not only a pleasure but also even more of symbolic value for him to present part of his life's work at the school where he studied and finally wrote his thesis on bed-load transport. The three presentations he gave during his very short visit were, therefore, completely devoted to the problem of sediment transport and the means of describing the processes involved in the view of his life long investigation of this utmost complex problem.

All three lectures were held in the main building of the Swiss Federal Institute of Technology, which overlooks Zurich from the right side of the river Limmat, and gives the impression, as Hans-Albert commented with a smile, it was the House of Commons. The lectures themselves were rather a slide show than a presentation. The slides were from Einstein's own famous collection on rivers, erosion sites of mudflows, debris, etc. He had photographed these pictures at many locations in all continents and with an extreme photographic sensitivity.

After the lecture, there was an astonishingly long discussion on several aspects of the shown examples, but then it was time to put the slides in proper order and store them in his old leather purse. Outside the building, it was rainy and so I offered him to share the umbrella with me, which he refused by saying that rain is so natural that it would be a pity not to wet our faces. It was at this occasion that he looked in my face asking me if I was disappointed by his lecture since he heard a lot of comments but none from me. My face must have been an open book to him, because I was disappointed. As a young theorist, I expected from Einstein's son an analysis of the possible theoretical approaches I was so eager to hear about, since I had worked the last months on sediment transport too and was fascinated by the diverse physical aspects that this phenomenon includes.

-You need not tell me your reasons, Hans-Albert helped me out of the situation, -but what you must tell me is what you expected and why you expected it. -It would have been the moment to apologize and telling him how much I was fascinated by all the study cases, but that I missed the theoretical aspects. I was blocked. Finally I had the answer -You Hans-Albert Einstein, -I said, -are known as the scientist who introduced statistics in a rigorous form to interpret and to analyze sediment transport. As one of the most complex processes in nature it was obvious to me from beginning that only a statistical description is appropriate to tackle the problem. Therefore, I expected that you comment, to which extent we can use decoupled variables, and perhaps give some hint which kind of coupling is needed to get a better description. -Young man, Einstein responded, as a theorist you should know what you are asking for; the mathematics of a description you seem to have in mind is for most of us common people by far too difficult, but I appreciate your enthusiasm and wish you all my best in your endeavor.

After a while I dared to ask what he thinks to be the correct approach to the problem.

-This was the topic of my lecture, young man, every event seems to be so extremely genuine and still the landscape is made of forms which can be put in order in a fairly general way. So there must be laws behind this fact, but they are hidden and I am seeking for them by studying a variety of sediment transport events. That means by finding the relations of appropriate parameters in single events to finally get the general laws behind the different relationships. To find these relations, I need a memory for storing the whole complexity, and this is my photo collection. I saw his argument but at that time it sounded somehow exotic to me.

Hans-Albert smiled and said, young man do you think we two have just solved the problem of sediment transport? -Oh yes, -I answered, -in the same sense as all problems of fluid dynamics are solved by the Navier–Stokes equation, however in a form that nobody can explain turbulence. It seemed that this was the right answer because Einstein started to laugh with an intensity only Einstein's are capable to laugh. The rest of the evening was more or less a social event, but when I got home I made a note on this conversation. Now more than a quarter of a century later I will, starting from this talk, develop some ideas, why in the research of sediment transport a paradigmatic change is needed. Starting from the last ideas of Einstein, sediment transport is not a phenomenon, which can be described by a unique equation; sediment transport stands for a series of mechanisms depending on a large number of parameters.

1.1.1 A Short Guide Through the Content of this Book

In Chap. 2 the classical theory of sediment transport will be discussed, and in Chap. 3 the more elaborated based on statistical concepts, together with some resulting questions, asking for new approaches and for a paradigmatic change. The formulated criticism of the existing theories will be elucidated in Chap. 5. Chapter 4 gives a description of the processes in accordance with theoretical principles from a more global viewpoint. This chapter is addressed to system analysts, such as geographers, who are searching for universal relations.

In complex systems, it is often of advantage to investigate only part of the system by treating some aspects of the problem. To do so, it is necessary to have a good knowledge of the scaling of all involved parameters. This is also an important tool for researcher, investigating the problem through numerical simulations. Some commented lists on scaling are given in Chap. 6.

Chapters 7 and 8 are devoted to basics needed to develop new concepts. In Chaps. 9–11, the physical aspects are the center of the description and in Chap. 12 the mathematical implications necessary for simulations is presented.

The common denominator of all is formed by the fundamentals, such as the definition of the phenomenon and the involved parameters as well as a series of methods. All these elements will be discussed in this chapter.

1.2 The Phenomenon and its Main Parameters

The word sediment in the title means that we restrict ourselves to the geophysical phenomenon of transport of solid material as encountered in rivers. Many of the results can, however, be used also for industrial processes etc. Vice versa, results found for example by investigating pure two-phase flows can be very useful, since for instance the mechanisms involved in fluidized beds are very similar to the suspension processes in rivers. With this restriction in mind, we start describing the phenomenon.

1.2.1 A Definition of the Sediment Transport

Sediment transport by a flow of a liquid or by wind presumes a source of material; here an erodable bed built up by nonadhesive single grains, e.g., heavy particles. This bed material can basically transported in four different ways: First, single grains are raised from the bed and transported as suspended particles until they get deposited and become part of the bed again. Second, they can be transported by rolling on the bed until they are stopped. Third, they can behave as a fluidized material moving as a two-phase flow. And fourth, they can be transported as material, once suspended, resting in suspension during the whole transport.

It is hard to distinguish these four well-defined states in practice. Therefore, it is common to call material a suspension when is not or only a short time embedded, and through its transport forms a suspension load. On the other hand, material which is only in motion for short-time intervals, e.g., in the near wall region, is called bed load. Depending on the volume-concentration of the sediment in motion, we speak of sediment transport for low concentration and as two-phase flow for higher concentration. The interface from one to the other regime often needs a finer classification. When not explicitly stated otherwise, we use the rough classification given earlier.

1.2.2 The Parameters

Any physical theory describes the interaction between variables, and it is therefore essential to introduce at least the main parameters of the system. Already, in doing so, we will recognize the complexity of the problem.

1.2.2.1 The Source and the Characterization of the Sediment

We suppose that the bed of a flow channel is covered by erodable, grainy, nonadhesive material, which is the source of the transported sediment. This limitation excludes weathering and abrasion processes, which are important for the formation and the shape of the sediment.

The limitation to nonadhesive sediments allows us to exclude electrostatic and electrolytic effects, as they appear, e.g., in the transport of clay. However, at certain places in the text, we will make some brief comments to provide a better understanding of the effects occurring in such suspensions.

The composition of the bed and mainly of its covering layer is continuously changing, and therefore the sediment source is depending on its history. We will not take account of this fact, because we will use the composition of the bed as an initial condition, the result will be a new composition.

The sediment material has a density ρ_s, where $\rho_s > \rho_f$, the density of the fluid, so is not floating. For wind driven transport, $\rho_s \gg \rho_f$, and in this case ρ_f can be neglected. The density ρ_s as well as the grain size d_s are statistical values and must be given as distribution functions. The variance of ρ_s is often very small, since the sediment originates from the same geological formation, and can therefore be replaced by a single value. However, in general both distributions correlate, a fact that makes their description rather difficult.

We see that the grain size and density distributions are main parameters for the transport. These quantities have to be determined, which can be done by a variety of methods. The most common procedures are the following: the distribution of the size can be evaluated by a sieve analysis that means the grain mixture will be classified by a cascade of sieves of different mesh sizes. By this method, histograms of the grain sizes are produced, which can be smoothed or transformed into a cumulative curve. This would be a unique classification, if the sediment would consist of spherical particles. In reality, the grains differ in shape and therefore the sieve would at best retain the grains with the smallest diameter being bigger than the mesh size. A method completely depending on hydrodynamic properties use sedimentation experiments. Density, size, and shape are involved in this process. The results for a uniform density are defined relatively to an equivalent sphere settling at the same velocity. A very similar definition is given by the nominal value, where also the volume of the particle has to be the same as for the sphere. For small particles, as they appear in suspensions, only optical methods are adequate to measure the grain diameter.

In textbooks, one usually finds plots of the grain distribution where a weight-function is shown as percentage of the grain sizes as they were evaluated by a sieve analysis.

If results from different sediment transport observations have to be compared, the grain size distribution has to be given in the form of correlations to other relevant parameters. A method to analyze such correlations is by evaluating the Joint Probability Density Function (JPDF). For a two-parameter correlation the JPDF can be visualized by a 2D diagram (Fig. 1.1).

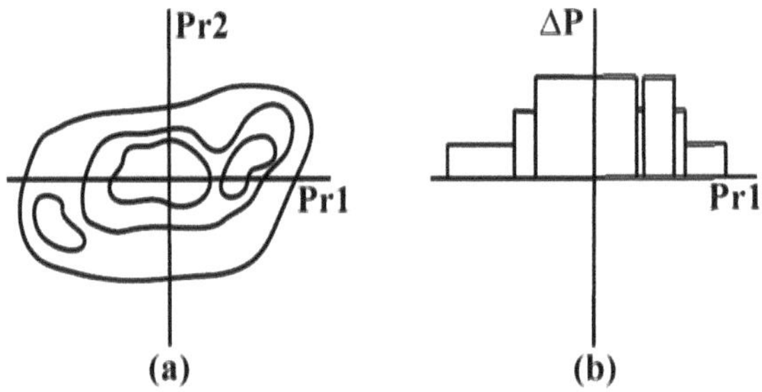

Fig. 1.1 (a) Representation of a JPDF for properties 1 and 2 as probabilities P_{r1} and P_{r2}. The origin of the coordinate system is given by $\left(\overline{P}_{r1}, \overline{P}_{r2}\right)$, the mean values of the probabilities of the two properties of interest. These are not necessarily the medians of the distributions. The curves are iso-probability contours, which show the probability of an event belonging to the category $\left(P_{r1} + P_{r1} + \Delta P_{r1}, P_{r2} + P_{r2} + \Delta P_{r2}\right)$. (b) A section through the JPDF for P_{r2} = constant represents a regular histogram of P_{r1} for that value of P_{r2}, here for $P_{r2} = 0$

The probability to find an event with combining P_{r1} and P_{r2} is

$$P\left(P_{r1}, P_{r2}\right) \geq 0 \tag{1.1}$$

usually standardized by

$$\int_{-\infty}^{+\infty}\int_{-\infty}^{+\infty} P\left(P_{r1}, P_{r2}\right) \mathrm{d}P_{r1}\, \mathrm{d}P_{r2} = 1 \tag{1.2}$$

The usual quantities correlated by JPDF's are the grain sizes d_s and the grain number density n_s, the grain surfaces $\mathrm{d}V_s$, the grain volumes V_s, the grain weight W_s, or some other parameters.

In several cases, these distributions become very simple, which leads to a simple histogram, as for example the most often published grain size distributions. Plotted on probability paper, it can be tested whether a grain mixture is composed of sets of normally distributed grains.

The JPDF contains the information on correlation, however, the interpretation becomes more elaborate. An overview of the relations one finds by cutting through the JPDF is shown in Fig. 1.1b. In this representation the probability P_{r1} reduces to the usual histogram for a constant P_{r2}.

The JPDF gives information on the modality of the distribution, e.g., in Fig. 1.1 we find a bimodal distribution. In this representation, the correlations show up in the width of the field enclosed by the iso-probability contours. The more the two properties correlate, the more the JPDF approaches a line with a probability distribution of ΔP in a histogram. In other words, a JPDF is recommended when an exact characterization of the sediment is needed.

JPDF are a general tool to represent probabilities of correlated variables, and it is often useful to standardize the parameters, e.g., by investigating the connection between the size and the form of the grains

$$P\left(d_s, \mathrm{d}V_s / \pi d_s^2\right); P\left(d_s, V_s / \frac{\pi}{6} d_s^3\right) \tag{1.3}$$

Good examples are spheroids, which are very suitable to characterize the sediment by shape, which allows differentiating between oblate and prolate grains. The oblate spheroid is formed by rotation of an ellipse around its short axis, whereas the prolate one by the rotation around the long axis. For an eccentricity, ε of the ellipse the surface and the volume for an oblate spheroid is given by

$$\begin{aligned} & a = d_s/2; b = \beta d_s/2 \quad \therefore \quad V_s = \frac{\pi}{6}\beta d_s^3, \mathrm{d}V_s = \frac{\pi}{4} d_s^2 \left(2 + \frac{\beta^2}{\varepsilon} \ln \frac{1+\varepsilon}{1-\varepsilon}\right) \\ & \wedge \quad \varepsilon = \frac{d_s}{2}\sqrt{1-\beta^2} \end{aligned} \tag{1.4}$$

and for prolate ones by

$$V_s = \frac{\pi}{6}\beta^2 d_s, \frac{\pi}{2} d_s^2 \left(\beta^2 + \frac{\beta}{\varepsilon} arc\sin\varepsilon\right) \tag{1.5}$$

Something very similar is introduced in the literature by the form-factor

$$c_{\mathrm{F}} = c / \sqrt{ab}$$

a, b, c are the axes' lengths of the adjoined ellipsoid, which are usually rather estimated than evaluated. c_{F} is used in transport equations as well as, e.g., for describing the settling velocity of a particle. From Eq. 1.4 the c_{F} for an oblate is equal to β, and for a prolate, Eq. 1.5, equal to $\beta\sqrt{\beta}$.

In cases in which the grain distribution is of minor importance, the distribution is given by a single value $d_{s(n)}$, where n stands for the upper limit in percentage of the summation curve. Therefore, $d_{s(50)}$ is the median of the grain distribution. More information is given by the gradation of the distribution function given by the deviation

$$\sigma_g = \left(\frac{d_{s(84)}}{d_{s(16)}} \right) \tag{1.6}$$

respectively the gradation coefficient

$$G_r = \frac{1}{2} \left(\frac{d_{s(84)}}{d_{s(50)}} + \frac{d_{s(50)}}{d_{s(16)}} \right) \tag{1.7}$$

The individual size classes of the grains have been standardized, and a selection can be found in Chap. 14 (Sect. 14.2.2.1). Typical grain size distributions of the well-known rivers can be found in the literature. An illustration of a bed, as found in hilly areas, is shown in Fig. 1.2.

Fig. 1.2 A top view of the surface layer of a bed of the river Buech in France close to the "porte du midi" in the scale 1:10

1.2.2.2 The Arrangement of the Sediment in the Bed

The arrangement of the grains in the surface layer of the bed is as important for the transport as the size distribution. Even sediment consisting of equal spheres can react differently in a flow if densely or loosely packed in this layer. In such a case, its porosity p_o, respectively the void fraction e of the sample, best describes the arrangement with

$$p_o = \frac{V_V}{V_V + V_s} = \frac{V_V}{V_t} = \frac{e}{1+e} \quad \wedge \quad V_V : \text{void volume} \tag{1.8}$$

$$e = \frac{V_V}{V_s} = \frac{p_o}{1-p_o} \tag{1.9}$$

For the loosest package of spheres p_o is 47.6%, and for the most dense package it is only 25.95%. For a grain distribution, the most dense package is given by its internal structure, as described by the so-called Fuller-curve (Fuller and Thompson, 1907), which is calculated by the passage through a sieve of size d_x in percent,

$$\%d_x = 100\left[\frac{d_x}{d_{s(100)}}\right]^{0.5} \tag{1.10}$$

with d_x, the mesh size of the sieve x, and $d_{s(100)}$ the biggest grains in the mixture.

Often a mean porosity is rather irrelevant since the bed composition differs from the layer beneath it. Usually the top layer is rougher than the one beneath. This is evident since the fine material subject to the outer flow erodes much easier, while the layer of finer material gives a better support from the bottom because of the higher friction it imposes onto the grain on top.

In consideration of this fact it is recommended to test the bed stability by exposing it to gravitational and friction forces only, that means without any contribution by the shear of a flow. One measures the natural slope given by the angle Φ, the angle of repose, which is the elevation angle against the horizontal, at which the bed material starts to slide. Φ is typically varying between 30° and 42°, and must therefore be more exactly evaluated for a given case.

1.2.2.3 The Fluids Their Sources and Their Characterizations

The wind draws its energy from atmospheric pressure gradients, and therefore depends on climatic conditions, while water flow draws its energy from the elevation of the drainage area collecting the rain and causing it to run downhill through the river system. To define the boundary conditions of the energy input, those who would like to treat sediment transport by wind under different climatic influences should consult Trenberth (1992). Similarly, for the rain input one has to consult the hydrological information of the drainage area under investigation, or one has to estimate these values based on hydrological tools.

As soon as a sediment transport starts, the flow is in its strict sense a two-phase flow, which means that the fluid properties start to deviate from a Newtonian fluid while a feedback loop between flow and material develops. We therefore restrict ourselves, at least for the moment, to characterizing the fluid independently of the sediment.

The fluids of interest are air and water, both of which are Newtonian fluids and their state is described by a series of material properties. The most important ones are the density ρ_f, the dynamical viscosity μ respectively the kinematic viscosity $\nu = \mu/\rho_f$, the temperature T, as well as the isotropic pressure p. The temperature is in most cases of minor importance for the sediment transport, although its influence is implicit by changing the values of other physical parameters as, e.g.,

$$\rho_f = \rho_f(T), \mu = \mu(T), \nu = \frac{\mu}{\rho_f} = \nu(T), p = p(T) \tag{1.11}$$

For example, the dependence of the (dynamical) viscosity on temperature can be estimated by using Arrhenius relation

$$\mu(T) = A\mathrm{e}^{-B/T} \tag{1.12}$$

with A and B typical constant values of a specific fluid, and by using the absolute temperature, T in Kelvin. In gases, such as air, p and ρ_f are temperature dependent, and the thermodynamic laws for a real gas reveal these relations. For water the compressibility can be neglected so it is treated as an incompressible medium and deviation from this assumption is only necessary if sound or pressure waves play an important role.

Usually the viscosity increases exponentially with p, with water under 30°C being a unique exception. Here, the viscosity first decreases until a minimum value is achieved at which the viscosity starts to increase exponentially similar to all the other fluids. Although the dependence on p can be described exponentially, the increase remains so small that it can be neglected.

The rheology of a Newtonian fluid is given by the linear relation between the shear-tensor $\underline{\underline{\sigma}}$ and the shear rate $\underline{\underline{\dot{\gamma}}}$ with the dynamical viscosity μ being the proportionality constant

$$\underline{\underline{\sigma}} = \mu\underline{\underline{\dot{\gamma}}} \wedge \quad \sigma_{ij} = p\delta_{ij} + d_{ij} \rightarrow p = -\frac{1}{3}\sigma_{ii} \tag{1.13}$$

and μ being independent of the shear rate $\underline{\underline{\dot{\gamma}}}$ or du_x/dz, respectively.

The first term to the right in the second equation stands for a fluid at rest, where all stresses are normal stresses. The pressure at a point in a moving fluid is defined by the mean normal stress with reversed sign and denoted by p for convenience. This is a purely mechanical definition of pressure without giving the relation between this mechanical quantity and the term pressure used in thermodynamics. The deviatoric stress tensor is denoted d_{ij}. Introducing the assumption that d_{ij} is in good approximation a linear function of the various components of the velocity gradient tensor for sufficiently small magnitudes of those components

$$
\begin{aligned}
& d_{ij} = A_{ijkl}\frac{\partial u_k}{\partial x_l} \\
& d_{ij} = A_{ijkl}e_{kl} - \frac{1}{2}A_{ijkl}\varepsilon_{klm}\omega_m \quad \wedge \quad e_{ij} = \frac{1}{2}\left(\frac{\partial u_i}{\partial x_j} + \frac{\partial u_j}{\partial x_i}\right) \\
& A_{ijkl} = \mu\delta_{ik}\delta_{jl} + \mu'\delta_{il}\delta_{jk} + \mu''\delta_{ij}\delta_{kl} \quad \wedge \quad A_{ijkl} = A_{jikl} \quad \therefore \quad \mu' = \mu \\
& A_{ijkl} = A_{ijlk} \quad \therefore \quad d_{ij} = 2\mu e_{ij} + \mu'' e_{kk}\delta_{ij} \\
& d_{ii} = (2\mu + 3\mu'')e_{ii} = 0 \quad \therefore \quad (2\mu + 3\mu'') = 0
\end{aligned} \tag{1.14}
$$

Choosing μ as the only independent scalar constant, we obtain the deviatoric stress tensor, given as a function of the rate of strain tensor e_{ij},

$$d_{ij} = 2\mu\left(e_{ij} - \frac{1}{3}e_{ll}\delta_{ij}\right) \quad \therefore \quad \sigma_{ij} = p\delta_{ij} + 2\mu\left(e_{ij} - \frac{1}{3}e_{kk}\delta_{ij}\right) \tag{1.15}$$

This formula was first given by Saint-Venant (1843) and Stokes (1845), see also Batchelor (1967). We use this form of description to give a better view of the rheology of a Newtonian fluid.

By choosing only one value for the viscosity, one does have to distinguish between extensional (=elongational) and shear viscosity anymore, where the extensional viscosity μ_E is given by the extensional stress divided by the rate of extension $\dot{\varepsilon}$, the change in extensional strain per unit time. For Newtonian fluids the relation is

$$\mu_{\mathrm{E}} = 3\mu \tag{1.16}$$

The ratio of the two viscosities is called the Trouton ratio T_R. It may differ from the Newtonian one and is characterizing the rheological state of the system,

$$T_{\mathrm{R}} = \frac{\mu_{\mathrm{E}}(\dot{\varepsilon})}{\mu(\dot{\gamma})} \xrightarrow{\text{Newtonian}} T_{\mathrm{RN}} = 3 \tag{1.17}$$

From this rheology, an equation of motion can be deduced for a known ρ_f and an external force-density $\underline{F}$ described by μ alone, where the total shear-tensor is given as in Eq. 1.15.

For a plane Couette flow as it is established in a shearing experiment, this deviatoric stress tensor reduces to the planar shear stress

$$\tau_{xz} = \mu\frac{\partial u_x}{\partial z} \tag{1.18}$$

and μ is independent of $\underline{\underline{\dot{\gamma}}}$.

1.2.2.4 Dynamical Description of the Flow

The equation of motion for a fluid, in its most fundamental form, is a relation equating the rate of change of momentum of a selected portion of fluid with the sum of all forces acting on that portion of fluid

$$\int_V \underline{u}\rho_f\, dV \quad \rightarrow \quad \int_V \frac{D\underline{u}}{Dt}\rho_f\, dV \quad \wedge \quad \frac{D}{Dt} = \frac{\partial}{\partial t} + u_j\frac{\partial}{\partial x_j}$$

$$\int_V \frac{Du_i}{Dt}\rho_f\, dV = \int_V F_i\rho_f\, dV + \int_V \frac{\partial \sigma_{ij}}{\partial x_j}dV \tag{1.19}$$

$$\therefore \quad \rho_f\frac{Du_i}{Dt} = \rho_f F_i + \frac{\partial \sigma_{ij}}{\partial x_j}$$

This differential equation giving the acceleration of the fluid in terms of the local volume force and stress tensor, is an equation of motion.

With Eqs. 1.15 and 1.14, Eq. 1.19 can be written as

$$\rho_f \frac{Du_i}{Dt} = \rho_f F_i - \frac{\partial p}{\partial x_i} + \frac{\partial}{\partial x_j}\left[2\mu\left(e_{ij} - \frac{1}{3}e_{ii}\delta_{ij}\right)\right] \tag{1.20}$$

This is called the Navier–Stokes equation; it is an integro-differential equation for the local motion of a unit volume element. The Navier–Stokes equation is the fluid mechanical equivalent to Newton's first law of motion of a solid body, which, however is only a differential equation.

If μ can be taken as independent of the location in the flow, Eq. 1.20 reduces to

$$\rho_f \frac{Du_i}{Dt} = \rho_f F_i - \frac{\partial p}{\partial x_i} + \mu\left(\frac{\partial^2 u_i}{\partial x_j \partial x_j} + \frac{1}{3}\frac{\partial e_{ii}}{\partial x_i}\right) \tag{1.21}$$

For incompressible fluids, the mass conservation equation, called continuity equation, is

$$\frac{\partial u_i}{\partial x_i} = 0 \quad \vee \quad (\nabla, \underline{u}) = 0 \tag{1.22}$$

and with Eq. 1.22 the Navier–Stokes equation gets its usual form

$$\frac{\partial u_i}{\partial t} + u_j \frac{\partial u_i}{\partial x_j} = -\frac{1}{\rho_f}\frac{\partial p}{\partial x_i} + \nu \frac{\partial^2 u_i}{\partial x_j \partial x_j} + F_i \quad \vee \quad \frac{D\underline{u}}{Dt} = -\frac{1}{\rho_f}\nabla p + \nu\nabla^2 \underline{u} + \underline{F} \tag{1.23}$$

This equation cannot be integrated in a closed form and therefore remains one of the big problems in the description of sediment transport. Some simplifications are unavoidable and will be given where appearing in the models.

As a last remark we mention that Eq. 1.23 describes laminar and turbulent flows. Since surface flows are practically always turbulent, we will restrict ourselves to that regime, the consequences will be given later in this book.

1.3 The Topography of a Drainage Area

Erosion and deposition by sediment transport transform the surface of the earth. Therefore, there exists a feedback loop between the sediment transport and the landscape. However, this coupling acts with some rare exceptions, on geological time scales, and is therefore not relevant for the posed question of the prediction of sediment transport on much shorter time scales. But those who simulate topographic evolutions have to consider such feedback mechanisms where the change of the geometry and the slope of a channel are the main parameters to be evaluated over long time scales. This is usually done by an extrapolation of the short-time scale results.

A drainage area is characterized by its boundaries, which are watersheds. From these edges, the water flows versus the lower grounds. Small channels merge to bigger and bigger ones and finally discharge in a single one. Arriving at an alluvial land area, where the slope of the river decreases, the geometry of the channel changes and the river starts to meander. If the slope of the terrain decreases even more, the channel starts to bifurcate and ends up in a typical delta like estuary. A typifying representation of such a drainage area is given in Fig. 1.3.

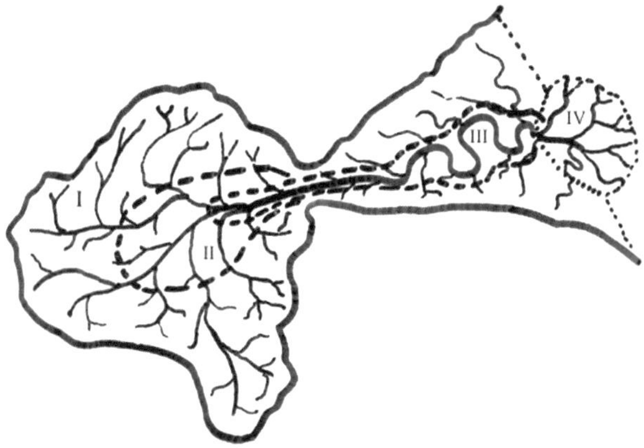

Fig. 1.3 A typifying drainage area as described in the text. Zone I stands for alpine regions with rivulets, zone II is the typical river area, in zone III a single stream is present occurring in two different channel geometries, which is the main case discussed in this text book, zone IV stands for the estuary area with a delta

This characterization is important for the simulation of sediment transport. Only in rare cases, one investigates the entire sediment transport from the spring to the estuary within a single model. Most investigations deal with the transport in a section only and use as the sediment transport rate through the limiting cross sections as boundary conditions. This allows composing the total system from modules if one is able to match the boundary conditions at the interfaces. Within this modular framework, two modules were extensively investigated in the past, namely the straight channel and the flow in curvature. Both geometries are especially suited for investigation in hydraulic laboratory tests.

A fluvial system shows up as a morphological structure as seen in Fig. 1.3, characterized by a network resembling a typical tree like structure. Such structures suggest that similarity laws may exist, which in a general manner combine geology and hydrology. It is assumed that under the same conditions equal or at least only statistically weakly diverging systems of a channel configuration can develop. The knowledge of such feedback mechanisms are not relevant for the sediment transport as such, however, it is relevant if one compares different systems. The more the characteristics of two systems depart from each other, the more problematic it becomes to compare even part of the channel system.

A method of quantitative characterization of such systems is achieved by analysis of its fractal geometry, whereby the network is considered as the sum of subnetworks starting from the first brook-network, all rivulets limited by their own watershed and occupying a given drainage area. To evaluate the fractal dimension, D_f, Mandelbrot (1977) found that one has to take the ratio of the length to the square root of the area. For a river system Hack (1957) did this by taking the ratio of the length L of the main river channel to its drainage area A_d, and found the relation,

$$(A_{\mathrm{d}})^{1/2} \propto (L)^{1/D_{\mathrm{f}}} \rightarrow D_{\mathrm{f}} > 1 \rightarrow D_{\mathrm{f}} \approx 1.2 \tag{1.24}$$

Instead of the length L, one can use for instance the length of the boundary around a drainage area or the cumulated length of all rivers belonging to a river-tree, for the last dimension Mandelbrot predicted the value $D_{\mathrm{f}} = 1$. All these dimensions allow classification of the different networks for comparison. Common characterizations are used in the work of Horton (1945) with the supplement of Strahler (1952), enlarged by Smart (1972). He introduced so-called inner and outer parameters. The outer regions have a boundary, which they have partly in common with the one of the total drainage area. As length scales, he chose the mean value of the outer and the inner branches of the system, l_{e} and l_{i}, together with their drainage areas a_{e} and a_{i}, respectively. The dimensionless values, which have to be evaluated first, are therefore

$$\lambda_{\mathrm{f}} = \frac{l_{\mathrm{e}}}{l_{\mathrm{i}}}; \alpha_{\mathrm{f}} = \frac{a_{\mathrm{e}}}{a_{\mathrm{i}}}; D_{\mathrm{fe}}^2 = \frac{l_{\mathrm{e}}^2}{a_{\mathrm{e}}}; D_{\mathrm{fi}}^2 = \frac{l_{\mathrm{i}}^2}{a_{\mathrm{i}}} \tag{1.25}$$

Coffman et al. (1972) proposed another characterization by postulating a relation between the number of bifurcations (knots) and the segments.

As indicated in Fig. 1.3, a fluvial network consists of zones which discern themselves mainly by their slope S, but also the sediment composition changes often drastically. The consequences are differences in the relative roughness $D_{\mathrm{s}} = d_{\mathrm{s}(100)}/H$, defined as the ratio of the diameter of the largest grains on the bed $d_{\mathrm{s}(100)}$ and the water depth H. Especially in mountainous terrain or in very shallow rivers, D_{s} can become larger than one, which means single stones dig through the water surface so that in a topological sense the channel is of higher topological order. Since the main interest is usually for rivers belonging to the zones II and III, which is equivalent to the assumption that the rivers have a certain size and a rather shallow, we assume an upper limit to be $D_{\mathrm{s}} \sim 0.1$. With this statement we conclude that there is no sharp criterion, which distinguishes between the zones of a network.

1.4 Modeling the Phenomenon

Sediment transport stands for an interaction of diverse parameters. To describe the entire process in all its complexity and to predict the erosion and deposition starting from boundary- and initial conditions as they are left by earlier transport events, is wishful thinking. Even if the whole process would be mathematically describable, neither the capacity of the computers nor the current solution algorithms would suffice to achieve this goal. Today not even the pure flow without sediment can be calculated. Therefore, the strategy must be to simplify the calculations. Some approaches will be shown later.

In the past, the modeling was mainly supported by laboratory investigations. The most accurate investigation would be to observe the system itself on the system scale during its evolution by measuring the relevant parameters, however this is not useful for the development of a prediction tool for a different setting as results cannot be transferred. The model, therefore, has to be transformed to smaller transferable scales in such a way that the result for the whole system is not changed. That means the process has to be described by similarity laws, which must remain valid in the model.

In a first step, the mechanisms have to be described in dimensionless parameters because in this form the variables become invariant. Very similar thoughts stand at the origin of the so-called dimensional analysis, which reduces the phenomenological information by parameter reduction to a representation, which only must be correct in its dimension. It is warned to interpret such results as physical laws or even worse to deduce from them any comprehension of physical facts. However, as a tool for a better understanding of functional structures or for modeling it is of great help. One can find a more complete description of the method in Barenblatt (1979).

Any physical relation between a_i, $(i = 1 \div n)$, parameters can be formulated as a function a of all parameters

$$a = f(a_1 \div a_n) \tag{1.26}$$

If we assume that a_i / a_k with $k < n$ have independent dimension, then $a_{k+1}, \ldots, a_n$ are reducible by the independent parameters to dimensionless numbers

$$\Pi_i = \frac{a_i}{a_1^{p_i} \cdots a_k^{r_i}} \qquad i = \frac{k+1}{n} \tag{1.27}$$

and with Eqs. 1.27 and 1.26 it can be written as

$$\Pi = \frac{f(a_1 \div a_n)}{a_1^{p} \cdots a_k^{r}} = F(a_1 \div a_k, \Pi_1 \div \Pi_{n-k}) \tag{1.28}$$

Since $\partial F / \partial a_i$ for $i = 1/k$ is zero the solution of Eq. 1.28 is

$$\Pi = \Phi(\Pi_1 \div \Pi_{n-k})$$

$$\vee \quad f(a_1 \div a_k, a_{k+1} \div a_n) = a_1^{p} \cdots a_k^{r} \Phi\left(\frac{a_{k+1}}{a_1^{p_{k+1}} \cdots a_k^{r_{k+1}}} \div \frac{a_n}{a_1^{p_n} \cdots a_k^{r_n}} \right) \tag{1.29}$$

known as the Buckingham Π-theorem.

Let us take for example the Navier–Stokes equation (Eq. 1.23) with its boundary condition for a pipe flow without an outer forcing. Then the flow is characterized by the parameters D, the inner diameter of the pipe, $\underline{u}$, ρ_f and μ. Since Eq. 1.23 is an integro-differential equation, a fixed relation exists between ∇p and $\underline{u}$, and p can be replaced in terms of $\underline{u}$, (with the same argument one could avoid $\underline{u}$). This relation is also the reason why ∇p is usually treated as an outer force, this, however, is a convention only. Therefore, we have four parameters showing the following dimensionless parameters

$$[D] = L; \quad [\underline{u}] = L/T; \quad [\rho_f] = M/L^3; \quad [\mu] = M/LT \tag{1.30}$$

Therefore, we have four parameters for the three independent dimensions and following the Π-theorem, the similarity relation results in a single dimensionless product Π_1. To evaluate this number one has first to write Eq. 1.23 in dimensionless form, by using the calibration values D and U, the mean velocity in the pipe, which results in the new dimensionless variables.

$$\hat{u} = \frac{u}{U}; \quad \hat{t} = \frac{tU}{D}; \quad \hat{x} = \frac{x}{D}; \quad \hat{p} = \frac{p - p_0}{\rho_f U^2} \tag{1.31}$$

with p_0 the representative value of the modified pressure in the flow. With this Eq. 1.23 in dimensionless form can be written as

$$\begin{aligned} &\frac{\partial \hat{u}_i}{\partial \hat{t}} + \hat{u}_j \frac{\partial \hat{u}_i}{\partial \hat{x}_j} = -\frac{\partial \hat{p}}{\partial \hat{x}_i} + \frac{1}{Re} \frac{\partial^2 \hat{u}_i}{\partial \hat{x}_j \partial \hat{x}_j} \quad \wedge \\ &Re = \frac{\rho_f LU}{\mu} = \frac{LU}{\nu} \end{aligned} \tag{1.32}$$

which contains only one reference number of the system, the Reynolds number, *Re*.

However, when the velocity distribution is influenced by the gravity force, e.g., if the flow has a free surface, the gravity has to be considered explicitly in Eq. 1.23 as an outer force ($\rho_f \underline{F} = \rho_f \underline{g}$) which produces an additional reference number

$$\mathrm{Fr} = \sqrt{\frac{U^2}{gL}} = \frac{U}{\sqrt{gL}} \tag{1.33}$$

the Froude number. In rivers this similarity relation is the dominant one, and in theory, a similarity model has to fulfill both reference numbers simultaneously, which is practically impossible. For the entire representation of the sediment transport, new parameters have to be added, and with every additional new parameter, a new reference number must be added while the hope for a similarity representation vanishes completely.

Before we introduce a higher series of parameters, we would like to briefly comment on the interpretation of reference numbers especially with respect to *Re*. Depending on which of the three values D, U, and ν one has chosen, *Re* physically represents, the natural basic values: a length measured in natural units, a velocity measured in natural units or a viscosity measured in the reciprocal natural units.

In alluvial systems it is assumed that gravity forces are dominating the viscous forces, for which physical models with a Froude similarity are required. This is mainly problematic for the investigation of flows with a high amount of suspension. With these remarks we hope to have cautioned the reader that he critically checks the results found by these methods, especially when a generalization of the results is suggested.

This description will end by an incomplete parameterization of the sediment transport to demonstrate the complexity of the phenomenon. The parameter function can take, for example, the form

$$f\left(\rho_s, \rho_f, d_s, u, H, \nu, q_s, q_f, g, w, u_{ic} \ldots, L_G \ldots, \rho_k, T, \mathrm{pH}, c_F, \sigma, \mathrm{etc.}\right) = \tag{1.34}$$

Besides the already known parameters, we additionally find material fluxes q_f of the fluid as well as of the contained sediment q_s, different critical velocities characterizing change in state, topographical lengths and curvatures given by radii, the electrolytic characteristic like pH value, the heat capacity, the surface tension, and additional parameters.

This shows that the channel geometry is only described very primitively, say we neglected a parameter for the exposition of the grains and with it the roughness of the bed, etc. This flood of parameters shows drastically that no similarity law can address all of them in a single model. However, the situation is not as bad as it seems, because several helpful simplifications can be made. For example, by compressing values of the same dimensions to dimensionless values as for instance the densities,

$$\rho' = \frac{\rho_s - \rho_f}{\rho_f} \quad \vee \quad \rho_r = \frac{\rho_s}{\rho_f} \tag{1.35}$$

However all these numbers appearing in the literature, shall not hide the fact that the right combination is unknown, and not detectable by a dimensional analysis. Nevertheless, some aspects can be examined in part, if the neglected parameters play a minor role in the physical process.

As important as the dimensional relations are the statistical quantities of the used parameters, that means their probability distributions.

These critical sentences were intentionally placed at the end of this chapter, because in the literature on sediment transport we often find dimensional arguments. This may be misleading in so far as the physics remain obscure and some experimental results remain over estimated.

2 The Classical Representation of the Sediment Transport

As mentioned in the introduction, currently we are still using the classical concept of relating bed shear stress to a particle transport rate. In this chapter, we discuss this commonly used representation. In practice, the transport formulated in such a form resembles the transport by bed-load.

There are dozens of equations treating the bed-load by the same fundamental concept. Their similarity law is given by the Froude number Eq. 1.33 and the transport is uniquely defined by the bed (or wall) shear stress τ_w. With this approach the total effect of the turbulent flow is combined into one parameter, the wall shear stress, but the relation between the status of the flow and the forces acting on the grains remains unresolved.

Assuming a so-called normal discharge, the wall shear stress is given by an equilibrium condition, which is used in the classical theories. The friction forces must be equal but opposite of the accelerating forces. Since the gravitational forces are continuously accelerating the flow, a friction force of the same amount has to be transferred to the bed by a continuous flux of momentum. This flux is represented either by τ_w or the wall shear velocity u_τ, respectively, which is given by:

$$u_\tau = \sqrt{\frac{\tau_w}{\rho_f}} \tag{2.1}$$

In an open channel flow with a slope S, the acceleration force is given by the weight component in flow direction as shown in Fig. 2.1

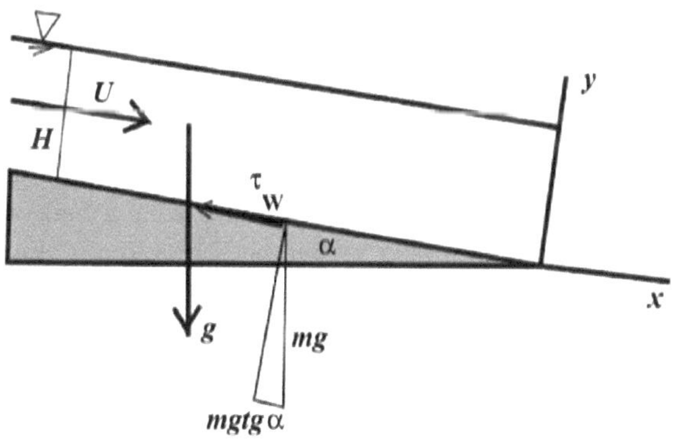

Fig. 2.1 A sketch showing the equilibrium state between the accelerating forces, which are gravitational body forces in flow direction given by $F_b = mg \tan \alpha$ and the corresponding friction forces representing the wall shear stress

The time averaged total shear stress τ_T at any height of the flow decreases as a linear function of the wall distance and vanishes at the height H, the surface level of the water; this is shown in Fig. 2.2. This linear decrease in τ_T does not mean that a smooth sliding process governs the momentum exchange from one layer to the neighboring one. On the contrary, the situation is extremely complex and the result of a hierarchy of flow structures responsible for the velocity profile.

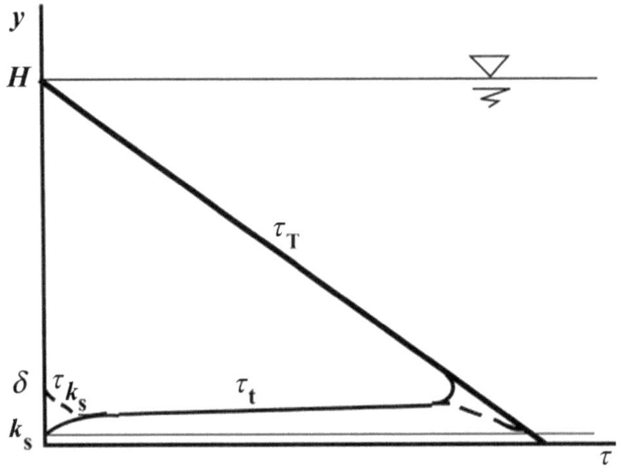

Fig. 2.2 The total shear stress τ_T, in a normal open channel flow is a linear function of the wall distance, and it is composed of different contributions. Above the height δ the momentum transfer is given by the turbulent shear stress τ_t only, whereas close to the wall the shear stress is given by a roughness term τ_{ks}, which is essentially a form drag, or a viscous term τ_v, depending whether the wall configuration is rough or smooth

The total shear stress is essentially composed of two components: the first originates from the turbulent momentum exchange, τ_t, which dominates the momentum flux above δ and defines the outer region of the flow. The second term is more dominant near the wall and comprises the viscous stress τ_v or stresses from roughness τ_{ks} depending on the smoothness of the bed.

We will explain the shear stress τ_{ks} originating from a bed of "equivalent sand roughness" (Nikuradse, 1933) in a later paragraph. The viscous contribution τ_v is very important for a flow over a smooth wall; however it is not relevant for a flow as it occurs in a natural river bed.

Of highest importance is the turbulent momentum exchange, which is also entirely reflected by τ_w. The classical approach describes the turbulent flow by splitting the relevant parameters into mean and fluctuating values according to the Reynolds decomposition.

2.1 The Representation of the Flow

It makes sense to discuss the implications on the physics of sediment transport following from this extremely simplified classical representation. First, we critically look at the modeling of turbulence together with the applied boundary conditions and will discuss the resulting flow description, especially with respect to the roughness of the bed and its influence on the flow.

2.1.1 The Reynolds Decomposition

For fixed initial- and boundary conditions, a turbulent flow exhibits an asymptotic behavior, which means that, although the flow states can differ momentarily, the long time average is represented by mean values of the flow variables. In general these mean values are formulated as,

$$\overline{\Phi}\left(\underline{x}\right) = \lim_{T \to \infty} \frac{1}{T} \int_0^T \phi\left(t, \underline{x}\right) \mathrm{d}t \tag{2.2}$$

where $\overline{\Phi}(\underline{x})$ is the temporal mean value of the function $\phi(\underline{x},t)$ and ϕ stands for any flow parameter influenced by turbulence. The momentary value of ϕ can therefore be expressed as

$$\phi\left(\underline{x},t\right) = \overline{\Phi}\left(\underline{x}\right) + \phi'\left(\underline{x},t\right) \tag{2.3}$$

the so-called Reynolds decomposition in which the fluctuating part is given by ϕ' with the identity

$$\overline{\phi'} \equiv 0 \tag{2.4}$$

Although this decomposition is always permitted mathematically, nature does not know about this arbitrary splitting. It is therefore problematic to make conclusions about the physical nature of the process from results found using Eq. 2.3. To elucidate this statement, we assume for example that the system possesses a periodic part because such information would get lost in the mentioned representation. Therefore, a series of new decompositions were proposed in the past to overcome this deficiency, e.g. as shown in Eq. 2.5,

$$\phi = \overline{\Phi} + \tilde{\phi} + \phi' \tag{2.5}$$

where $\tilde{\phi}$ is the periodic part of the system. However, if the fluctuating part is composed of other deterministic contributions, such as by structural elements with a quasi-periodicity, one is not able to discern those from the turbulent fluctuations without further assumptions. Additionally, if the Reynolds decomposition is also used to investigate quantities, such as concentrations of suspensions, we have to also show that in such cases an asymptotic description is valid and makes physical sense.

> *What is the most appropriate decomposition for the different relevant parameters encountered in the sediment transport?*

In general terms, we are confronted with the question of how a process will be divided into temporal sequences without influencing the resulting flow quantities.

As a first approach, the time sequences used must be long enough so that the autocorrelation of the investigated function truly vanishes. The nonvanishing part in an autocorrelation representation is the periodic part of the signal.

Due to Eq. 2.4, the mean values of the mixed products are also identical to zero,

$$\overline{\overline{\Phi}\phi'} \equiv 0 \tag{2.6}$$

but the products of the fluctuations are not equal to zero.

$$\overline{\phi_i'\phi_j'} \neq 0 \tag{2.7}$$

Also the mean values of quantities of the form shown in Eq. 2.7 can approach an asymptotic estimate, and many theories of sediment transport make use of that fact. However, what has been said for the fluctuating parameters is even more important for the fluctuations of the higher order products.

In other words, whenever a parameter is used in a decomposed form, we have to investigate its physical meaning before the results can be generalized. This becomes even more important if spatial, rather than temporal, decompositions are used. For the moment, we will assume that the fluctuations are distributed homogeneously in space. The classical theories rely on this assumption. Later we will discuss deviations from this assumption.

2.1.2 The Flow Equations

With the Reynolds decomposition given by Eq. 2.3 the Navier–Stokes equation (Eq. 1.23) for the mean flow can be written in mean and fluctuating terms as:

$$\frac{\partial \overline{u}_i}{\partial t} + \overline{u}_j \frac{\partial \overline{u}_i}{\partial x_j} = -\frac{1}{\rho_f}\frac{\partial \overline{p}}{\partial x_i} + \frac{\partial}{\partial x_j}\left(\nu \frac{\partial \overline{u}_i}{\partial x_j} - \overline{u_i' u_j'}\right) + F_i \tag{2.8}$$

the term, which relates the mean motion and turbulence, is:

$$\overline{u_i' u_j'} \tag{2.9}$$

the tensor of the so-called Reynolds stresses.

The next simplification is the reduction to a so-called one-dimensional flow, that means the mean velocity is parallel to the river slope and has therefore only one direction,

$$\overline{u} = \underline{U}(z) = \left(U_x(z), 0, 0\right) \quad \wedge \quad \underline{U} = (U, V, W) \tag{2.10}$$

Introducing a Cartesian coordinate system *x, y,* and *z* in flow, lateral and bed normal direction, respectively, and assuming that the only external force is gravitational we see that only one of the nine terms of tensor (Eq. 2.9), namely

$$\overline{u_x' u_z'}(z,t) \neq 0 \tag{2.11}$$

is relevant and therefore Eq. 2.8 can be simplified as follows:

$$\begin{aligned} 0 &= -\frac{1}{\rho_f}\frac{\partial \overline{p}}{\partial x} + \nu \frac{\partial^2 U(z)}{\partial z^2} - \frac{\partial}{\partial z}\overline{u_x' u_z'} + g_x \\ 0 &= -\frac{1}{\rho_f}\frac{\partial \overline{p}}{\partial z} + g_z \end{aligned} \tag{2.12}$$

The continuity equation for an incompressible fluid (Eq. 1.22) becomes

$$\frac{\partial \overline{u}_i}{\partial x_i} = \frac{\partial u_i'}{\partial x_i} = 0 \quad \wedge \quad \frac{\mathrm{D}\rho_f}{\mathrm{D}t} = 0 \tag{2.13}$$

By using Eq. 2.1 and knowing that $\tau_w \approx \tau_t$ we can derive

$$\tau_w \approx \rho_f \overline{u'_x u'_z} \quad \therefore \quad u_\tau = \sqrt{\frac{\tau_w}{\rho_f}} \approx \sqrt{\frac{\overline{u'_x u'}}{\rho_f}} \tag{2.14}$$

The effects of the simplifications shown yield the result that the entire influence of the turbulence can be combined into a single parameter τ_w or u_τ.

For a quick estimate of the friction velocity, one can use the experimentally found approximate relation

$$U = 10 u_\tau \tag{2.15}$$

2.1.3 The Velocity Distribution and the Influence of the Roughness

Flow over smooth walls: Let us start with the no-slip condition at the wall given by

$$\underline{u}_w \equiv 0 \quad \wedge \quad \frac{\partial \underline{u}}{\partial x} = \frac{\partial \underline{u}}{\partial y} = 0, \frac{\partial \underline{u}}{\partial z} \neq 0 \tag{2.16}$$

which is the reason why the flow cannot be described by a single length scale anymore, as was possible for the free flow condition. Now there exists a restriction in z-direction. In case of a 1D free flow, one could assign the length scale l in x-direction where l was defined by the mean velocity gradient and could be interpreted as the length scale of the energy containing vortices. It is observed that

$$\frac{\partial U}{\partial x} = 0\left(\frac{U}{L}\right), \quad \frac{\partial U}{\partial z} = 0\left(\frac{U}{l}\right) \quad \therefore \frac{l}{L} \ll 1 \tag{2.17}$$

Since according to Eq. 2.16 the velocity at the wall is zero, the viscosity must define an additional length scale in the layer close to the wall

$$\frac{\nu}{u_\tau}; \quad \tau_w = \rho_f u_\tau^2 = \mu \left.\frac{\partial U}{\partial z}\right|_w \tag{2.18}$$

Therefore in a thin region close to the wall, we have a system with two length scales, however outside, as soon as the criterion

$$\delta = z \gg \frac{u_\tau}{\nu} \tag{2.19}$$

is fulfilled, the viscous influence can be neglected. Even for rather small Reynolds numbers δ is reached at about $z^+ \approx 50$ and is increasing with Re. Here we have introduced the so-called viscous (or wall) units, which are based on u_τ and ν as the basic scaling parameters,

$$z^+ = \frac{z u_\tau}{\nu}, \quad U^+ = \frac{U}{u_\tau} \tag{2.20}$$

The viscous length scale z^+ and the friction velocity u_τ itself allow defining the flow regions near to the wall, where the flow remains invariant when scaled with these dimensionless parameters. However, the outer part of the flow will not scale universally when scaled by viscous units. The usual treatment is to use the results of the boundary layer flow as it develops on a smooth plate exposed to a uniform parallel flow. This is obviously something different from the boundary layer flow over the bed of an open

channel flow. Nevertheless, the results are somewhat useful since it was shown that at least the structures developing in the wall-near zone are very similar. This we will explain in more detail when we introduce the concept of coherent structures.
When U_s is representative of a velocity of the outer flow

$$U_s = U(x, +\infty) \tag{2.21}$$

where the fluid stresses are governed by turbulent momentum exchange, then the velocity defect

$$U_s - U = O(u_\tau), \quad U = O(U_s), \quad \frac{\partial U}{\partial y} = O\left(\frac{u_\tau}{l}\right); \quad \frac{u_\tau}{U_s} \ll 1 \tag{2.22}$$

It becomes evident that the two different length scales stand for the velocity field and the velocity gradients. The experimentally determined value for the velocity ratio was found to be 1/10 for moderate Reynolds numbers and slowly decreasing to 1/30 for high Reynolds numbers.

Using the continuity Eq. 2.13 the vertical flow velocity W can be estimated if u_τ, l, and L are introduced as scaling parameters, where L is the distance in flow direction from the edge of the plate,

$$W = O\left(\frac{u_\tau l}{L}\right) \tag{2.23}$$

Using this vertical velocity, one can evaluate the mean lateral vorticity

$$\Omega_y = \frac{\partial W}{\partial x} - \frac{\partial U}{\partial z}, \quad \frac{\partial W}{\partial x} = O\left(\frac{u_\tau l}{L^2}\right), \quad \frac{\partial U}{\partial z} = O\left(\frac{u_\tau}{l}\right),$$

$$\frac{\partial W / \partial x}{\partial U / \partial z} = O\left(\frac{l}{L}\right)^2 \quad \therefore \quad \Omega_y \approx -\frac{\partial U}{\partial z} \tag{2.24}$$

$$\underline{\Omega} = \left(\frac{\partial W}{\partial y} - \frac{\partial V}{\partial z}, \frac{\partial U}{\partial z} - \frac{\partial W}{\partial x}, \frac{\partial V}{\partial x} - \frac{\partial U}{\partial y}\right) \quad \vee \quad \Omega_i = \varepsilon_{ijk} \frac{\partial \overline{u}_k}{\partial x_j}$$

with,

$$\left.\frac{\partial U}{\partial z}\right|_{\mathrm{w}} \rightarrow = \frac{u_\tau^2}{\nu} = \text{const}; \quad \left.\frac{\partial W}{\partial x}\right|_{\mathrm{w}} \rightarrow 0 \tag{2.25}$$

one gets:

$$\int_0^\infty \Omega_y \mathrm{d}z \approx -\int_0^\infty \frac{\partial U}{\partial z} \mathrm{d}z = -U(x, \infty) = -U_s \tag{2.26}$$

That means, the mean lateral vorticity is independent of the shape of the cross section and is a constant. This vorticity is created at the surface of the plate and distributes with the growing boundary layer through diffusion. When it reaches the free water surface, the boundary layer engulfs the whole flow field and therefore boundary layer theory can be applied to the open channel flow.

For incompressible flows with no-slip boundary condition it can be shown that the following relation holds

$$\frac{\partial}{\partial x} \int_0^\infty U\left(U_s - U\right) \mathrm{d}z = \nu \left.\frac{\partial U}{\partial z}\right|_{\mathrm{w}} = u_\tau^2 \tag{2.27}$$

and momentum is transferred continuously into the wall.

In the outer boundary layer, where l is the scaling parameter, or $z \gg \nu/u_\tau$, a self similarity can be assumed and with the velocity scaling parameter u_τ as follows:

$$U_s - U = u_\tau F\left(\frac{z}{l}\right) \tag{2.28}$$

and on the other hand close to the wall where the viscosity is dominant we find,

$$\frac{U}{u_\tau} = f\left(\frac{zu_\tau}{\nu}\right) = f(z^+) \tag{2.29}$$

Both layers can overlap and the zone where both relations Eqs. 2.28 and 2.29 are valid is called the inertial sublayer. From Eqs. 2.28 and 2.29 follows:

$$\frac{\partial U}{\partial z} = -u_\tau \frac{F'}{l} = u_\tau^2 \frac{f'}{\nu} \tag{2.30}$$

and it follows from Eq. 2.30

$$\begin{aligned} &\frac{U - U_s}{u_\tau} = \left(\frac{1}{\kappa}\right) \ln\left(\frac{z}{l}\right) + b, \quad \frac{U}{u_\tau} = \left(\frac{1}{\kappa}\right) \ln z^+ + a \\ &\therefore \quad \frac{U_s}{u_\tau} = \left(\frac{1}{\kappa}\right) \ln Re_l + a - b \quad \wedge \quad Re_l = \frac{u_\tau l}{\nu} \end{aligned} \tag{2.31}$$

the well-known logarithmic velocity profile with κ the von Karman constant, where a and b are constants which have to be determined by experiments. This law has been verified experimentally, it is often also called the logarithmic friction law.

In addition to its role in determining the velocity profile, the resistance plays an important role in sediment transport. The resistance coefficient describes the friction on the bed and it is given by

$$C_D = \frac{\tau_W}{\frac{1}{2}\rho_f U_s^2} = \frac{\rho_f u_\tau^2}{\frac{1}{2}\rho_f U_s^2} = 2\left(\frac{u_\tau}{U_s}\right)^2 \quad \therefore \quad \frac{u_\tau}{U_s} = \sqrt{\frac{C_D}{2}} \tag{2.32}$$

The stress in the fluid layer nearest to the wall is viscous dominated; therefore this layer is called the viscous sublayer. Its velocity profile is given by

$$\begin{aligned} &\tau_T \approx \tau_v = \mu \frac{\partial U}{\partial z} \quad \wedge \quad \tau = \rho_f u_\tau^2 \\ &\therefore \ (\textstyle\int) \rightarrow \frac{U}{u_\tau} = \frac{u_\tau z}{\nu} + c; \quad \left(\frac{U}{u_\tau} = 0\right)_{z=0} \quad \therefore \ c = 0 \\ &\frac{U}{u_\tau} = z^+, \quad z^+ \leq 5 \end{aligned} \tag{2.33}$$

Here we must emphasize, that the flow in the viscous sublayer is not laminar. The flow velocities remain fluctuating, primarily in the wall parallel directions.

The layer in between the two layers described by Eqs. 2.31 and 2.33 is called the buffer zone and its upper height is at about $z^+ \approx 50$. The area where viscous and turbulent influences are of about the same order is at about $z^+ \approx 12$. These layers and their scaling are shown in Fig. 2.3.

With these scaling laws the dominant Reynolds stress component $< u'_x u'_z >$ for the four flow layers are

$$\begin{aligned}
&\langle u'_x u'_z \rangle \approx 0 && z^+ \leq 5 \\
&\langle u'_x u'_z \rangle = \nu \frac{\partial U}{\partial z} - u_\tau^2 && 5 \leq z^+ \approx 50 \\
&\langle u'_x u'_z \rangle = u_\tau^2 \left(\frac{1}{\kappa z^+} - 1 \right) && 50 \approx z^+ \leq 2 - 300 \\
&\langle u'_x u'_z \rangle = \frac{1}{\kappa} \left(\frac{z}{l} - \frac{\delta}{l} \right) && z^+ \geq 300
\end{aligned} \tag{2.34}$$

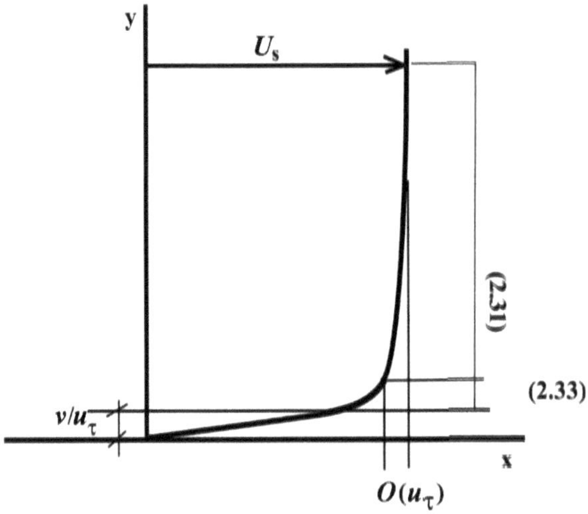

Fig. 2.3 Schematic diagram of the velocity profile of a turbulent boundary layer flow with its diverse length and velocity scales. The explanation can be found in the text, especially the definition of k_s by Eq. 2.40

2.1.3.1 Over Rough walls

First, we introduce the roughness conditions as they are used in classical transport models and later we will discuss the properties of a rough wall in detail. The main difference between smooth and rough walls is how the momentum transferred to the wall. On a rough wall, the transfer of momentum occurs due to differences in the pressure forces arising from flow separations at the roughness elements and it is no longer a result of the viscous shear gradients.

Separations can occur at single roughness elements or at bed-forms, both events contribute to τ_w, and it became customary to use a splitting of the wall shear stress into two components. The classical concepts generally use the total wall shear stress without introducing an additional length scale for the bed-forms. The new scaling laws must depend on the roughness elements, which are given by z/ε and cannot depend on viscous forces any more.

In the outer region of a turbulent flow, the Reynolds stress components are responsible for the total momentum exchange. This also holds for flow over rough walls, although a new length scale ε has to be used. The resulting velocity profiles are, as derived for the smooth case logarithmic analogs to Eq. 2.31

$$\frac{U}{u_\tau} = \frac{1}{\kappa}\ln z^+ + 5.5 \qquad \{\nu, u_\tau\}$$
$$\kappa = 0.4 \tag{2.35}$$
$$\frac{U}{u_\tau} = \frac{1}{\kappa}\ln\frac{z}{\varepsilon} + 8.5 \qquad \{u_\tau, \varepsilon\}$$

It seems simple to introduce an effective roughness ε. However, ε stands for an elevation, which can be measured on a machine tooled surface. Transcribing this concept to a river bed, one has to take into account the size distribution, the exposition, and last but not least the arrangement in the bed surface layer (Chap. 1, Sect. 1.2.2.2). For a uniform size distribution ε approaches a fixed value proportional to d_s,

$$\varepsilon \propto d_s \quad \therefore \quad \varepsilon = c d_s \wedge \varepsilon < d_s \tag{2.36}$$

However if the bed has a grain size distribution, it is not appropriate to use a single ε in Eq. 2.35. One has to formulate a mean $\langle\varepsilon\rangle$, which needs calibration through experiment. Usually this is done by flow experiments through a pipe with an inner diameter D, since in such a configuration the friction force can easily be measured by the pressure loss over a given pipe length L

$$-\frac{\Delta p}{\rho g} = \lambda \frac{L}{D}\frac{U^2}{2g}, \qquad \lambda = \lambda\left(\mathrm{Re}, \frac{\langle\varepsilon\rangle}{D}\right), \qquad \mathrm{Re} = \frac{UD}{\nu} \tag{2.37}$$

and the friction factor λ can be evaluated. The relation between l and $\langle\varepsilon\rangle$ was given by Moody (1944) and his results, compiled in the Moody-diagram, can be found in practically every textbook (Chap. 14, Appendix 14.3.2). A first calibration is achieved by measuring a laminar flow in a pipe with completely smooth walls, the result is

$$\lambda = \frac{64}{\mathrm{Re}} \tag{2.38}$$

For turbulent flows in so-called sand-rough pipes the friction factor becomes (Schlichting, 1958)

$$\lambda = 8\frac{u_\tau^2}{U^2} \tag{2.39}$$

The concept of an equivalent sand-roughness k_s was introduced by Nikuradse (1933). He evaluated a virtual particle diameter from particles glued as a monolayer to the inner wall of a pipe, which could be matched in resistance to the riverbed under discussion. Therefore k_s is not a measure for the exposition, and k_s can be replaced in good approximation by d_s, respectively its mean value $\langle d_s\rangle$.

$$k_s = O(d_s) \tag{2.40}$$

It is evident that such a definition makes sense only if the differences within a size distributions are not too big. Since the size distribution is so important, one finds all kinds of definitions in the literature depending on the case under study. One often-used definition is for example,

$$k_s \approx d_{s(65)} \tag{2.41}$$

With k_s instead of ε respectively $\langle\varepsilon\rangle$ in Eq. 2.35, three flow regimes can be defined. The regime in which the wall near zone is still dominated by viscous forces is called hydraulically smooth, because the roughness is too small to act as elevations producing a flow separation. For the hydraulically smooth case, the grains are embedded in the viscous sublayer Eq. 2.42a. From Eq. 2.23, it follows that the thickness of the viscous sublayer decreases with increasing u_τ. Therefore, when d_s remains constant its value scaled by viscous units increases with u_τ or the flow velocity, respectively. For larger velocities, the condition of a hydraulically smooth surface cannot be fulfilled any more and even a very fine surface becomes rough in a dynamical sense. Therefore it is important to always check whether the criterion is fulfilled when one assumes hydraulically smooth conditions.

If the grains belong to the category of Eq. 2.42b then both momentum exchange mechanisms are present and for even bigger grains only the separation mechanism is relevant Eq. 242c.

$$\begin{aligned} &0 \le k_s^+ \le 5 && \lambda = f(\mathrm{Re}) && (a)\\ &5 \le k_s^+ \le 70 && \lambda = f(\mathrm{Re}, k_s/D) && (b)\\ &k_s^+ \ge 70 && \lambda = f(k_s/D) && (c) \end{aligned} \tag{2.42}$$

The change in the velocity profile is closely related to these dependencies of the grain size. An extreme example is high roughness case in which the entire buffer zone vanishes and the logarithmic profile reaches the bed. This is the so-called hydraulically rough wall condition, and it is this kind of flow for which the constant in Eq. 2.35 holds. For the case, Eq. 2.42b which covers the transition from smooth to rough walls, the constant varies and many approximations exist depending also on the height at which the fully turbulent profile starts.

For an open channel flow, D has to be replaced by an adequate length scale. The standard is to replace D by the hydraulic diameter d_H,

$$\begin{aligned} &d_H = 4\frac{A}{\mathrm{Per}}, \qquad A = \frac{\pi D^2}{4} \text{ respectively } = BH; \quad \mathrm{Per} = \pi D \text{ respectively} = B + 2H\\ &d_H = \frac{4BH}{B+2H} \end{aligned} \tag{2.43}$$

for a rectangular channel with B being the width of the channel and Per standing for the wetted perimeter. For many channels $B >> H$ and therefore d_H can often be set $d_H \approx 2H$. For a circular pipe, this definition of the hydraulic diameter results in $d_h = D$. In the literature, there exist several competing definitions for the roughness, which can be transformed from one to the other, but we will not go into this detail at this point.

The roughness conditions are relevant for the velocity distribution but they do not appear in the sediment transport formulas given by the classical models. Therefore we will focus on this issue in the framework of a new description. The questions to be answered are:

Can we treat sediment transport using a mean roughness, or do we have to investigate the transport locally with respect to roughness elements?

What are the consequences of the assumption of as fixed bed? In other words, can we neglect the feedback from the transport on the roughness?

How would a change of roughness influence the bed?

2.2 The Classical Bed-Load Theories

As already mentioned, the classical theories make use of a simplification by which the influence of turbulence on the transport can be described by the wall shear stress. This simplification starts from Eq. 2.12 and makes use of the fact that under normal discharge conditions no pressure gradient is present and that the total shear stress is given by a linear relation of the height at which one investigates the shear stress

$$\begin{aligned}
&\tau_{\mathrm{T}} = \tau_{\mathrm{w}}\left(\frac{H-z}{z}\right) \approx \tau_{\mathrm{t}} = \rho_{\mathrm{f}}\overline{u_x' u_z'} \quad \therefore \quad \frac{\partial \overline{u_x' u_z'}}{\partial z} \approx \frac{1}{\rho_{\mathrm{f}}}\frac{\tau_{\mathrm{w}}}{H} \\
&\text{Eq. (2.35)} \rightarrow \frac{\partial^2 U}{\partial z^2} = -\frac{u_\tau}{\kappa}\frac{1}{z^2} \quad \therefore \\
&\text{Eq. (2.12)} \rightarrow \frac{\mu u_\tau}{\rho_{\mathrm{f}}\kappa}\frac{1}{z^2} + \frac{1}{\rho_{\mathrm{f}}}\frac{\tau_{\mathrm{w}}}{H} = g_x = gtg\alpha \wedge \frac{\mu u_\tau}{\rho_{\mathrm{f}}\kappa}\frac{1}{z^2} \rightarrow 0\,;\, z > k_{\mathrm{s}} \\
&\frac{\tau_{\mathrm{w}}}{H} = \rho_{\mathrm{f}} gtg\alpha \quad \vee \quad \tau_{\mathrm{w}} = H\rho_{\mathrm{f}} gtg\alpha
\end{aligned} \tag{2.44}$$

The total wall shear stress is just compensating the gravity forces acting on the fluid.

2.2.1 Du Bois' Description of the Bed-Load

The first mathematical description of the bed-load was probably the one by Du Bois (1879). He derived his formula by assuming that bed-load is in fact the gliding of sediment layers of the size of one grain diameter forced by the wall shear stress at the bed surface. Assuming a linear decreasing shear stress down into the bed until the layer is reached, which cannot be moved anymore. Nowadays we know that this is not the case, and the only grains lying in the top layer are participating in the grain motion. However, when the bed-load becomes high, the particles are transported in a two-phase flow condition, and for this regime Du Bois' approximation of the form

$$q_{\mathrm{bl}} = \Psi_{\mathrm{s}}\tau_{\mathrm{w}}\left(\tau_{\mathrm{w}} - \tau_{\mathrm{c}}\right) \tag{2.45}$$

works well.

For a model like this, one can only describe the bed-load transport per time interval through a standardized cross section. The decisive advantage of this model is that, for the first time, τ_{w} was introduced as the essential parameter governing the transport. In addition, a drag stress was introduced given by the difference between the wall shear stress and the empirically evaluated shear stress at which grains start to move. The factor Ψ_{s}, which contains for example the influence of the grain size distribution, is also a quantity to be determined empirically. This formula is still used for estimation, although its physics is inadequate.

2.2.2 Meyer-Peter's and Similar Descriptions of Bed-Load

In Eq. 1.34, we indicated that the sediment transport phenomenon is the result of many interacting parameters, and one or the other parameter can be replaced by a function of some other ones. One example is the wall shear stress, which depends directly on the slope and the bed roughness. In other words, by using such relations one can formulate a series of new equations without changing the basic concept. Using dimensional analysis, one can construct a dimensionally correct potential function of the parameters used, and so a series of equations can be derived. However, all have to be calibrated by laboratory or field measurements. All these formulas are empirical ones and very useful when a prediction has to be made for a river with similar conditions as the one used to calibrate the formula. Since many of such formulas exist, one can choose the most appropriate one for the case under investigation. With this remark, we already made clear that for many practical cases one could find the proper classical formula. However, doing so, one uses a recipe, which tells little about what is really going on physically.

The most commonly used formulas are the one of Schoklitsch (1934, 1943, 1950) and the one of Meyer-Peter and Müller (1948, 1949).

2.2.2.1 Schoklitsch

$$q_{\mathrm{bl}} = 7\cdot 10^3 d_{\mathrm{s}}^{-1/2} S^{2/3}\left(Q - Q_{\mathrm{c}}\right);\quad Q_{\mathrm{c}} = \frac{1.944\cdot 10^{-5} d_{\mathrm{s}}}{S^{4/3}};\quad d_{\mathrm{s}}\,[\mathrm{mm}] \tag{2.46}$$

with Q being the discharge, and Q_{c}, the critical discharge defining the beginning of the transport. One can see that in the basic interpretation, his relation replaces the wall shear stress by the total discharge. In addition, Eq. 2.46 is not dimensionless, d_{s} has to be inserted in mm. Schoklitsch revised his formula, after calibration at some other rivers (Aare and Donau), to the final form

$$\begin{gathered} q_{\mathrm{bl}} = 2500 S^{3/2}\left(Q - Q_{\mathrm{c}}\right);\quad Q_{\mathrm{c}} = 0.26\rho'^{5/3}\frac{d_{\mathrm{s}}^{3/2}}{S^{7/6}} \\ \wedge\ \rho' = \frac{\rho_{\mathrm{s}} - \rho_{\mathrm{f}}}{\rho_{\mathrm{f}}}\quad \mathit{and}\quad \wedge\quad d_{\mathrm{s}} = d_{\mathrm{s}(40)}\,[\mathrm{m}] \end{gathered} \tag{2.47}$$

Again, the equation is not dimensionless and all values are given in the kms-System.

2.2.2.2 Meyer-Peter and Müller

In elaborated experiments the two authors developed an equation for coarser material with a variation in the densities of the grains.

$$\frac{R'S_{\mathrm{r}}}{\rho' d_{\mathrm{m}}} = \frac{Q_{\mathrm{bl}}}{Q}\left(\frac{k'_{\mathrm{s}}}{k_{\mathrm{r}}}\right)^{3/2}\frac{HS}{\rho' d_{\mathrm{m}}} = 0.047 + 0.25\left(\frac{\gamma_{\mathrm{f}}}{g}\right)^{1/3}\frac{q''^{2/3}_{\mathrm{bl}}}{\left(\gamma_{\mathrm{s}} - \gamma_{\mathrm{f}}\right) d_{\mathrm{m}}} \tag{2.48}$$

In Eq. 2.48, we see how the complexity is represented by a series of interlaced functions. It is an implicit formula where q''_{bl} stands for the weight of bed-load transported under water. Q_{bl} is the discharge responsible for the transport. The concept is that only a part of the velocity profile is responsible for the transport and a new optional parameter appears is given by a relative hydraulic radius R'

$$R' = H\frac{Q_{\mathrm{bl}}}{Q} \tag{2.49}$$

This concept also introduced a so-called roughness slope S_{r}

$$S_{\mathrm{r}} = \left(\frac{k'_{\mathrm{s}}}{k_{\mathrm{r}}}\right)^{3/2} S \quad \wedge \quad k'_{\mathrm{s}}: \text{Strickler}; \quad k_{\mathrm{r}} = \frac{26}{\left(d_{90}\right)^{\frac{1}{6}}}, \quad \left[d_{90}\right]\text{in}\left[\mathrm{m}\right]$$
$$d_{\mathrm{m}} = \frac{\sum d_{\mathrm{s}}\Delta p_{\mathrm{s}}}{100} \tag{2.50}$$

with Δp_{x} being the percentages of the respective grain fractions.

Again, the equation is not truly dimensional and contains a complicated relation of different relevant particle diameters. Furthermore instead of densities, specific weights are used. Equation 2.48 is so complex that several authors simplified it later on since the database was so scarce at the time that it made sense to use it again. Chien (1956) reformulated Eq. 2.48 in the following form

$$\frac{q_{\mathrm{bl}}}{\sqrt{(\gamma'-1)gd_{\mathrm{s}}}} = 8\left(\tau_{\mathrm{w}} - \tau_{\mathrm{c}}\right)^{3/2} \quad \wedge \quad \gamma' = \frac{\gamma_{\mathrm{s}}}{\gamma} \tag{2.51}$$

and we find again the basic format as in Eq. 2.45. Another transformation of this equation makes use of a Froude number etc.

2.2.2.3 Shields

Based on these classical thoughts, Shields (1936) took the condition of a critical wall shear stress more seriously, by carefully investigating the incipient motion and a critical value associated with it. His equation is

$$q_{\mathrm{bl}} = 10\frac{QS\left(\tau_{\mathrm{w}} - \tau_{\mathrm{c}}\right)}{\rho'\left(\gamma_{\mathrm{s}} - \gamma_{\mathrm{f}}\right)d_{\mathrm{s}}} \quad \wedge \quad \rho' = \frac{\rho_{\mathrm{s}} - \rho_{\mathrm{f}}}{\rho_{\mathrm{f}}} \tag{2.52}$$

Instead of the Froude representation of Chien it is often advantageous to use a Froude number based on viscous units, that means in relation to single grains,

$$\mathrm{Fr}_{\mathrm{v}} = \sqrt{\frac{u_{\tau}^2}{\rho' g d_{\mathrm{s}}}} \tag{2.53}$$

This thought is appropriately addressed in the Shields representation, which too is based on the reaction of single grains. With this expression, often a dimensionless transport characteristic is formulated

$$G_{\mathrm{sv}} = \frac{q_{\mathrm{sv}}}{\mathrm{Fr}_{\mathrm{v}}} \quad \wedge \quad q_{\mathrm{sv}} = \frac{q_{\mathrm{s}}}{u_{\tau}d_{\mathrm{s}}} \tag{2.54}$$

and with this Eq. 2.52 can also be written

$$G_{\mathrm{sv}} = 10\frac{U}{u_{\tau}}\left(\mathrm{Fr}_{\mathrm{v}} - \mathrm{Fr}_{\mathrm{vc}}\right) \tag{2.55}$$

2.2.2.4 A General Remark and Other Formulas

All these equations have an empirical part and therefore a calibrating component, and in addition they contain a series of parameters, which can be used for fitting. If one uses those in systems with similar boundary conditions the results can be quite good. Here are some additional references: Vollmers and Pernecker (1965) tried to minimize well-known descriptions and formulated the most simple equation

$$G_{\mathrm{sv}} = 25\mathrm{Fr}_{\mathrm{v}} - 1 \tag{2.56}$$

Engelund and Hansen (1967) introduced an energy argument by using the squares of the velocities

$$G_{\mathrm{sv}} = 0.05\left(\frac{U}{u_{\tau}}\right)^{2} \mathrm{Fr}_{\mathrm{v}} \tag{2.57}$$

Yang (1973) introduced a unit stream power US together with the settling velocity w_{s} of the grains. With w_{s}, a parameter was introduced that belonged to a new concept since with it a transport mechanism entered in more detail. The equation is rather complicated and the reader has to consult the original paper if he wants to use this description. Vanoni (1977) published additional equations of this kind, and the most elaborated is the one of Zanke (1982), where more modern approaches like a force balance on single grains are considered. All equations are missing a part that relates the transport to the turbulent regime of the flow. In so far has entered only through the wall shear stress in an averaged form. Turbulence was thought to be of minor importance, but when suspension loads had to be treated, it became evident that these theories were too simple minded.

3 Turbulence and the Statistical Aspects of the Sediment Transport

Since the Reynolds numbers in rivers are of the order of 10^6–10^7, turbulence is unavoidable and omnipresent. One of the main properties of a turbulent flow under identical initial- and boundary conditions is that the flow is asymptotic, which means they are identical in the mean flow characteristics. Assuming a potential flow, this can be used for turbulent flows to evaluate some integral properties. For the sediment transport, however this approach is rather problematic since the turbulent fluctuations in the flow field are interacting with the single grains. The scales of the particles with respect to the turbulent fluctuations determine to which extent these interactions become relevant.

So far we introduced turbulence just by the mean value of the most significant Reynolds stress component Eq. 2.11. This reduction is too simple, and it is also obvious that turbulence has to be treated in a statistical form. Therefore, the skill of combining turbulence with a theory of sediment transport is finding the middle way between a statistical and a deterministic description. This condenses to the question to which extent we have to study micromechanical processes in a mechanistic form and when can we determine the result by a statistical treatment.

In this chapter, we will discuss the fundamentals of a statistical approach as well as already existing statistical theories.

3.1 The Incipient Motion

Through the introduction of a critical wall shear stress τ_c, one already used the concept of a threshold level needed to move a grain that is embedded in the top layer of a bed. The wall shear stress generally is an averaged value, which is a constant for the given flow. However, this mean value usually is not large enough to move a single grain. Therefore, there must be events in the flow, which locally cause a much higher force on the bed. It is clear that such forces result from the turbulent fluctuations. Measurements show that these fluctuations cause variations in local shear stress by an order of magnitude of the mean. The maximum instantaneous Reynolds stress is about 16 stronger than the mean value

$$\left(u'_x u'_z\right)_{\max} \gg 16\overline{u'_x u'_z} \tag{3.1}$$

Using this statement, Gessler (1965) explained the armoring process of beds. It is important to recognize that one can define a wall shear using the mean Reynolds stress, however this cannot be a local instantaneous value since the area on which the velocity fluctuation acts is unknown. This deficiency can be formulated through the following question.

What are the relations between the Reynolds stresses and the forces of the flow acting on the bed?

Defining the threshold value for the incipient motion is just as complex as describing the configuration of the grains on the bed surface. Therefore, it has been investigated by many authors through numerous projects in the last decades. In practice, this task necessitates a very large database because every sediment transport formula needs this critical threshold value in one form or the other.

A series of authors tried to derive formulas for this value. For example, one series can be found in Zanke (1982). Most of the data we use today were evaluated by Shields (1936) using uniform grains of different size of a noncohesive material. He found that for an incipient motion, the Reynolds number and the Froude number based on viscous parameters

$$Re_v = \frac{u_\tau d_s}{\nu} \quad \vee \quad Re_{ks} = \frac{u_\tau k_s}{\nu} \quad \wedge \quad \mathrm{Fr}_v = \sqrt{\frac{u_\tau^2}{\rho' g d_s}} \tag{3.2}$$

are correlated as shown in Fig. 3.1, and with Re_v taking turbulence into account by appointing a value with a small scattering to the critical value.

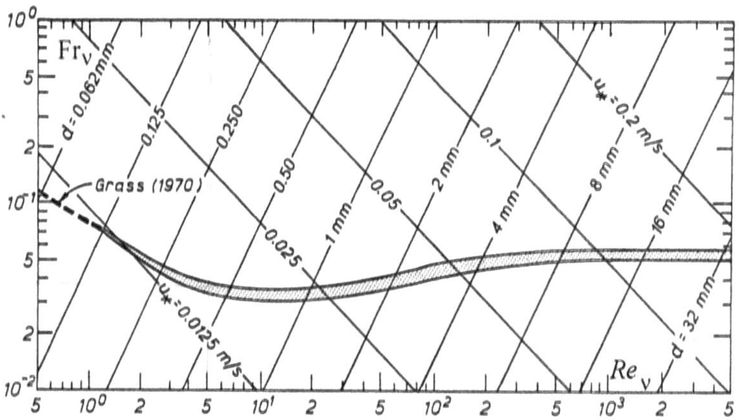

Fig. 3.1 The Shields diagram for the incipient motion as often reproduced with supplementary measurements by Yalin and Karahan (1979). Particles above the critical gray zone are in motion. In this plot alternative notations are used, which are often found in the literature ($\mathrm{Fr}_* = \mathrm{Fr}_v$; $Re_* = Re_v$; $d = d_s$; $u_* = u_\tau$)

Bonnefille (1963) replaced the Froude number by a sedimentological grain diameter D^* and showed that the Reynolds number is a pure function of this parameter

$$Re_v = f(D^*), \qquad D^* = \left(\frac{Re_v^2}{\mathrm{Fr}_v}\right)^{1/3} = \left(\frac{\rho' g}{\nu}\right)^{1/3} d_s \tag{3.3}$$

More elaborate representations are given by Vollmers and Pernecker (1967) who also incorporated adhesive material; in such a case it could be shown that D^* depends on the investigated material and can be described approximately by

$$D^* = c\,\mathrm{Re}_v^n \tag{3.4}$$

For adhesion- and cohesion-free fine material, the values are $c = 2.15$ and $n = 1$. Further information can be found in Yalin (1972), Bogardi (1974) and Graf (1971). A helpful list is given in Table 3.1, with Φ being the angle of repose, discussed in Chap. 1 (Sect. 1.2.2.2).

Table 3.1 Angle of repose

$d_{fs} = d_s \left(\frac{(\gamma'-1)g}{\nu^2} \right)$	τ_c
<0.3	$0.5\tan\Phi$
$0.3<d_{fs}<19$	$0.25d\ d_{fs}^{-0.6} tg\Phi$
$19< d_{fs}<50$	$0.013\ d_{fs}^{0.4} tg\Phi$
$d_{fs} >50$	$0.06 tg\Phi$

Since these critical wall shear stresses were evaluated with uniform grain sizes, the question remains how to formulate critical values for a mixed grain distribution. Julien (1995) gave the following approximation

$$tg\Phi = \frac{d_{s1}}{\sqrt{(d_{s1}+d_{s2})^2 - 2d_{s1}^2}} = \sqrt{\frac{1}{(1+d_{s2}/d_{s1})^2 - 2}} \tag{3.5}$$

The material is assumed to consist of two classes of spherical particles, those of class 2 sitting in top of the one of class 1, which means a grain of class 2 sits on top of four grains of class 1. For $d_{s1} = d_{s2}$ Φ is 35.3°, it decreases when $d_{s2} > d_{s1}$ and increases for fine particles $d_{s2} < d_{s1}$ until $d_{s2} = 0.41 d_{s1}$. More detail on the angle of repose for granular material is presented in Simons (1957).

Naturally, the smallest grains are the ones which are the first to be transported; therefore, it is possible that the transport stops after this fine partition has been transported away. We will come back to this case later. The Shields results are normally used for the fraction of the smallest grains. This is however incorrect since the fine material is deposited preferentially behind larger grains. The surface layer becomes rougher and is finally composed of bigger grains, whereas the layer beneath becomes finer in grain size. The result of such a process can be an incipient rolling transport of bigger grains, although they would not move due to the roughness of the bed (Julien et al., 1993).

In the case of graded sediment mixtures, the incipient motion of a particle of size d_{si} in a mixture of average grain size $d_{s(50)}$ is given by the critical Shields parameter

$$d_{si} \in \text{Sed}(d_{50}), \quad \tau_{w^*ci} = \frac{\tau_{wci}}{(\gamma_s - \gamma_m) d_{si}} \tag{3.6}$$

for the fraction i with reference to the critical dimensionless shear stress

$$\tau_{w^*c50} \cong 0.083 \tag{3.7}$$

at which the grain size $d_{s(50)}$ starts to move. The relationship of $\tau_{w^*ci} / \tau_{wc(50)}$ and the grain size ratio $d_{si} / d_{s(50)}$ are shown in Fig. 3.2. The data are based on literature of Tsujimoto (1992) and Garde and Ranga Raju (1985) and show in a double logarithmic representation a straight line with some scatter at the extremes.

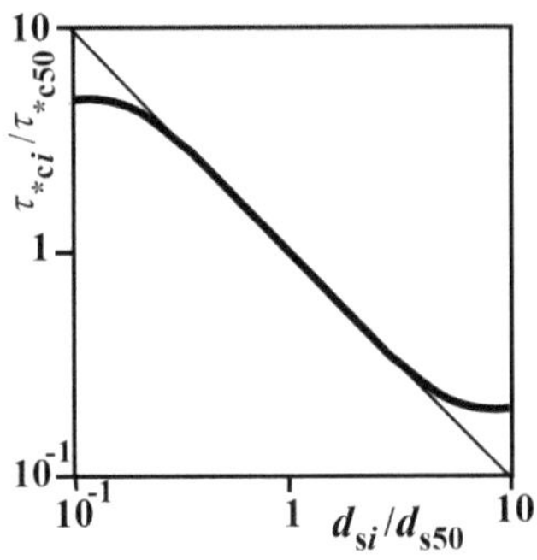

Fig. 3.2 Critical wall shear stress for sediment mixtures

The critical wall shear stress τ_{c12} for a particle resting on a bed of particles of diameter d_{s1} can be compared with the critical value τ_{c1} for uniform grains of size d_{s1}, and from Eq. 3.5 we find

$$\frac{\tau_{c21}}{\tau_{c1}} = \frac{d_{s2}}{d_{s1}}\sqrt{\frac{2}{\left(1+d_{s2}/d_{s1}\right)^{2}-2}} \tag{3.8}$$

If one plots the situation for $d_{s2} = d_{si}$ and $d_{s1} = d_{s1(50)}$ in Fig. 3.2, one recognizes that the particles of diameter $d_{s2} > d_{s1}$ are almost as mobile as particles of the same size, $d_{s2} = d_{s1}$.

Within this frame, one can formulate an equal concept of mobility, with respect to the median given by

$$\tau_{*ci}d_{si} = \tau_{*c50}d_{50} \tag{3.9}$$

3.1.1 The Forces Acting on a Single Grain

After these more empirical considerations, we turn to a more mechanistic view by which the forces and angular moments acting on a single grain embedded in the surface layer of the bed are investigated in case we know the arrangement. Then we can calculate the criteria for the forces and angular moments which are necessary to move the grain by the flow (Fig. 3.3a and b).

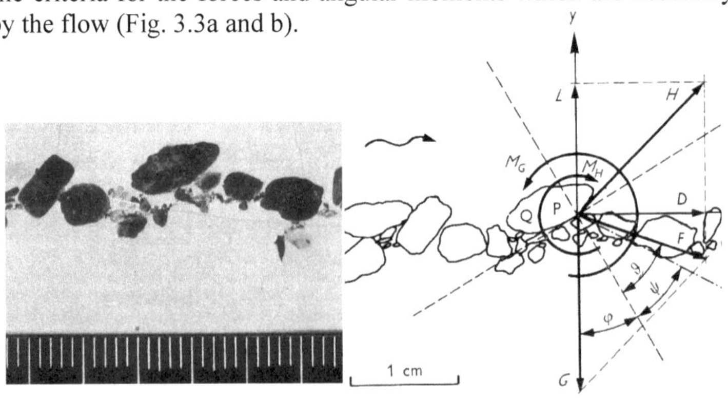

Fig. 3.3 (a) A cut through a stable bed configuration produced by the self-organizing process of the transport, a photographic view, from Günther (1971). (b) The representation of the forces and the angular moments acting on the chosen grain

Forces independent of the flow are the weight $\underline{G}_s$ and the hydrostatic buoyancy $\underline{L}_f$. Since these two forces are acting in opposite direction, it is a standard procedure to comprise them in a single force, the weight of the submerged body, given in the text by $\underline{G}$.

The hydrodynamical force $\underline{H}$ has to overcome $\underline{G}$ and coupled with it the friction force $\underline{F}$. By splitting $\underline{H}$ into a horizontal drag force $\underline{D}$, and a vertical lift force $\underline{L}$, we can formulate the destabilization by two different motions. First by rolling, and second by gliding, these two types of initial motion are ending up in a saltation. For both case, the crucial parameter is the elevation of the highest point at which the grain reposes since the grain has to turn around this point due to angular momentum or has to glide tangentially over this point of support, which is at a gliding angle φ to the vertical.

On the grain, two forces and angular momentum pairs are acting ($\underline{G}$,$\underline{M}_G$) and ($\underline{H}$,$\underline{M}_H$), and the equilibrium remains undisturbed, if one of the two criteria (Eqs. 3.10 and 3.11) is fulfilled.

The rolling condition is defined as

$$\underline{F} = \underline{H} + \underline{G}, \qquad M_G + M_H = M_F < 0 \tag{3.10}$$

where the angular momentum $\underline{M}$ is defined as positive in the counter clockwise direction and where the angular momentum due to friction was neglected. When the point of contact P is shifted toward Q, then the roll instability increases.

The gliding criterion is

$$(\underline{F}, \underline{n}_t) = Fn_t \cos\vartheta, \quad \vartheta > \psi, \quad \psi \approx \Phi \text{ Fig.3.1b} \tag{3.11}$$

with ϑ the friction angle and $\underline{n}_t$ the unit vector normal on the tangent $\underline{t}$.

Since the grains usually have convex surfaces, P has the largest horizontal distance from the center of gravity so that the stabilizing moment due to weight becomes extreme. Also the tangential plane at P has the highest inclination φ so that the reactive forces opposing movement are maximal. To formulate the equilibrium conditions, we consider the 2D flat case. This simplification disregards a certain jamming effect between the concave seat and the grain so that the grain is less stable in the 2D case.

One finds several approximations in the literature for Eq. 3.10, e.g., Raudkivi (1982) for uniform spherical particles with a high exposure

$$\pi \frac{d_s^3}{6} g(\rho_s - \rho_f) \frac{d_s}{2} \sin\alpha = \beta \rho u_\tau^2 \pi \frac{d_s^2}{4} \frac{d_s}{2} \cos\alpha \wedge \alpha = \varphi + \psi \tag{3.12a}$$

$$\tau_{wc} A = \rho u_\tau^2 \pi \frac{d_s^2}{4}$$

$$\therefore \quad u_{\tau c} = \sqrt{\frac{2}{3} \frac{tg\alpha}{\beta}} \sqrt{g\rho' d_s} = B\sqrt{g\rho' d_s} \tag{3.12b}$$

with β being a parameter accounting for the exposure. With a logarithmic velocity profile the critical value $u_{\tau c}$ at height z becomes

$$u_{\tau c} = 5.75 B \sqrt{g\rho' d} \ln \frac{z}{z'} \tag{3.13}$$

where z' is a fictitious origin of the logarithmic profile given by the extrapolation of the profile to its intersection with the vertical axis z. We will discuss this problem later.

Essential for the whole is the experimental evaluation of $\underline{H}$ because it accounts for the instantaneous flow separation on the grain.

Often it is assumed that the grains are exposed to a parallel flow or a laminar boundary layer flow condition. Such approximations show that L scaled by G is

$$\frac{L}{G} \geq \frac{tg\alpha}{1+tg\alpha} = 0.73 \vee \frac{L}{G} \approx 0.7 + 1 \tag{3.14}$$

if ($\varphi + \psi = \alpha$) is smaller than 45°. That means to move a grain a relatively large force produced by the low pressure of the separating flow must occur in a turbulent flow (Müller et al., 1971).

From this estimate it became evident that a consideration of a high critical shear stress as introduced by Gessler (1965) is not sufficient. It is necessary to find a relation between the instantaneous momentum acting on the grain and the pressure difference Δp acting on the particles. The pressure field has a spatial extension, and therefore a 2D or a 1D representation of the flow cannot describe the required forces. What is needed is a pressure distribution at the bed. Measurements of this kind are very rare and mainly limited to smooth surfaces, where the pressure fluctuations are the results of coherent structures, which we will discuss later. These structures are related to the vorticity production as it was described, e.g., by Eckelmann et al. (1977), a regime for which the wall pressure distribution was measured by Emmerling (1973) and Dinkelacker et al. (1977) who showed that between p and τ_w the following relation exists

$$\sqrt{\overline{p'^2}} = p' = \Delta p \approx 3\tau_w, \qquad p'_{max} \approx 18\tau_w \tag{3.15}$$

For an incipient motion, the lift force can be set equal to the weight and we find

$$\left(\pi \frac{d_s^2}{4}\right) 18\tau_w = g\rho'\pi \frac{d_s^3}{6} \quad \therefore \quad \frac{\tau_w}{g\rho' d_s} = Fr_v = 0.037 \tag{3.16}$$

a value, which is in good agreement with the critical value of the Shields curve. This is not surprising since Shields evaluated his curve using uniform spherical particles.

3.2 Statistical Bed-Load Models

Based on the classical concepts together with a closer view on the incipient motion, a series of statistical models are derived. Here, however, we discuss only those which were important for the further progress.

3.2.1 A First Statistical Formula

Kalinske (1947) tried to introduce the turbulent fluctuations instead of a mean wall shear stress. In his formula, he distinguished between grains of different specific weight. His transport formula is given as

$$\tilde{q}_s = \gamma_s A_V d_s^3 \overline{v}_s N = \alpha_s A_s \gamma_s d_s \overline{v}_s \tag{3.17}$$

with $\tilde{q}_s$ being the transport per unit time and width,

$$\tilde{q}_s = q_s \gamma_s p_o \quad \wedge \quad p_o = \frac{V_s}{V} \tag{3.18}$$

with the porosity p_o given by V_s the volume of the solid without voids, and V the bulk volume with voids. A_V is a volume factor of the grains, N the number of grains of a unity area of the bed in motion, A_s is the percentage of the bed covered by this class of grains, $\bar{u}_s$ the mean velocity of a transported grain, and α a grain factor. Not having enough data, Kalinske fixed the drag as $U/u_\tau = 11$, and assumed that at these conditions 35% of the grains on the bed is in motion. In addition, he assumed that the turbulent velocity fluctuations have a Gaussian distribution. With all these restriction he finally finds

$$\tilde{q}_s = 2.5 u_\tau \gamma_s d_s f\left(\frac{\tau_c}{\tau_w}\right) \tag{3.19}$$

The experimental input is considerable, and Kalinske has provided a plot of τ_c / τ_w against $\tilde{q}_s / \gamma_s u_\tau d_s$.

3.2.2 Einstein's Bed-Load Formula

A paradigm change became evident as the turbulence had to be incorporated in a more explicit form. It was just not enough to take care of the turbulent fluctuations by a mean wall shear stress. It is the merit of Einstein (1942, 1950) who realized this gap in the transport theories and started incorporating turbulence. However because of the lack of a good turbulence theory at that time, he introduced turbulence indirectly through statistical laws based on observations rather than a theoretical deduction. In this form, his transport equations are purely probabilistic and are therefore different from the representation through statistical mechanics used in the deduction of thermodynamical laws. The state is described by the statistic occurrence of individual interactions, which have to be defined by statistical rules. Einstein's law is defined by observations and given by the following rules:

1. The probability that hydrodynamic forces move a single grain on the bed depends on ρ_s, d_s, grain form, and the instantaneous flow condition of the near-wall flow field. The arrangement and the history of their deposition can be neglected.
2. The particle moves when the resulting forces acting on the grain are strong enough. In his opinion, this is the case when $\underline{L}$ is larger than $\underline{G}$, a result based on the investigation by Einstein and El Samni (1949) ($\underline{L} \approx \underline{G}$)
3. A particle in motion can be deposited again on the bed if at that location $L < G$. However, Einstein makes the assumption that the deposition probability is everywhere the same, an assumption which provoked a controversial discussion.
4. The transport distance is independent of the flow conditions and for spherical material approximately $100d_s$. This is an empirical result and has to be discussed later in more detail since it allows some conclusions on the structure of the flow field.

These rules postulate a spatial unit for the transport of 100 grain diameters. It is a homogeneous theory, and feedback mechanisms are only feasible if one is introducing the influence of a deposition on the flow field in this area.

Based on these assumptions, Einstein defines the probability P that a grain becomes removed from the bed as

$$\frac{P}{1-P} = A_{\mathrm{E}} \Phi_{\mathrm{s}} \tag{3.20}$$

which is a very compact formulation. Since Einstein's formula is an interlaced con-truction, it can only be understood when all the implicitly involved functions are discussed. Einstein calls Φ_{s} "the intensity of the sediment transport." He postulated that this function could be used as criterion to investigate the dynamical similarity of two rivers exhibiting bed-load transport. It is given by

$$\Phi_{\mathrm{s}} = \frac{\tilde{q}_{\mathrm{rs}}}{\rho_{\mathrm{s}} g} \sqrt{\frac{\rho_{\mathrm{f}}}{(\rho_{\mathrm{s}} - \rho_{\mathrm{f}})}} \sqrt{\frac{1}{g d_{\mathrm{s}}^3}} = G_{\mathrm{sv}} \mathrm{Fr}_{\mathrm{v}}^{3/2} \quad \wedge \quad \text{Eqs. 2.54 and 2.53}$$
$$\wedge \quad \tilde{q}_{\mathrm{rs}} = \tilde{q}_{\mathrm{s}} \left(\frac{\rho_{\mathrm{s}} - \rho_{\mathrm{f}}}{\rho_{\mathrm{s}}} \right) \tag{3.21}$$

In this function, the entire dynamical contribution of the intensity of the sediment transport is hidden in $\tilde{q}_{\mathrm{s}}$, and therefore Einstein had to introduce an additional function Ψ_{s} which he called the flow intensity and which has the following form for uniform grain size

$$\Psi_{\mathrm{s}} = \frac{\rho' d_{\mathrm{s}}}{R' S} = \frac{1}{\mathrm{Fr}_{\mathrm{v}}} \tag{3.22}$$

From Eq. 3.21 it is evident that

$$\Phi_{\mathrm{s}} = f(\Psi_{\mathrm{s}}) \quad \vee \quad \Phi_{\mathrm{s*}} = f(\Psi_{\mathrm{s*}}) \quad \wedge \quad \Phi_{\mathrm{s*}} = \left(\frac{i_{\mathrm{B}}}{i_{\mathrm{b}}} \right) \Phi_{\mathrm{s}} \tag{3.23}$$

where the second form applies to mixtures, where i_{B} stands for the fraction of the grains i in the grain distribution and i_{b} for the fraction of the same material in the bed surface, and

$$\Psi_{\mathrm{s*}} = \xi Y \left(\frac{\beta}{\beta_x} \right)^2 \Psi_{\mathrm{s}} \tag{3.24}$$

The real stochastic element, however, was introduced by the probability of a grain being removed from the bed. Einstein formulates this probability by a new function containing several new parameters among them the ratio of L/G_{r}, where G_{r} is the relative weight

$$P = 1 - \frac{1}{\sqrt{\pi}} \int_{-B_{\mathrm{s}} \Psi_{\mathrm{s}} - 1/\eta_0}^{B_{\mathrm{s}} \Psi_{\mathrm{s}} - 1/\eta_0} \mathrm{e}^{-t'^2} \mathrm{d}t' \tag{3.25}$$

He postulated a Gaussian distribution with the constants B_s and η_0. With Eq. 3.25 inserted into Eq. 3.21 and making use of Eq. 3.22, one finds

$$1-\frac{1}{\sqrt{\pi}}\int_{-B_s\Psi_s-1/\eta_0}^{B_s\Psi_s-1/\eta_0} e^{-t'^2}\,dt' = \frac{A_E\Phi_s}{1+A_E\Phi_s}$$
$$\wedge\quad A_E = 43.5;\quad B_s = 0.143;\quad \eta_0 = 1/2 \tag{3.26}$$

Because of Eq. 3.23, this rather complicated expression is usually given by a graphical representation of Ψ_s versus Φ_s.

With Eq. 3.24 Einstein introduced a new idea, which goes beyond the purely statistical description of turbulence. It includes statistics of grain positions, and it is this supplement, which bridges between the classical and the modern representations. The so-called hiding factor ξ is revolutionary because it states that in addition to the turbulent fluctuations the position with respect to the neighboring grains is responsible for the transport. In other words, the separation was introduced. The statistical representation of this wake positioning needs a complex description, by an interlaced series of dependencies, which have to be inserted into Eq. 3.24

$$\begin{aligned}
&\xi = \frac{d_s}{X} \quad \rightarrow \quad \text{given often in graphical form}\\
&X = 0.77\Delta \quad \wedge \quad \Delta/\delta' > 1.8\\
&X = 1.39\delta' \quad \wedge \quad \Delta/\delta' < 1.8\\
&\delta' = \frac{11.5\nu}{u_\tau}\\
&\Delta = \frac{d_{si}}{x} \quad \wedge \quad x = f\left(\frac{d_{si}}{\delta'}\right) \quad \text{given in graphical form}\\
&Y = f\left(\frac{d_{s65}}{\delta'}\right) \quad \text{given in graphical form}\\
&\beta = \ln 10.6 \quad \beta_x = \ln\left(\frac{10.6X}{\Delta}\right)
\end{aligned} \tag{3.27}$$

Einstein expanded this bed-load equation later to a complete sediment transport formula including the suspended part of the transport.

One parameter in Einstein's representation has to be stressed prominently, it is δ a characteristic thickness for the near-wall zone of the flow, something like a boundary layer thickness. We will see later that the value $\delta'^+ = 11.5$ has a physical meaning, which Einstein did not know at his time, however, the good values confirms the high quality of his calibration measurements.

Although his representation is to a high extent semiempirical one needs several datasets, which are given in the Appendix 13.4. For fine material, Einstein's formula was very predictive, however, also complicated, and several authors tried to simplify it. Brown (1950), e.g., showed that using the most available datasets a simple relation could be formulated

$$\Phi = 40\left(\frac{1}{\Psi}\right)^3 \tag{3.28}$$

or by Vollmers and Pernecker (1965)

$$\frac{\tilde{q}_{rs}}{\rho_s g} = 25\mathrm{Fr}_v\left(\mathrm{Fr}_v - 1\right) \quad \vee \quad G_{sv} = 25\mathrm{Fr}_v - 1 \tag{3.29}$$

To the same category of simplifications belongs the Eq. 2.57 of Engelund and Hansen (1967)

$$2\left(\frac{u_\tau}{\overline{U}}\right)^2 \Phi = 0.1\mathrm{Fr}_v^{5/2} \quad \wedge \quad \Phi = \frac{\tilde{q}_s}{\rho_s g}\sqrt{\frac{1}{\rho' g d_s^3}} \rightarrow Eq.2.57 \tag{3.30}$$

or Bagnold for the transport in air

$$\Phi = AB\mathrm{Fr}_v^{1/2}\left(\mathrm{Fr}_v - \mathrm{Fr}_{vc}\right) \tag{3.31}$$

where A and B have to be found empirically.

3.2.3 The Concept of Grass

Since not enough datasets of the pressure distribution on the bed were available, Grass (1970) proposed using a statistical description of the forces on a single grain getting it into motion. He used two probabilities. One for the distribution of the forces and a second one for the forces needed to put a grain in motion. In the tradition of the classical equations he used τ_w and τ_{wc} as the probabilistic parameters, where τ_{wc} represents an individually needed force. This force can be described by a wall shear stress, which is highly correlated with the grain diameter. However, using a distribution of the wall shear stress as a result of turbulent fluctuations is very problematic since a mean value such as $\langle\tau_w\rangle$, has no fluctuations. In other words, τ_w has to be understood as a much more local or instantaneous parameter. Since we assume a homogeneous bed, the new formulation stands for the conversion

$$\left\langle u_x' u_z' \right\rangle \approx \tau_w \Rightarrow P\left(u_x' u_z'\right) \tag{3.32}$$

and uses the same thoughts as Gessler (1965). But even in this form it remains problematic since not the value $u_x' u_z'$ moves the grain but rather a kind of integral value of this stress over the exposed grain area. Such a probabilistic concept would need at least an additional parameter. The pressure $p(\underline{x}, t)$ is not a direct function of $u_x' u_z'$, and neither is a force acting on the grain, for which the interaction time would be needed also.

However, the concept of Grass is very instructive and perceptive since it is evident that incipient motion occurs where the two probability distributions overlap (Fig. 3.4).

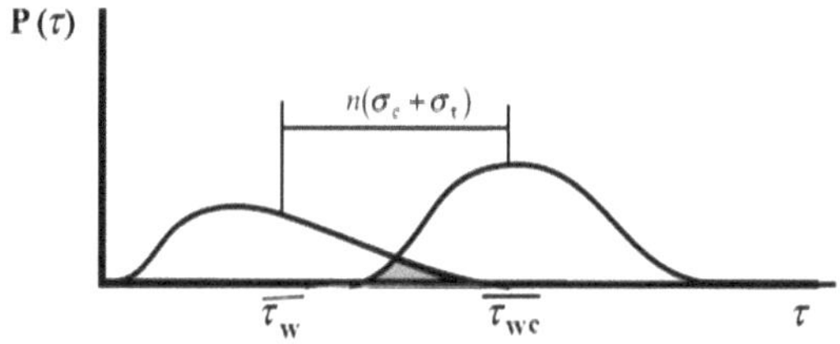

Fig. 3.4 Probability distributions of the wall shear stress τ_w and τ_{wc} due to the flow and the distribution of the resistance of the grains against wall shear stresses

The characteristic values of the probability distributions are their deviations, here σ_τ and σ_c, and they are used to characterize the distance of the two mean values

$$\overline{\tau}_{wc} - \overline{\tau}_w = n(\sigma_c + \sigma_\tau) \qquad (3.33)$$

From measurements, the mean and the deviations are known as

$$\frac{\sigma_\tau}{\overline{\overline{\tau}}_w} = 0.4, \quad \frac{\sigma_c}{\overline{\overline{\tau}}_{wc}} = 0.3 \quad \therefore \quad \overline{\tau}_w = \overline{\tau}_{wc}\frac{1-0.3n}{1+0.4n} \qquad (3.34)$$

Therefore, it is possible to compare the description of Grass and Shields. It is found that for $n = 0.625$ the two are equivalent.

One of the advantages of Grass' formulation is the possibility of recognizing the change of $\overline{\tau}_{wc}$ due to a distribution in the grain size. Very low transport exists for $n = 1$, which defines an additional stability criterion based on a very low degree of overlap of the two probability distributions

$$n = 1 \qquad (3.35)$$

With this representation, we come to an end of the "classical theories," and our interest must now be to find theories in which turbulence is introduced in a more modern form. Our goal is to receive a theory based on the local pressure distribution on the bed.

3.2.4. The Concept of Coherent Structures

The situation changed drastically as it was recognized that turbulent flows contain so-called coherent structures. In other words, a turbulent regime is at least partially composed of flow structures, which can be described in a deterministic form. Therefore the statistical description becomes simplified since now it can be given in the form of a statistic of coherent structures. Such a description introduces a new change in the paradigm that turbulence can only be formulated as a statistical quality. This statement will have to be replaced.

The theory of coherent structure applies to the deterministic turbulence elements in the boundary layer flow over smooth walls, for further detail Holmes et al. (1966) is recommended. Since we know that the boundary layer flow is strongly related to transport, it became evident that the results of the new theory must be relevant for the sediment transport. This has been recognized, and a series of new models were formulated. However, often they could not be dynamically confirmed, or good models were interpreted wrongly. Therefore, we have to introduce the concept of coherent structures more carefully. Most of the properties of coherent structures can be found in Robinson (1989), Kline and Robinson (1990), which are only two publications out of a series published on this matter in the late 1980s and early 1990s.

The main idea is that the viscous sublayer of a boundary layer flow grows in time until an instability process ends up in a burst which releases concentrated vorticity to the outer turbulent flow. It is the mechanism by which the turbulent flow is fed with vorticity, which can be produced at the wall of the flow only. After the burst the viscous layer recovers, a layer appears very near the wall growing again. In other words, the flow is fed periodically with vorticity mainly through the form of vortices. This idea was postulated by Einstein and Li (1958a), long before coherent structures became fashionable. Since the destabilizing process always needs about the same time, the bursting became quasi-periodic and the released vortices became very similar, therefore the name coherent structures.

In fact such elements were detected experimentally, and launched an intense discussion, not so much on their existence but on their dynamical relevance for the turbulent flow. However, since such structures can be separated from the background turbulence, the temptation is strong also to construct models which are dynamically not consistent, in other words they do not fulfill the Navier–Stokes equation, which always has to be demanded. Therefore, we restrict ourselves to models that are dynamically consistent.

Depending from the method of investigation, two classes of coherent structures were found

1. Longitudinal vortices in flow direction, slightly raised and connected at the forward end by a weak lateral vortex bow.
2. A series of ordered vortices, which due to their form are called Λ- or hairpin-vortices organized in turbulent spots. They first were visualized and documented by Head and Bandyopadhyay (1981).

Since both forms exist, they must be related by a mechanism. Several models were postulated, but none of them was able to satisfy completely. However, numerical simulations helped a lot to formulate consistent pictures of what goes on in the near-wall region. A first finding was that the longitudinal vortices could not be explained by a nonlinear development of a Tollmien-Schlichting wave (Klebanoff et al., 1962). A description compatible with Tollmien-Schlichting waves was found by Benney and Lin (1960) and Benney (1961) by superposing on the primary wave a 3D one. Such a system produces the longitudinal vortices as they have been observed. In competition with this model Holmes et al. (1996) propagate the idea that the longitudinal vortices are structures as found in Langmuir cells. The mean velocity profile produces lateral vorticity Eq. 2.24, whereas the vorticity lines can in good approximation assumed to be transported as material lines, and they are disturbed by the turbulent fluctuations. In Lagrangean representation the convectional velocity in x-direction has a gradient in z-direction, therefore the lines raised by the disturbances are transported more rapidly than those turning toward the wall. By this process the vorticity lines get tilted and elongated and a Ω_x component results. The stretching due to the velocity gradient leads to further intensification (Fig 3.5). These are the vortices producing the rolls searched for. However in contradiction to most descriptions, they rarely form symmetric vortex pairs (Adrian, 2000).

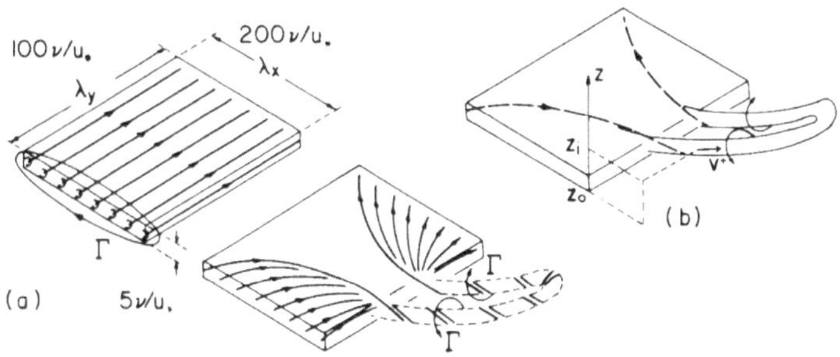

Fig. 3.5 Schematic representation how a Λ-vortex forms out of lateral near-wall vorticity lines, and how they are stretched by the flow field

The roll-model for the coherent structures was introduced first by Blackwelder and Eckelmann (1979) based on the dataset of Blackwelder and Kaplan (1976). The numerical simulation by Leonard (1980) helped introducing this model as a standard description for long time. This model is very similar to the one of Holmes et al., however, differs in the breakdown process. Blackwelder and Eckelmann thought that a vortex embedded on one side in the boundary layer transports fluid of high velocity toward the bed and on the other side pumps low velocity fluid into the outer flow. This fluid of low velocity must first be accelerated. In this phase of the process the fluid of low velocity forms zones, the so-called low speed streaks, which become unstable. This can best be seen in a (x, z)-plane through these zones, where we recognize that the velocity profile has an inflexion point, and is therefore instable. The stimulated instability ends up in an eruptive local breakdown. This process is called a burst, which will occur in a quasi-periodic interval because the rolls reform after such an eruption. Therefore, a mean period of the bursts can be determined, given by the interval time T_B of the bursts, which can be thought of as an element of the near-wall intermittency of the turbulent flow. This model was improved by Holmes et al. based on the results found by Hamilton et al. (1995). The latter found that the velocity profile has inflexion points on both sides. These inflexions produce a secondary instability, which after Lundbladh et al. (1994) stimulate the first disturbance. A view of this process is shown in Fig. 3.6.

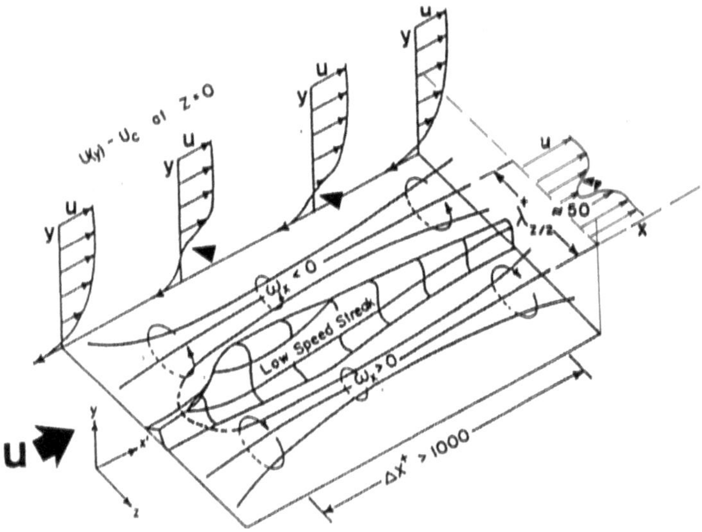

Fig. 3.6 A schematic explanation how burst events are produced due to instability processes on low-speed streaks, produced by a system of longitudinal rolls. The explanation is given in the text. Adapted from Blackwelder and Eckelmann (1979)

These investigations yield

$$T_{\mathrm{B}} = 3.6 \cdot 10^{2} \frac{\nu}{u_{\tau}^{2}} \qquad C_{\mathrm{D}} = 7.2 \cdot 10^{2} \frac{\nu}{T_{\mathrm{B}} U_{\mathrm{s}}^{2}} \tag{3.36}$$

and Holmes et al. pointed out that T_{B} and C_{D} were inversely proportional, which elucidates the observation that change in the turbulent structures can lead to drag reduction.

In competition to this roll-model, Perry postulated a process of local vortex separation at the bed. With this concept a very difficult problem was approached. Because of the nonslip condition of Eq. 2.16 we get

$$\Omega_{z}\big|_{\mathrm{w}} \equiv 0 \tag{3.37}$$

Therefore no vortex line can end at the wall. This fact was discussed at length by Lighthill (1963) because Eq. 2.47 contradicts one of the laws of Helmholtz, by which a vortex line must be closed or end up on a wall. Viscosity plays a major roll and it was Lim et al. (1980) who could show—using a linearized Navier–Stokes equation–that vortex lines can detach from the wall in the form of a viscous tornado, or in Lighthill's terminology in a critical point with complex eigenvalue solutions. The mechanism is reproduced in the Fig. 3.7.

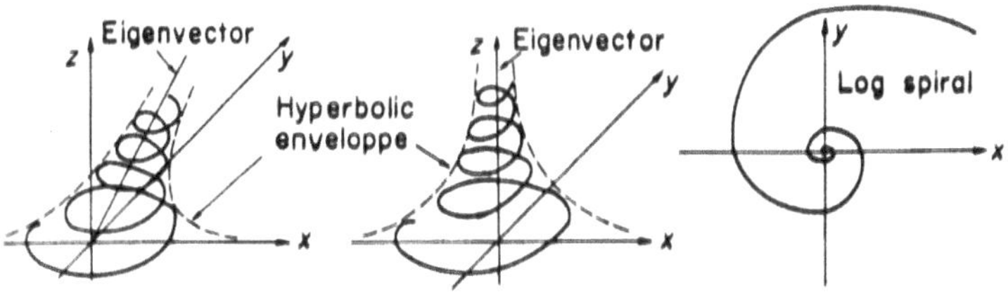

Fig. 3.7 The trajectories of a flow around a critical point with complex eigenvalues brought into its canonical form

The initialization of a viscous tornado is also due to the turbulent fluctuations, however the near-wall vorticity starts to organize in vortex lines of spiral form. They are limited by an equilibrium between stretching and diffusion of the vorticity, see also Gyr (1985). The Λ-vortex model has to be combined with the viscous tornado model such that some vortex lines of the Λ-vortex attach on the bed in a critical point and thus produce a stretched vortex loop very similar to the one postulated by Helmholtz, only that in the new formulation these vortices are compatible with the nonslip condition. Fluid and vorticity are thus pumped into the "head" of the forming vortex tube and will be released by the instability process destroying the Λ-vortex in a burst. We will come back to this mechanism later when discussing new sediment transport mechanisms based on coherent structures. Here, however, we will only discuss some preliminary models based on structures also called coherent but of completely different origin.

This confusing situation is the result of a high similarity in the footprints of the two classes of structures. The coherent structures as discussed earlier are located in the area of the loop pumping low-speed fluid in z-direction whereas at the outside of the vortex fast fluid is pumped toward the wall (-z-direction). In the extreme case of a breakdown, the outflow is violent and bundled and called an ejection whereas the down flows are called sweeps. These events have to be extracted from the flow signal, and special pattern recognition programs, usually conditional sampling methods, do this. One of the most popular one is the averaging by the variable-interval-time-averaging method (VITA) that first was introduced by Blackwelder and Kaplan (1976). The most simple classification of the structural events as found in a 2D open channel flow can be given in form of a quadrant splitting, or decomposition (Wallace et al., 1972; and Willmarth and Lu, 1972). By this method one plots the instantaneous velocity fluctuation vector $\underline{u}' = \left(u_x', u_z'\right)$ in a $\left(u_x', u_z'\right)$-coordinate plane. In the simplest form, an event is defined as a signal given by the vector (u_x', u_z') from the moment it enters a quadrant until it leaves this coordinate sector (Table 3.2). By this definition one also defines an event time, or duration, T_e. The transport of momentum by such an event is given by

$$I_e = \int_{T_e} u_x' u_z' \mathrm{d}t \tag{3.38}$$

and we can define a relative intermittency by

$$\gamma(T_i)_{\text{rel}} = \frac{\sum_{j=1}^{n} T_{ij}}{\sum_{j=1}^{n}\sum_{i=1}^{4} T_{ij}} = \frac{\sum_{j=1}^{n} T_{ij}}{T} \tag{3.39}$$

i is the index of the quadrant and j the one of the n particular events. The total summation is equal to the total time of observation. For smooth boundary conditions at $y^+ > 20$, this intermittency value is approximately a constant. In Fig. 3.8 one finds a definition of the parameters.

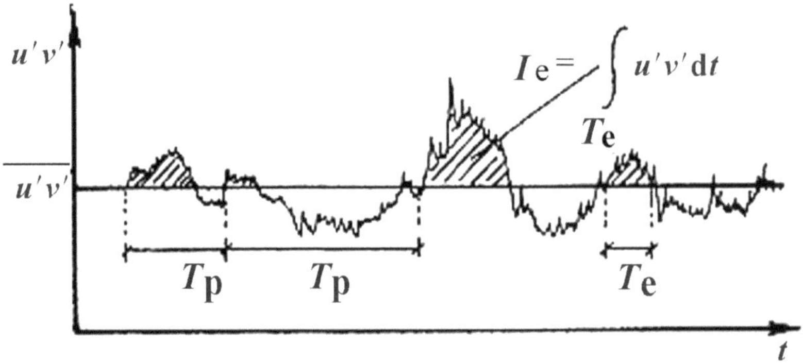

Fig. 3.8 A definition sketch for T_e, I_e, and T_P the time of a burst sequence called a period

If the residence time of the velocity fluctuation vector (u_x', u_z') is plotted as a residence probability, corresponding to a JPDF function as shown in Fig. 1.1, additional results can be drawn from such a representation, as we will see later. Such a representation can always be produced whether coherent structures exist or not, and the JPDF does not even react much whether we have or do not have these coherent structures. So one finds that the 2nd quadrant (ejections), and the 4th quadrant (sweeps) contribute each 70% to the negative momentum transfer and the 40% surplus is compensated by 40% positive momentum transferred in quadrants 1 and 3 (Table 3.2).

Table 3.2 The quadrant definitions

The boundaries of the quadrants	Contributions to the momentum transfer (%)	Name of the events
$u_x' \geq 0$; $u_z' \geq 0$	+20%	Inflexion
$u_x' \leq 0$; $u_z' \geq 0$	–70%	Ejection
$u_x' \leq 0$; $u_z' \leq 0$	+20%	Inflexion
$u_x' \geq 0$; $u_z' \leq 0$	–70%	Sweep

This representation is dangerous insofar as the contributions by the 2nd and the 4th quadrant are also called “coherent structures,” which is the origin of a lot of misunderstanding in the literature. Ejection and sweep are the notation for the instability structures of the near-wall flow close to a smooth wall. However because the quadrant method can always be used, even when no coherent structure are present, this classification is independent of the model of coherent structures.

Especially the fully developed turbulent flow has so-called large-scale structures, which are different from the described instability structures. They can also be classified by the quadrant method, and since they are sheared by the mean velocity profile, the JPDF is very similar. The confusion becomes even greater when the concept of the instability process is also used for rough walls. The structures found under such conditions by the quadrant decomposition should by no means be called ejections and sweeps but events of the 2nd and 4th quadrant.

These quadrant events are mainly large structures, and therefore the concept as it is used now is fairly problematic since these elements differ from the real coherent structures and cannot be extrapolated from those. It is evident that the new view did not enlarge our knowledge and the concept is used rather to explain the intermittent sediment transport than the sediment transport in a quantitative form. This is the reason why a more detailed representation of the transport due to turbulence based on coherent structures will be introduced in later chapters in which new concepts are discussed.

3.3 Transport in Suspension

Since the grains usually are heavier than the fluid, they always have the tendency to settle, and the fact that they remain suspended in the fluid is the result of the turbulent flow. There exist enough flow areas with a local upward flow by which the fine material is transported into the flow. In comparison with the bed-load transport the transported grain is completely surrounded by fluid and in balance with the local flow conditions. By definition only those particles are treated as suspended, which during their whole transport time or at least during most of this time are in suspension, and therefore not in contact with the bed. In other words, a sharp definition is not possible because there are always grains that are at rest on the bed surface but once in motion can belong completely to the suspended part of the grain distribution. It is evident that the suspended material must be rather fine, which needs to be defined. For scaling two measures based on the grain size are used, and both numbers must be of order one to define suspended material: the Reynolds number based on the grain size $\mathrm{Re_d}$, as a characteristic of the dynamical interrelation, and the Stokes number St based on the static motion of a falling grain, as a measure of viscous forces, in relation to the gravity of the submerged grains.

$$\mathrm{St} = \frac{u_{\mathrm{v}}\nu}{L^2}\frac{1}{g\rho'} \quad \vee \quad \mathrm{St} = \frac{u_{\mathrm{v}}\nu}{d_{\mathrm{s}}^2}\frac{1}{g\rho'} \quad ; \quad Re_{\mathrm{d}} = \frac{u_{\mathrm{v}}d_{\mathrm{s}}}{\nu} \tag{3.40}$$

with u_{v} the asymptotic settling velocity and L a length parameter usually set to the diameter of the grain. For

$$\mathrm{St} = O(1) \quad \wedge \quad Re_{\mathrm{d}} = O(1) \tag{3.41}$$

we speak of suspended material, which moves like in honey.

3.3.1 A Suspended Particle in a Turbulent Flow

Due to inertia, a grain moves relative to the surrounding fluid with a relative velocity u_r

$$\underline{u}_r = \underline{u}_f - \underline{u}_s \tag{3.42}$$

If the fluid and the grain move parallel, as is the case in a settling process, the drag of the particle is given in Rayleigh's formulation as

$$\underline{F}_D = \zeta A\left(\frac{\rho_f \underline{u}_r |\underline{u}_r|}{2}\right) \tag{3.43}$$

with A the area of the largest cross section of the grain, and ζ its drag coefficient, which has to be known.

In the stationary case, such as encountered in sedimentation in a fluid at rest we find for $\rho_s > \rho_f$

$$\begin{aligned} &\underline{F} = -(\underline{G} - \underline{L}) \quad \therefore \quad u_r = u_v \\ &G - L = g\rho_f \frac{d_s^3 \pi}{6}\rho' \quad (\text{Eq.3.17}) \quad \frac{u_r |u_r|}{g d_s} = -\frac{4}{3}\rho' \frac{1}{\zeta} \\ &\therefore \quad u_v = \sqrt{\frac{4\rho' d_s g}{3\zeta}} \quad \therefore \quad \text{St} = 2\nu\sqrt{\frac{1}{\sqrt{3g\rho' d_s^3 \zeta}}} \end{aligned} \tag{3.44}$$

For suspensions we can assume that the form of the grain plays a minor role, we can in good approximation assume that all grains would be spheres. For a sphere we find the drag coefficient from Stokes' law,

$$\zeta = \frac{24}{Re_d} \tag{3.45}$$

For the Stokes domain (St $\approx$ 1) we find

$$\frac{u_r d_s}{\nu} = -\frac{Ar}{18}, \quad Ar \equiv \frac{d_s^3 g}{\nu^2}\rho' \equiv Ga\rho', \quad Ga \equiv \frac{d_s^3 g}{\nu^2}, \quad Fr \equiv \frac{Re_d^2}{Ga} \tag{3.46}$$

with Ar the Archimedes number and Ga the Galilei number.

Let us stay with the spherical form but introduce more complicated flow conditions. The flow surrounding a particle has to fulfill the Navier–Stokes equation with its initial and boundary conditions, and since the Navier–Stokes equation is an integro-differential equation, we have to evaluate the pressure field exerted on the grain on the condition that the flow field at the surface of the grain remains steady. By applying the rotational operator to the Navier–Stokes equation, we can eliminate the pressure term and the problem remains to calculate the vorticity given by the vorticity diffusion equation. Another way to find a solution is by calculating the pressure field on the grain. This method is necessary if the flow separates at the grain at higher Reynolds numbers. In that case the flow field changes drastically with Re_d since the wake area depends on this number, and one can observe a characteristic wake-sledge. The frictional fraction of the forces on a grain decreases with increasing Re_d, and the opposite is the case for the pressure forces. For a creeping flow as it is discussed here the frictional part takes only care of two-thirds of the drag forces whereas the pressure difference accounts for one-third. If all inertial forces are neglected we come back to Eq. 3.45 the Stokes drag law, which is only fulfilled for

$$Re_d \le 0.1 \tag{3.47}$$

Oseen (1910) used linearized inertia terms and could expand the formulation of the drag; a full solution of this kind was given by Goldstein (1929)

$$\begin{aligned}
\zeta &= \frac{24}{Re_d}\left(1+\frac{3}{16}Re_d\right) && (\text{Oseen}) \quad Re_d \le 1 \\
\zeta &= \frac{24}{Re_d}\left(1+\frac{3}{16}Re_d - \frac{19}{1280}Re_d^2 + \frac{71}{20480}Re_d^3 - \ldots\right) && (\text{Goldstein}) \quad Re_d \le 1
\end{aligned} \tag{3.48}$$

For higher values of Re_d, the drag coefficient could not be calculated and was evaluated experimentally. Based on these results a series of approximation formulas were derived. At higher Reynolds numbers the drag coefficient still decreases, however much more moderately, and reaches a minimum at

$$\begin{aligned}
Re_d &= 4\times10^3 \rightarrow \zeta = 0.4 \\
Re_d &= \frac{4\times10^3}{3\cdot10^5} \rightarrow \zeta \approx \text{const.} = 0.44
\end{aligned} \tag{3.49}$$

and changes drastically at

$$Re_d = 3\times10^5 \rightarrow \zeta = 0.07 \tag{3.50}$$

This drastic reduction is the result of the transition of the boundary layer flow from the laminar to the turbulent regime. Due to its higher momentum transport the turbulent boundary layer is capable to reach a region further to the back of the grain and therefore separates much later, resulting in a smaller wake. The domain in which ζ is quasi-constant (Eq. 3.49) is called the Newtonian domain. In the undercritical domain the formula of Kaskas (1964) can be used in good approximation

$$\zeta = \frac{24}{Re_d} + \frac{4}{Re_d^{1/2}} + 0.4 \quad (\text{Kaskas}) \quad Re_d \le Re_{dc} \tag{3.51}$$

Torobin and Gauvin (1959, 1960, 1961) discuss 478 papers dealing with the drag of spheres, and Garner et al. (1959) have added another series on this matter. The results are summarized in Fig. 3.9.

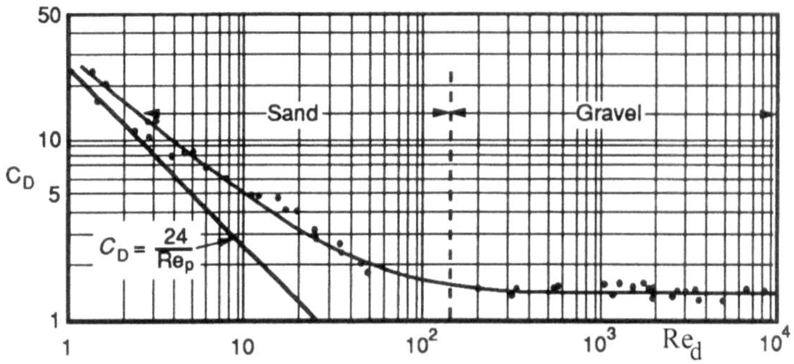

Fig. 3.9 The drag coefficient as function of Re_d of spherical sand grains, with C_D given by Stokes law

With these drag coefficients one usually calculates the steady-state settling velocity, however, in a turbulent flow the instationary behavior of the particle motion has to be accounted for.

$$m_s \frac{d\underline{u}_r'}{dt} = \underline{G} - \underline{L} + \underline{F}, \quad m_s = \frac{\pi \rho_s d_s^3}{6} \tag{3.52}$$

∴

$$\frac{1}{g}\frac{d\underline{u}_r'}{dt} = \left(1 - \frac{\rho_f}{\rho_s}\right) + \frac{3}{4}\frac{\rho_f}{\rho_s}\left(\frac{\underline{u}_r' |\underline{u}_r'|}{g d_s}\right)\zeta_b, \qquad \underline{u}_r' = \underline{u} - \underline{u}_s' \tag{3.53}$$

The grains feel the flow by its fluctuating force field as well as the constant gravity force. This results in inertial forces, which have to be in equilibrium with the sum of the external forces and thus allows formulating an equation of motion for the grains.

Usually ζ_b is larger than ζ, the reason being that by accelerating the particles, one has to accelerate an additional mass of fluid too. In air this can be neglected as well as the buoyancy term, and it is recognized that the suspension flow in water and air must differ. The added mass concept was introduced by Bessel as Torobin and Gauvin (1959, 1960) refer. The term can also be transformed into an additional volume.

$$m_b = m_s + \tilde{m}_f, \quad m_b = \rho_s V_s + \rho_f \tilde{V}, \quad m_b = V_s\left(\rho_s + \alpha\rho_f\right), \quad \tilde{V} = \alpha V_s$$

$$m_b = \frac{\pi d_s^3}{6}\left(\rho_s + \alpha\rho_f\right) \tag{3.54}$$

α can be set in most cases to 0.5 (Tollmien, 1938).

If one restricts the drag as to the part given by Stokes law, the differential equation for a particle becomes

$$\frac{1}{g}\frac{du_r'}{dt} = \frac{1-\rho_r}{1+\alpha\rho_r} + 18\frac{1-\rho_r}{1+\alpha\rho_r}\frac{\nu u_r'}{g d_s^2}, \quad \frac{\rho_f}{\rho_s} = \rho_r \tag{3.55}$$

and the traveled distance h_r' resulting from transport with u_r' can be found by the integration over time

$$h_{\mathrm{r}}' = \int_0^t u_{\mathrm{r}}' dt \rightarrow (3.55) \rightarrow |h_{\mathrm{r}}'| = -\frac{1}{18}\frac{|u_{\mathrm{r}}'| d_{\mathrm{s}}^2}{\nu \rho_{\mathrm{r}}}(1+\rho_{\mathrm{r}})\left[\frac{u_{\mathrm{r}}'}{u_{\mathrm{r}}} + \ln\left(1 - \frac{u_{\mathrm{r}}'}{u_{\mathrm{r}}}\right)\right] \tag{3.56}$$

In gases these distances must be taken into account as they may become larger.

The general differential equation for a particle motion can now be formulated

$$x: \quad m_{\mathrm{b}} \frac{du_{\mathrm{s}x}'}{dt} = F_x \qquad z: \quad m_{\mathrm{b}} \frac{du_{\mathrm{s}z}'}{dt} = G - AL + F_z \tag{3.57}$$

and the resistance is

$$\underline{F} = \zeta \frac{\pi d_{\mathrm{s}}^2}{4} \frac{\rho_{\mathrm{f}} \underline{u}_{\mathrm{r}}' |\underline{u}_{\mathrm{r}}'|}{2}, \quad F_i = \zeta \frac{\pi d_{\mathrm{s}}^2}{4} \frac{\rho_{\mathrm{f}} |\underline{u}_{\mathrm{r}}'|}{2} u_{\mathrm{r}i}'$$
$$\wedge \quad \cos\gamma = \frac{u_{\mathrm{r}x}'}{u_{\mathrm{r}}'}, \quad \sin\gamma = \frac{u_{rz}'}{u_r'} \tag{3.58}$$

Equation 3.58 inserted into Eq. 3.57 yields the differential equations

$$m_{\mathrm{b}} \frac{du_{\mathrm{s}x}'}{dt} = \zeta \frac{\pi d_{\mathrm{s}}^2}{4} \frac{\rho_{\mathrm{f}} |\underline{u}_{\mathrm{r}}'|}{2} u_{\mathrm{r}x}'$$
$$m_{\mathrm{b}} \frac{du_{\mathrm{s}z}'}{dt} = \zeta \frac{\pi d_{\mathrm{s}}^2}{4} \frac{\rho_{\mathrm{f}} |\underline{u}_{\mathrm{r}}'|}{2} u_{\mathrm{r}z}' + g \frac{\pi \rho_{\mathrm{f}} d_{\mathrm{s}}^3}{6} \rho' \tag{3.59}$$

with

$$u_{\mathrm{s}i}' = \frac{dx_i}{dt}, \quad \frac{du_{\mathrm{s}i}'}{dt} = \frac{d^2 x_i}{dt^2} \tag{3.60}$$

With these we formulate the differential equations for the particle trajectories in the two main directions (x, z) as found by Grave (1967) using a nondimensional formulation

$$x: \quad \frac{d^2 x^*}{dt^{*2}} - 18\left(u_x^* - \frac{dx^*}{dt^*}\right) \bullet$$
$$\left\{1 + \frac{1}{6}\left[\left(u_x^* - \frac{dx^*}{dt^*}\right)^2 + \left(u_z^* - \frac{dz^*}{dt^*}\right)^2\right]^{1/4} + \frac{1}{60}\left[\left(u_x^* - \frac{dx^*}{dt^*}\right)^2 + \left(u_z^* - \frac{dz^*}{dt^*}\right)^2\right]^{1/2}\right\}$$
$$z: \quad \frac{d^2 z^*}{dt^{*2}} = 18\left(u_z^* - \frac{dz^*}{dt^*}\right) \bullet$$
$$\left\{1 + \frac{1}{6}\left[\left(u_x^* - \frac{dx^*}{dt^*}\right)^2 + \left(u_z^* - \frac{dz^*}{dt^*}\right)^2\right]^{1/4} + \frac{1}{60}\left[\left(u_x^* - \frac{dx^*}{dt^*}\right)^2 + \left(u_z^* - \frac{dz^*}{dt^*}\right)^2\right]^{1/2}\right\} + \mathrm{Ar} \tag{3.61}$$
$$x_i^* \equiv \frac{x_i / d_{\mathrm{s}}}{\left(\frac{\rho_{\mathrm{s}}}{\rho_{\mathrm{f}}} + \alpha\right)}, \quad t^* \equiv \frac{\nu t d_{\mathrm{s}}^2}{\left(\frac{\rho_{\mathrm{s}}}{\rho_{\mathrm{f}}} + \alpha\right)}, \quad u_i^* \equiv \frac{u_i d_{\mathrm{s}}}{\nu}, \quad \mathrm{Ar} \equiv \frac{g d_{\mathrm{s}}^3}{\nu^2}\left(\frac{\rho_{\mathrm{s}}}{\rho_{\mathrm{f}}} + 1\right)$$

These are two coupled nonlinear inhomogeneous differential equations, and the curved brackets represent the terms differing from Stokes law. Neglecting those, we find for the equation of motion for particles following Stokes' law Eq. 3.45

$$x: \quad \frac{\mathrm{d}^2 x^*}{\mathrm{d}t^{*2}} = 18\left(u_x^* - \frac{\mathrm{d}x^*}{\mathrm{d}t}\right)$$
$$z: \quad \frac{\mathrm{d}^2 z^*}{\mathrm{d}t^{*2}} = 18\left(u_z^* - \frac{\mathrm{d}z^*}{\mathrm{d}t}\right) + \mathrm{Ar} \tag{3.62}$$

which are not coupled and can be integrated in a close form.

Beginning with Eq. 3.62, one can construct a series of simplified equations depending on assumptions made for the disturbances, e.g., considering vertical velocity fluctuations only ($u^*_{xy} = 0$). Similarly only horizontal fluctuations can be considered ($u^*_z = 0$).

In analogy to the bed load the transport of suspension must be formulated in a statistical form. However since the grain is exposed to the local field, it would be necessary to describe the field by a statistic of the turbulent flow field decomposed into structures that means in a nonlocal way. The calculations of single grain motions are therefore only a remedy for analyzing complicated cases and are used in the literature most exclusively for the evaluation of the settling velocity.

Here, however, we go a step further by investigating grains in curved flow fields because this representation is essential for calculating the movement of particles in vortices. Assuming a flow represented by a 2D field given in polar coordinates we are searching for the velocity components of the particles due to this field.

$$u = f_1(r,\varphi), \quad v = f_2(r,\varphi) \;\Rightarrow\; u'_{\mathrm{sr}} = \dot{r}, \quad u'_{\mathrm{s}\varphi} = r\dot{\varphi} \tag{3.63}$$

The inertial forces are calculated from the accelerations

$$a_{\mathrm{r}} = \ddot{r} - r\dot{\varphi}, a_{\varphi} = r\ddot{\varphi} + 2\dot{r}\dot{\varphi}, \Rightarrow I_{\mathrm{r}} = m_{\mathrm{s}}\left(\ddot{r} - r\dot{\varphi}\right), \quad I_{\varphi} = m_{\mathrm{s}}\left(r\ddot{\varphi} + 2\dot{r}\dot{\varphi}\right) \tag{3.64}$$

The drag force was defined by Eq. 3.58 and now has to be transformed into polar coordinates with the pertinent relative velocities

$$u'_{\mathrm{rr}} = u - \dot{r}, u'_{\mathrm{r}\varphi} = v - r\dot{\varphi}, \quad \left|\underline{u}'_{\mathrm{r}}\right| = \sqrt{(u-\dot{r})^2 + (v - r\dot{\varphi})^2} \tag{3.65}$$

Together with the components of the resulting $\underline{G} - \underline{L}$-force,

$$\underline{G} - \underline{L} = \left(-m_{\mathrm{s}}\left(1 - \frac{\rho_{\mathrm{f}}}{\rho_{\mathrm{s}}}\right) g \sin\varphi, -m_{\mathrm{s}}\left(1 - \frac{\rho_{\mathrm{f}}}{\rho_{\mathrm{s}}}\right) g \cos\varphi\right) \tag{3.66}$$

the equations of motion and $\rho_{\mathrm{f}}/\rho_{\mathrm{s}} = \rho_{\mathrm{r}}$ we arrive at

$$\ddot{r} - r\dot{\varphi}^2 = \frac{3}{4}\zeta\rho_{\mathrm{r}}\frac{\left|\underline{u}'_{\mathrm{r}}\right|}{d_{\mathrm{s}}}(u - \dot{r}) - g\frac{\rho'}{\rho_{\mathrm{r}}}\sin\varphi$$
$$r\ddot{\varphi} + 2\dot{r}\dot{\varphi} = \frac{3}{4}\zeta\rho_{\mathrm{r}}\frac{\left|\underline{u}'_{\mathrm{r}}\right|}{d_{\mathrm{s}}}(v - r\dot{\varphi}) - g\frac{\rho'}{\rho_{\mathrm{r}}}\cos\varphi \tag{3.67}$$

Equation 3.67 again is a coupled system of two non-linear inhomogeneous differential equations, which can be solved only numerically.

3.3.2 Particle Swarms

The sedimentation of particles as well as the sediment transport cannot be treated without incorporating the swarm behavior of the suspended particles since the particles will never be homogeneously distributed in the fluid. The drag wakes of the particles, e.g., in a settling process, have a feedback on the flow field. In a fluid at rest, the wakes of the settling particles are the only source of the start up of motion, which one observes.

The sedimentation, however, is not discussed in this book and we, therefore, treat this process only briefly. The simplest possible setup for investigating the swarm behavior starts looking at the interaction of a pair of proximate particles. The literature treating this simple configuration is respectable, and by no means closed, which shows how difficult it will be to describe a swarm flow analytically (Stimson and Jefferey, 1926).

In a next step, the suspension composed of two fractions of suspended material was treated. Due to the complexity of this problem, swarms are mainly investigated as a two-phase flow system with variable density and viscosity using rheological approaches. In the frame of self-organization, we will come back to such aspects later.

3.3.3 The Transport of Suspension as Described by a Continuum Mechanical Model

A grain in suspension has once been entrained into the flow from the bed. The mechanism of removal is in principle the same as for the incipient motion of grains moving as bed load. Therefore, it seems adequate to describe the incipient motion of a suspending grain again by the wall shear stress or $u_{\tau c}$, although the critical value for bed-load transport is lower than for entrainment. For rolling of the grains it must be demanded that they are not injected into the outer flow. Formally this is described by the empirically evaluated value $u_{\tau l}$,

$$U_{\tau l} > u_{\tau c} \tag{3.68}$$

Once the grains are in suspension a flow force must be present to compensate the weight. A grain surrounded by fluid fulfills the nonslip condition and acts as a solid body. If the flow field acting on the grain is subject to a shear gradient, than the volume of the grain can be thought of as a displaced fluid volume with the vorticity given by the shear.

The flow around the grain acts as if the grain would rotate—which it usually does with a certain delay with respect to the shear—resulting in a lift force perpendicular to the plane of (u_s, Ω_{grain}) as described by the Magnus effect on the grain (Magnus, 1853). Therefore, the grain moves away from the bed into the flow however this force is usually small with respect to the turbulent lift forces. In a laminar flow however, the Magnus-forces are the dominant ones. If particles are exposed to quasi-laminar local flow fields like in flow structures as eddies, we have to take this effect into account.

Turbulence is produced at the walls and "diffuses" turbulently into the flow thereby decreasing in intensity with increasing wall distance, a fact already encountered in the description of the velocity profile Eq. 2.31. Therefore, the gravity forces become relatively more important, which results in a concentration gradient of the suspended material. One important part of this model, however, is that the suspended grains remain more or less in suspension.

The entrainment: The entrainment of material is described the same way as the incipient motion for bed-load material (Raudkivi, 1982) and classified as follows:

$$\begin{aligned} &6 > u_{\mathrm{v}} / u_{\tau} > 2 && \text{bed load} \\ &2 > u_{\mathrm{v}} / u_{\tau} > 0.6 && \text{saltation} \\ &0.85 > u_{\mathrm{v}} / u_{\tau} > 0 && \text{suspension} \end{aligned} \tag{3.69}$$

where the asymptotic settling-velocity u_{v} was used for classification. These values can relate to the critical value $u_{\tau 1}$, however, the values found differed considerably, and Zanke (1982) proposed to use his value from (1976) as reasonable mean

$$\begin{aligned} &u_{\tau 1} = 0.25 u_{\mathrm{v}} && \text{Engelund } (1965) \\ &u_{\tau 1} = 0.812 u_{\mathrm{v}} && \text{Bagnold } (1966) \\ &u_{\tau 1} = 0.4 u_{\mathrm{v}} && \text{Zanke } (1976) \\ &\frac{u_{\tau 1}}{u_{\tau c}} = 2.2 && \text{Zanke } (1982) \end{aligned} \tag{3.70}$$

In other words, we conclude that the suspension flow is far from being understood properly. There are two reasons for this lack of knowledge. The first is identical with the main deficiency found for the bed load, the missing knowledge about the pressure distribution on the bed and the grains when suspended. In several experiments it was found that the pressure distribution inside the bed has practically no influence on the bed load; however, this need not be the case for very small particles, which could be destabilized by pressure gradients in the bed. There is a flow into and out of the bed. Therefore the question is:

> *What is the influence of a pressure gradient inside a bed on the entrainment process, producing a suspension?*

A diffusion model for the suspension load: For the evaluation of the suspension load a series of methods are known. The most popular ones are the diffusion- and the energy model, and model of mixed form of stochastic representation. The diffusion model is, however, the best known of all.

This representation for the volume concentration $\dot{c}_{\mathrm{V}}$ starts from a mass balance for mass in the suspension load in turbulent flow considering outer sources and sinks. This can be summarized in a continuity equation

$$\frac{\partial c_{\mathrm{V}}}{\partial t} + \operatorname{div} q_{\mathrm{su}} = \dot{c}_{\mathrm{V}} \; ; \quad c_{\mathrm{V}} = \frac{V_{\mathrm{s}}}{V_{\mathrm{t}}} \tag{3.71}$$

In other words, it is assumed that c_{V} changes according to the suspension flux q_{su} through the surface of a unit volume.

The suspension flux is composed of three contributions. First, the advective one describing the suspension transport by the mean flow field. Second, the transport by the Brownian motion, a molecular diffusion process given by Fick's diffusion law in which the flux is proportional to the gradient of the concentration. In a laminar flow, this would be the only additional term and could be described by the molecular diffusion coefficient *D*. The third contribution comes from the turbulent velocity fluctuations. This contribution can be described in analogy to the Brownian motion by a kind of Fick's law, however the diffusion coefficient is not a scalar anymore but in a homogeneous turbulence a diffusion vector $\underline{\varepsilon}$,

$$q_{\text{su}i} = u_i c_{\text{V}} - \left(D + \varepsilon_i\right)\frac{\partial c_{\text{V}}}{\partial x_i} \tag{3.72}$$

The minus sign of the bracketed term is a convention and ensures that the net mass transport is from the higher to the lower concentration. By inserting Eq. 3.71 into Eq. 3.72, we get

$$\frac{\partial c_{\text{V}}}{\partial t} + \frac{\partial u_i c_{\text{V}}}{\partial x_i} = \dot{c}_{\text{V}} + \left(D + \varepsilon_i\right)\frac{\partial^2 c_{\text{V}}}{\partial x_i^2} \tag{3.73}$$

in which we use the Einstein summation convention, summing over terms of the same index unless specified otherwise. Equation 3.73 is a linear equation in c_{V} and can be solved in a closed form, together with the continuity equation for the fluid (Eq. 1.22), which still holds for low concentrations. In Eq. 3.73, the velocity u_{i} can be moved out of the differential in the advective term. In addition, in a homogeneous isotropic turbulent flow where $\varepsilon >> D$ the bracket ($\varepsilon + D$) can be replaced by a scalar turbulent diffusion coefficient D_{t}, which has to be evaluated experimentally or by a turbulence theory. D_{t} is not a material property but a characteristic value for the status of the turbulent flow field. With these simplifications Eq. 3.73 becomes

$$\frac{\partial c_{\text{V}}}{\partial t} + u_i \frac{\partial c_{\text{V}}}{\partial x_i} = D_t \frac{\partial^2 c_{\text{V}}}{\partial x_i \partial x_i} \tag{3.74}$$

Here D_{t} is a simplified version of the second order diffusion tensor D_{ij}.

If the Reynolds decomposition Eq. 2.3 is applied not only on *u* but now also to c_{V} and then inserted in Eq. 3.74 by averaging, one gets the relation between the two fluctuating parameters and D_{ij}

$$\overline{c_V' u_i'} = -D_{ij} \frac{\partial \overline{c_{\text{V}}}}{\partial x_j} \tag{3.75}$$

and in place of Eq. 3.74 respectively for Eq. 3.73 we get the complete averaged diffusion equation for the suspension concentration

$$\frac{\partial \overline{c_{\text{V}}}}{\partial t} + \overline{u}\frac{\partial \overline{c_{\text{V}}}}{\partial x_i} = -\frac{\partial}{\partial x_i}\left(D_{ij}\frac{\partial \overline{c_{\text{V}}}}{\partial x_j} + D\frac{\partial \overline{c_{\text{V}}}}{\partial x_i}\right) = -D\frac{\partial^2 \overline{c_{\text{V}}}}{\partial x_i^2} - \frac{\partial}{\partial x_i}\overline{u_i' c_{\text{V}}'} \tag{3.76}$$

Here D_{ij} plays the same role as the Reynolds shear stress does in the turbulent equation of motion, and Eq. 3.76 is undefined as long as D_{ij} cannot be given as a function of the mean values of $\overline{u_i' c_{\text{V}}'}$. We encounter again a closure problem. In addition, one has to remark that Eq. 3.76 is not the most general one since it was assumed that the variance of the spreading of the concentration is linear in time (Fischer et al., (1979). A certain confirmation of this hypothesis is the result of the turbulent diffusion theory published by Taylor (1921). He found for the mean variance in Lagrangian coordinates for a stationary homogeneous case

$$\begin{aligned}&\langle x^2\rangle = 2\langle u_x'^2\rangle \int_{\tau=0}^{t} \tau R_x(\tau)\,\mathrm{d}\tau\\ &R_x\big((t+\tau)-t\big) = R_x(\tau) = \frac{\langle u_x'(t)\,u_x'(t+\tau)\rangle}{\langle u_x'^2\rangle};\quad R_x(0)=1,\quad R(\infty)=0\end{aligned} \tag{3.77}$$

The Lagrangian autocorrelation R_x cannot be predicted theoretically, however it was evaluated empirically, and Eq. 3.77 reduces to

$$\begin{aligned}&\langle x^2\rangle = \langle u_x'^2\rangle t^2 && R_x = 1,\quad x = 0\\ &\langle x^2\rangle = 2\langle u_x'\rangle \int_{\tau=0}^{\infty} R_x(\tau)\,\mathrm{d}\tau + \text{const.} && R_x = 0,\quad x \to \infty\end{aligned} \tag{3.78}$$

The integral represents the Lagrangian integral timescale and is a measure for the time needed that a particle forgets its initial velocity. This is the time in which the variance of a cloud grows linearly with time and quadratic with the turbulence intensity and therefore complies with Fick's representation.

In Eqs. 3.71 and 72, q_s was one of the quantities later replaced by c_V, and by doing this we replaced the suspension with a passive tracer solution. In other words, instead of grains one could also use molecules in solution. However, such a simplification holds only as long as the particles are cotransported with the flow. The description used is therefore restricted to the cases with particles of St < 1 (Eq. 3.40). For particles with a Stokes number larger than 1, the transport equation has to be modified by incorporating the relative velocities.

The mixing of waste is of high interest but is not treated within the frame of this book, therefore concerning this problem we refer the reader to the book of Rutherford (1994). Another ecological problem not treated is the adhesion of harmful waste on the suspended particles, which act as a sink for such problematic solutes.

The source of the suspension is the bed as long as it has enough suspendable material in its top layer. In this case, with a known distribution of concentration in bed distance, the complete sediment transport can be calculated by an integration of the flux through a cross section, with the assumption that height to width ratio is small, so that the influence of the sideboards can be neglected.

In the stationary case, Eq. 3.76 simplifies to the following form when the mean velocity profile has an x-component only

$$\bar{u}\frac{\partial c_V}{\partial z} = \frac{\partial}{\partial z}\overline{u_z' c_V'},\qquad \text{(Eq.3.75)},\quad \bar{u} = \left(0,0,\overline{u_x}(z)\right) \tag{3.79}$$

and an even stronger simplification can be achieved by using Eq. 3.73 without any assumption of the mean velocity profile as

$$u_z \frac{\partial c_V}{\partial z} = \varepsilon_z \frac{\partial^2 c_V}{\partial z^2} \tag{3.80}$$

Equation 3.80 has to be integrated over z. The turbulent fluctuations do not produce a mean vertical velocity, Eq. 3.79, and the particles settle with a velocity u_z given by the velocity u_v. The result is

$$u_v c_V + \varepsilon_z \frac{\partial c_V}{\partial z} = A, \quad c_V \Big|_{(z=H)} = 0 \therefore A = 0 \tag{3.81}$$

and in case of a constant c_V and ε_z, Eq. 3.81 can be integrated by a separation of variables

$$c_V = c_{V0} e^{-u_v z/\varepsilon_z} \tag{3.82}$$

The distribution of the concentration is determined by the exponent of Eq. 3.82,

$$\begin{aligned} &u_v z > \varepsilon_z \rightarrow \text{smooth distribution} \\ &u_v z < \varepsilon_z \rightarrow \text{concentrations just above the bed} \end{aligned} \tag{3.83}$$

Usually one encounters the second condition.

For a suspended particle, the sedimentation velocity can be calculated for a fluid at rest. The gravity forces have to be in equilibrium with the frictional forces for a particle of low St at its surface, a problem treated on a variety of conditions summarized in Torobin and Gauvin (1959) and for fine material in Happel and Brenner (1965). For spherical particles it can be found that

$$\begin{aligned} &\zeta \pi \frac{d_s^2}{4} \frac{\rho_f u_v^2}{2} = \pi \frac{d_s^3}{6} g(\rho_s - \rho_f) \quad \therefore \quad u_v^2 = \frac{4}{3} \frac{1}{\zeta} g d_s \rho' \\ &\wedge \quad \text{St} = 1 \rightarrow \zeta = \frac{24}{Re_d} \quad \text{(Eq. 3.19)} \quad \therefore \quad u_v = \frac{1}{18} \frac{g \rho' d_s^2}{\nu} \end{aligned} \tag{3.84}$$

The result of Eq. 3.82 was found using a series of assumptions, and therefore it is necessary to make adjustments. In particular, ε_z is a parameter given by the flow field and not a material constant. This situation is very similar to one of the Reynolds shear stress tensor describing the momentum transfer (Eq. 2.14). Therefore, it is appropriate to draw an analogy between the momentum transfer and the concentration transport. For the momentum exchange, we start with a mixing length theory as described by Prandtl, using a mixing length l_m.

$$\tau_{ij} = -\rho_f \overline{u_i' u_j'} \ \text{Eq. 2.14} \tag{3.85a}$$

$$\begin{aligned} &\overline{|u_x'|} \approx \overline{|u_z'|} \approx l_m \frac{dU_x}{dz} \therefore \overline{u_x' u_z'} \approx l_m^2 \left(\frac{dU_x}{dz}\right)^2 \\ &\tau_{zx} = \rho_f l_m^2 \left(\frac{dU_x}{dz}\right)^2 = \rho_f \varepsilon_m \left(\frac{dU_x}{dz}\right) \quad \wedge \quad \varepsilon_m = l_m^2 \left(\frac{dU_x}{dz}\right) \end{aligned} \tag{3.85b}$$

This is the representation of the shear stress in x-direction by using an eddy-viscosity or turbulent viscosity, ε_m, for the vertical motion.

With a flow as shown in Fig. 2.1 with its proper shear stress distribution given by Fig. 2.2, we get from Eq. 3.85

$$\tau_{zx} = \rho_f g (H - z) \sin\alpha \tag{3.86}$$

and

$$l_m \approx z \rightarrow \text{Eq.2.35} \quad \therefore \quad l_m = \kappa z \tag{3.87}$$

With this relation one can formulate an analogy between ε_z of the diffusion and ε_m of the momentum transfer introduced by their ratio

$$\varepsilon_z = \beta_s \varepsilon_m = \beta_s \frac{\tau}{\rho_f} \frac{dz}{dU_x} \quad \wedge \quad \beta_s = \frac{\varepsilon_z}{\varepsilon_m} \tag{3.88}$$

Inserting Eqs. 2.35 and 3.85 into Eq. 3.88 results in

$$\varepsilon_z = \beta_s \kappa u_\tau \frac{z}{H}(H - z) \tag{3.89}$$

Equation 3.89 can now be inserted into Eq. 3.81 and integrated by separation of variables to yield

$$\frac{c_V}{c_{Va}} = \left(\frac{H - z}{z} \frac{a}{H - a}\right)^{u_v / \beta_s \kappa u_\tau} = \left(\frac{H - z}{z} \frac{a}{H - a}\right)^{\text{Ro}} \quad \wedge \quad \text{Ro} = \frac{u_v}{\beta_s \kappa u_\tau} \tag{3.90}$$

Herein c_{Va} is the concentration of the suspension on the height $z = a$, and the Rouse number Ro is the ratio of the velocity of sedimentation versus the hydrodynamic characteristic of the turbulent flow.
With

$$\begin{aligned} &\beta_s \cong 1, (\text{Vanoni 1977}); \quad \kappa = 0.4 \quad \therefore \quad \text{Ro} \rightarrow \frac{u_v}{u_\tau} \\ &\text{we see that for} \\ &\text{St} << 1 \quad \therefore \quad u_v << u_\tau \quad \therefore \quad \text{Ro} << 1 \\ &\text{and for} \\ &\text{St} > 1 \quad \therefore \quad u_v > u_\tau \quad \therefore \quad \text{Ro} \rightarrow \infty \end{aligned} \tag{3.91}$$

and from the representation Fig. 3.10, we conclude that for very small particles the suspension tends to be homogeneously distributed, whereas for larger particles the suspension stays concentrated near the bed.

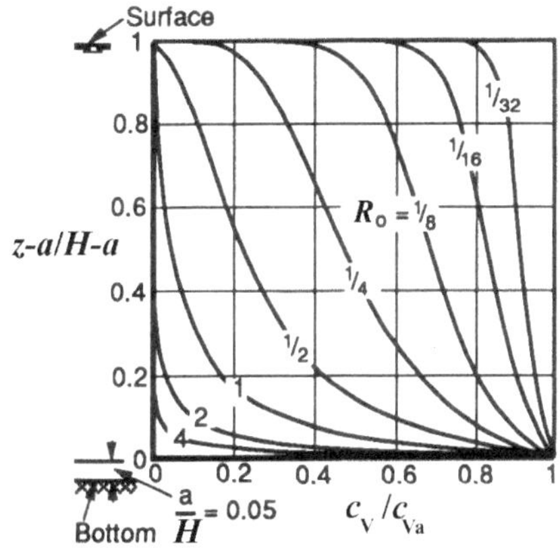

Fig. 3.10 Distribution of the suspensions given by Eq. 3.90 parameterized for different Ro with a given value a ($a/H = 0.05$) (Einstein, 1950)

With the diverse assumptions made to simplify the situation, we need to discuss the problems arising from these simplifications. Let us formulate three questions related to this deficiency:

1. What is the roll of the inertia for larger and heavier particles? What influence has the particle distribution on the distribution in the concentration?
2. How does the concentration distribution affect κ or ε_m, which are characterizing the turbulent flow? This question has to be treated in the framework of a rheological representation.
3. In which way does the bed form alter as a function of the concentration distribution? This question will be treated in the framework of the self-organization of bed forms.

Grain distribution and inertial effects: In the discussion of the motion of a single particle in Chap. 3 (Sect. 3.3.1), we showed the relevance of the relative velocity of a grain in a fluid due to inertial forces, and it was found that grains with St > 1 are the ones which depart from a coflowing state (pure advection). If the suspension is dilute enough, it can be assumed that the concentration can be evaluated by a superposition of the different fractions as given in Chap. 3 (Sect. 3.3.3.1). For $u_{\tau 1}$ a characteristic value can be obtained from a rough classification as given by Eq. 3.70.

Zanke (1979) argues that for the stationary case

$$u_\tau < u_{\tau 1} \rightarrow \varepsilon_s = 0 \tag{3.92}$$

that means the grains cannot be entrained anymore by the turbulent flow and therefore β_s as given by Eq. 3.88 should be replaced by a new value taking care of the effective entrainment given by Eq. 3.70

$$\beta_s = \frac{u_\tau - u_{\tau 1}}{u_\tau} \tag{3.93}$$

Consequently, the Rouse number has to be modified too

$$\tilde{\mathrm{R}}\mathrm{o} = \frac{u_{\mathrm{v}}}{\kappa\left(u_\tau - u_{\tau 1}\right)} \tag{3.94}$$

Colby and Hembree (1955) and Nordin and Dempster (1963) went even further by recommending an empirical formula for calculating $\tilde{\mathrm{R}}\mathrm{o}$ for the diverse grain size classes

$$\tilde{\mathrm{R}}\mathrm{o} = \frac{u_{\mathrm{v}}^{0.7}}{\kappa u_\tau} \tag{3.95}$$

Equation (3.95) shows the importance of the settling velocity Eq. 3.84 for the suspensions, and since u_{v} is a function of the viscosity, it indirectly depends on the temperature (Eqs. 1.11 and 1.12). This is the reason why the bed-load behaviour is practically independent of the temperature whereas the suspension reacts to changes in T. A special problem of inertial effects is the behavior of particles in vortical fields. A first introduction was given in Chap. 3 (Sect. 3.3.1, Eqs. 3.63–3.67). For a detailed study of such problems, we refer to the book of Ungarish (1993).

Influence of the concentration: In the introductional text to the flow of suspensions, we assumed that all relevant flow parameters are independent of c_{V} the volume concentration. What happens when this is not anymore a valid assumption? This question is to be answered by a theory incorporating the rheology for a mixture of highly concentrated suspensions. Such a task is very complicated and it became common practice to patch the whole complexity of the rheology into one parameter and treat the fluid as a Newtonian one. The parameter chosen for this purpose was the v. Kàrman constant κ, which then was set to be concentration dependent. κ was chosen because its empirical evaluation is very simple as long as the velocity profile remains logarithmic,

$$\kappa = \frac{U - U_s}{u_\tau} \ln y^+ \quad \text{(Eq.2.31)} \quad \text{respectively} \quad \kappa = \frac{U - U_s}{u_\tau} \ln \frac{y}{H} \tag{3.96}$$

and κ decreases with increasing concentration.

For a Newtonian flow, κ has been taken as constant because the momentum exchange is thought to be universal. However, since the turbulent structures of high concentration suspensions change also in a statistical sense, κ must change too. This corresponds to a change in rheology. With this remark, we point out that the relation between κ and the turbulent structures belongs to the theory of turbulence, which is a still unsolved problem. A first attempt to find a relation was undertaken by Perry and Chong (1982) based on a theory of coherent structures, which allows giving a theoretical description of κ based on the present structures. The resulting logarithmic profile showed a relation between the mean velocity profile and the turbulent shear stresses. Such a description would allow making predictions, if we knew how the different turbulent velocity scales are influenced by the suspensions.

> *How is the concentration related to the structure of a turbulent flow? A result, would also explain several drag reducing effects occurring in a suspension flow. This is a question belonging to the complex theme of the influence of the rheology on the transport.*

The most evident influence of the concentration is the effect on the local density of the mixture

$$\rho_{\mathrm{m}} = \rho_{\mathrm{f}}\left(1+\rho' c_{\mathrm{V}}\right) \tag{3.97}$$

At normal flow condition the slope of the energy line is equal the slope of the bed (Fig. 2.1) and given by the shear stress as described in Fig. 2.2. The density at a given level z can be calculated

$$\tau = \rho_{\mathrm{m}} g z S_{\mathrm{e}} \quad \wedge \quad S_{\mathrm{e}} = S_{\mathrm{b}} \quad \therefore \quad \rho_{\mathrm{m}} = \rho_{\mathrm{f}}\left(1+\frac{\rho'}{H-z}\right)\int_z^H c_{\mathrm{V}}\mathrm{d}z \tag{3.98}$$

In this description, u_τ is independent of the flow details, and u_{v} at constant κ would be a function of the Richardson number Ri,

$$\mathrm{Ri} = \frac{\dfrac{g}{\rho_{\mathrm{m}}}\dfrac{\partial \rho_{\mathrm{m}}}{\partial z}}{\left(\dfrac{\partial u}{\partial z}\right)^2} \rightarrow \frac{gz^2\dfrac{\partial \rho_{\mathrm{m}}}{\partial z}}{\rho_{\mathrm{f}} U^2} \tag{3.99}$$

$$\therefore \varepsilon_{\mathrm{m}} = \varepsilon_{\mathrm{m0}}\left(1-a\mathrm{Ri}\right)$$

u is the local velocity and the z coordinate is taken vertical to the stratification. Most often the eddy viscosity is written as a function of Ri in the form as shown in Eq. 3.99 with $\varepsilon_{\mathrm{m0}}$ and a constant that can be tuned to a particular situation. More particularly with the grain diameter ε_{m} also changes (Coleman, 1970). Zanke (1979) used this already implicitly (Eq. 3.93), by modifying β_{s}. Probably, this is the result of centrifugal forces acting on vortices, which we have to study in more detail. Coleman found experimentally that for

$$z/H = 0.2/1 \rightarrow \varepsilon_{\mathrm{m}}/u_\tau H = \text{const.} \tag{3.100}$$

The integration of Eq. 3.81 with the integration constant A and with Eq. 3.88 can now be redone, and we get

$$\frac{c_{\mathrm{V}}}{c_{\mathrm{V}a}} = \mathrm{e}^{\left(-\frac{1}{A}\frac{u_{\mathrm{v}}}{u_\tau}\right)} \quad \therefore \quad c_{\mathrm{V}} = c_{\mathrm{V}a} z^{-u_{\mathrm{v}}/\beta u_\tau} \quad \wedge \quad z > a \tag{3.101}$$

A proper continuum mechanical description does not allow using these parameters without describing their relations (e.g., $\nu = \nu(c)$ and $\rho = \rho(c)$). A two-phase flow has to be described by its deformation differing from the one of a Newtonian fluid. This needs a new rheology. According to Bingham, the flow of the given fluid in mathematical form is given by its deformation, a definition introduced (1929) by the American Society of Rheology. A good introduction to rheology of suspensions and solutions is found in Barnes et al. (1989). Rheological problems encountered in sediment transport will be discussed later.

Evaluation of the suspension load by an energy model: Bagnold (1962) and Vlugter (1962) proposed to evaluate the suspension load by equilibrating the energy needed to

hold the grains in suspension and the energy needed to compensate the energy dissipation. Since the higher weight of the fluid is due to the suspended particles, the situation as shown in Fig. 2.1 has to be replaced by an analogous one.
Work has to be done and this is usually represented by the difference in the stream power ΔP, given by

$$\begin{aligned}&(\rho_s-\rho_f)c_V V(u_v\cos\alpha-u_{su}\sin\alpha)g=\Delta P\\ \therefore\ &\frac{(\rho_s-\rho_f)}{\rho_s}\rho_s c_V V(u_v\cos\alpha-u_{su}\sin\alpha)g=\Delta P\\ \therefore\ &\Delta P=0\rightarrow u_v=u_{su}tg\alpha\end{aligned}\tag{3.102}$$

with u_{su} the outer flow velocity of the fluid suspension mixture. For normal flow conditions, we have equilibrium between the input of energy and its dissipation, therefore we get

$$\left[V_s(\rho_s-\rho_f)+(V-V_s)\rho_f\right]gSu_{su}=V_s(\rho_s-\rho_f)gu_v\cos\alpha+V\rho_f gSu_{su}\left(\frac{u_{su}^2}{u^2}\right)$$

$$\therefore V_s\left[(\rho_s-2\rho_f)Su_{su}-(\rho_s-\rho_f)u_v\cos\alpha\right]=V\left[\rho_f Su_{su}\left(\frac{u_s^2}{u^2}\right)-\rho_f Su_{su}\right]=V\rho_f Su_{su}\left[\left(\frac{u_{su}^2}{u^2}\right)-1\right]$$

$$\frac{V_s}{V}=c_V=\frac{\rho_f Su_{su}\left[\left(\frac{u_{su}^2}{u^2}\right)-1\right]}{\left[(\rho_s-2\rho_f)Su_{su}-(\rho_s-\rho_f)u_v\cos\alpha\right]}=\frac{\frac{\rho_f}{\rho_s}\left[\left(\frac{u_{su}^2}{u^2}\right)-1\right]}{\left(1-\frac{2\rho_f}{\rho_s}\right)-\left(1-\frac{\rho_f}{\rho_s}\right)\frac{u_v\cos\alpha}{u_{su}S}}\tag{3.103}$$

where the dissipation with and without suspension are considered by the ratio u_{su}^2/u^2.

Here, the value c_V is a mean over the total height, since the energy equation was used in its global form and not as a local relation. This simplifies the calculation of the suspension load. Significant information is lost however, and this is the reason why the diffusion model is preferred, although an experimentally evaluated calibration value c_{Va}, has to be introduced. If the bed is the only source of sediment, this is a reasonable approach. For quick estimation the energy method is a better choice. The densities are known, and therefore Eq. 3.113 reduces to an equation of three velocities. u_v can also be derived and therefore is a known parameter, and Eq. 3.103 reduces to

$$u_{su}^2=u^2 f(u_{su},c_V)\tag{3.104}$$

With a variety of assumptions, this equation can be solved approximately, and one recognizes that u_{su} and u differ only by 1%. In the equilibrium state, it is more appropriate to insert Eq. 3.102 into Eq. 3.103, and we find

$$c_{\mathrm{V}} = \frac{\rho_{\mathrm{f}}\left[\dfrac{u_{\mathrm{V}}^{2}}{u^{2}tg^{2}(\alpha)} - 1\right]}{(\rho_{\mathrm{s}} - 2\rho_{\mathrm{f}}) - (\rho_{\mathrm{s}} - \rho_{\mathrm{f}})ctg\alpha} \quad \wedge \quad S \approx \cos\alpha \tag{3.105}$$

Herein all values are known or can be estimated.

3.3.4 The Integration of the Suspended Load

When c_{V} is known the total suspension load can be calculated by a simple integration over the cross section

$$\begin{aligned} q_{\mathrm{su}} &= \int_{a}^{H} c_{\mathrm{V}} u_{x} \mathrm{d}z \quad \wedge \quad z > a \\ Q_{\mathrm{su}} &= \int_{A} q_{\mathrm{su}} \mathrm{d}A \\ L_{\mathrm{su12}} &= \int_{t_1}^{t_2} Q_{\mathrm{su}} \mathrm{d}t \end{aligned} \tag{3.106}$$

with q_{su} being the suspension flux at elevation z, Q_{su} the suspension mass transported over the entire cross-section A in a unit time, and L_{su12} the load transported during a time span Δt. The lower limit a of the integration corresponds usually to the roughness height (Fig. 3.10). The main remaining problem is the evaluation of the concentration at $z = a$. Formulase of this kind have been published by a variety of researchers and brought into a graphical form (Einstein, 1950; Brooks, 1956 in Raudkivi, 1976).

3.4 The Total Sediment Transport

In Chap. 3 (Sects. 3.2 and 3.3), the two transport forms by bed load and suspension were described and the goal was to elucidate the mechanisms involved. This representation would have to be supplemented by a discussion on the transport capacity, since the bed as a source can be depleted or the input into a control volume can differ from the output, which means the system is not in equilibrium. We will come back to this problem when we investigate the saturation concentration. Here we join the two transport forms, and the simplest ansatz would be a superposition

$$Q_{\mathrm{st}} = Q_{\mathrm{sb}} + Q_{\mathrm{su}} \tag{3.107}$$

The reality is not as simple because there is no clear criterion for the two states. Part of the bed load moves by saltation and does not remain in the surface layer of the bed for long time, and on the other hand a respectable part of the material usually is in suspension and can settle on the bed if it gets into the wake region of a bigger particle. It is evident that such a complex system needs empirical inputs. However, such inputs depend on the measurement techniques used, and it is therefore not surprising to find a whole variety of formulas, depending on the parameters which were taken to be the important ones in the different evaluations, e.g., the grain distribution etc. We will present some of the newer results later in the chapter devoted to the newer aspects in sediment transport.

One easily understands that the formulas give better results when the grain distribution becomes more homogeneous and when the transport form is restricted to only one transport type. The more classical equations, which will be discussed here, combine all the physics into the already mentioned three velocities u_τ, $u_{\tau c}$ and u_v.

Zanke (1982) used this approach for a dimensional analysis. He assumes that the transport is given by a horizontal and a vertical force and q_{st} being proportional to a power law of the ratio of these two forces.

$$q_{st} \propto \left(\frac{F_H}{F_V}\right)^{\alpha_1} \quad \wedge \quad F_H = F - F_c \quad \vee \quad F_V = G_A \quad \wedge \quad G_A = G - L$$

$$q_{st} \propto \left(\frac{F - F_c}{G_A}\right)^{\alpha_1} \quad \wedge \quad F = c_1 \frac{\rho_f}{2} u_\tau^2 A_e \quad \vee \quad F_c = c_2 \frac{\rho_f}{2} u_{\tau c}^2 A_e \tag{3.108}$$

$$G_A \propto F_V = c_3 \frac{\rho_f}{2} u_v^2 A_e \quad \therefore \quad q_{sb} \propto \left(\frac{c_1 u_\tau^2 - c_2 u_{\tau c}^2}{c_3 u_v^2}\right)$$

with the drag coefficients c_1 and c_2 and A_e the exposed area. With constant u_τ Eq. 3.108 becomes

$$c_1, c_2 \propto c_3 \quad \therefore \quad q_s \propto \left(\frac{u_\tau^2 - u_{\tau c}^2}{u_v^2}\right)^{\alpha_1} = F_{sb}^{*\alpha_1} \tag{3.109}$$

With this equation, the sediment load can be represented as a function of the following variables

$$\phi\left(q_{sb}, \rho_s, \rho_f, \nu, d_s, g, F_{sb}^*\right) = 0 \tag{3.110}$$

to which the Buckingham π-theorem is applied (Eq. 1.29). Doing this the following dimensionless numbers can be evaluated

$$\pi_1 = \frac{q_{sb}}{\nu} = q^*, \quad \pi_2 = \frac{\rho_s - \rho_f}{\rho_f} = \rho', \quad \pi_3 = \frac{u_\tau^2 - u_{\tau c}^2}{u_v^2} = F_{sb}^* \text{ (Eq.3.119)}$$

$$\pi_4 = \left(\frac{g}{\nu^2}\right)^{1/3} d_s, \quad \pi_{2,4} = \left(\frac{\rho' g}{\nu^2}\right)^{1/3} d_s = D^* \text{ (Eq.3.3)} \tag{3.111}$$

With Eq. 3.111 F in Eq. 3.110 reduces to the form

$$\phi\left(q_s^*, D^*, F_{sb}^*\right) = 0 \tag{3.112}$$

Zanke consequently postulated a bed-load formula of the kind,

$$q_s^* = F_{sb}^{*\alpha_1} D^{*\alpha_2} a \tag{3.113}$$

Equation 3.113 is an empirical one and has to be evaluated by measurements.

We repeat for the suspension load

$$\phi\left(q_{ss}, \rho_s, \rho_f, \nu, d_s, g, u_\tau, u_v, u_{\tau l}, T\right) = 0 \tag{3.114}$$

with additional variables which are a critical entrainment-shear velocity given by Eq. 3.70 and the temperature, which influences the viscosity. As a reference value ν_0 the viscosity at $T = 0°C$ is used, and the additional dimensionless numbers are

$$\pi_5 = \frac{u_\tau^2 - u_{\tau l}^2}{u_v^2} = F_{su}^*, \quad \pi_6 = \left(\frac{\nu}{\nu_0 - \nu}\right)^{1/4} = T^* \tag{3.115}$$

and the transport equation in this form becomes

$$q_{su} = F_{su}^{*\beta_1} F_{sb}^{*\beta_2} T^* b \tag{3.116}$$

For the total load Zanke continues with the dimensional analysis using the empirically evaluated parameters

$$\alpha_2 = 2\alpha_1 = 4 \quad \vee \quad a = 6.36 \cdot 10^{-4} \frac{1}{e} \quad \wedge \quad e \approx 0.7 \text{ (Eq.1.9)} \tag{3.117}$$

in Eq. 3.113, and we get

$$\begin{aligned} q_{sb} &= \frac{1}{e} 6.36 \times 10^{-4} \left(\frac{u_\tau^2 - u_{\tau c}^2}{u_v^2}\right)^2 D^{*4} \nu \\ q_{su} &= \frac{1}{e} 6.36 \times 10^{-5} \frac{H}{H_d} \frac{\left(u_\tau^2 - u_{\tau c}^2\right)\left(u_\tau^2 - u_{\tau l}^2\right)}{u_v^4} D^{*4} \nu \left(\frac{\nu}{\nu_0 - \nu}\right)^{1/4} \end{aligned} \tag{3.118}$$

and the total load by superposition as given by Eq. 3.107.

Einstein (1950) tried to incorporate the overlapping of bed and suspension load by treating the transport up to the height $a = 2d_s$ separately. By using his Eq. 3.20 respectively Eq. 3.23 for q_s, and by using Eq. 3.106 and the diffusion model Eq. 3.90 with the integration over the defined area it follows that

$$\begin{aligned} q_{st} &= q_{sb} + \int c_a \frac{u_\tau}{\kappa} \left[\frac{H - z}{z} \frac{a}{H - a}\right]^{\text{Ro}} \ln \frac{30z}{d_s} dz \\ &\wedge \quad c_a = \frac{q_{sb}}{a u_{x,z=a}} \quad \vee \wedge \quad u_{x,z=a} = \frac{u_\tau}{\kappa} \ln \frac{30z}{d_s} = 4.09 \frac{u_\tau}{\kappa} \end{aligned} \tag{3.119}$$

By introducing a series of dimensionless parameters, he found for the total load

$$\begin{aligned} q_{st} &= q_{sb} \left[1 + I_1 \ln \frac{30H}{d_s} + I_2\right] \\ &\wedge I_1 = 0.216 \frac{E^{\text{Ro}-1}}{(1-E)^{\text{Ro}}} \int_E^1 \left[\frac{1 - z^*}{z^*}\right]^{\text{Ro}} dz^* \\ &\vee I_2 = 0.216 \frac{E^{\text{Ro}-1}}{(1-E)^{\text{Ro}}} \int_E^1 \left[\frac{1 - z^*}{z^*}\right]^{\text{Ro}} \ln z^* dz^* \\ &\wedge \quad z^* = z / H \quad \vee \quad E = 2d_s / H \end{aligned} \tag{3.120}$$

All functions involved can be evaluated numerically or be found in a series of graphs existing in the literature. This is by far the most complete sediment transport equation existing at the moment. Together with some corrections for the grain size distribution, the hiding and grain-drag also is the most universal one. Results are good as long as the part of the bed-load contribution is large, which means $u_\tau > 2u_V$.

It is still debated whether it is not better to introduce a stream power

$$P = \tau U = \rho u_\tau^2 U \tag{3.121}$$

instead of the dominant parameter u_τ which was shown by Eq. 3.121 to be an equivalent representation. The idea behind this description is to become independent from any fluctuation. In the same sense the correction of Colby (1964b) in Einstein's formula must be understood because Einstein's formula gives bad results for too fine material, probably because Einstein did not integrate the temperature dependence of the viscosity for the suspension load. Colby too used the superposition representation and only corrected the transport through the suspensions using the new parameter set (U, H, d_{50}, T, c_{Vfine}). The resulting equation is

$$\hat{q}_{\text{su}} = \left[1 + \left(c_1 c_2 - 1\right) c_3\right] q_{\text{su}} \tag{3.122}$$

$$\wedge \quad c_1 = 1 \rightarrow \left(T = 15.5°\text{C}\right) \quad \vee \quad c_2 = 1 \rightarrow d_{10} \approx 0 \quad \vee \quad c_3 = 1 \rightarrow d_{50} \approx 0.2 - 0.3mm$$

where the constants were empirically evaluated.

Simons et al. (1981) developed a power law from Meyer Peter's Eq. 2.46 and Einstein's representation of the suspension load based on empirical data given as

$$q_{\text{ss}} = c_1 H^{c_2} U^{c_3} \tag{3.123}$$

The values c_1, c_2 and c_3 are available as graphical representation in the literature.

Bagnold (1966) used the energy model to develop Eq. 3.105, which for high transport rates gives fairly good results. But Bagnold uses the superposition of the two types of transport too

$$q_{\text{st}} = q_{\text{sb}} + q_{\text{su}} = \frac{\tau_{\text{W}} U}{G-1}\left(e_{\text{B}} - 0.01\frac{U}{u_{\text{v}}}\right) \quad \wedge \quad 0.2 < e_{\text{B}} < 0.3 \tag{3.124}$$

In this form, the sediment transport is represented by the stream power as defined in Eq. 3.121, and the total sediment transport is given by a constant, the so-called Bagnold coefficient. This representation is one of the simplest and therefore appropriate for estimations. Yang (1973) went a step further by introducing a description based on the concentration by weight,

$$c_{\text{W}} = \frac{c_{\text{V}} G}{1 + \left(G-1\right) c_{\text{V}}} \tag{3.125}$$

for the contribution of the suspension to the energy dissipation in the energy model. It is this part in the dissipation, given by $S\left(u_s^2 / u^2\right)$ in Eq. 3.103, which he changed. By doing so he recognized that the sediment transport must be split into two parts:

Transport of the fine sand and transport of gravel. He also used his results to formulate transport criteria based on the mean velocities

$$\begin{aligned} \frac{u_{\text{s}}}{u_{\text{v}}} &= \frac{2.5}{\ln\left(u_\tau d_{\text{s}} / \nu\right) - 0.06} && \wedge \quad 1.2 < u_\tau d_{\text{s}} / \nu < 70 \\ \frac{u_{\text{s}}}{u_{\text{v}}} &= 2.05 && \wedge \quad u_\tau d_{\text{s}} / \nu \geq 70 \end{aligned} \tag{3.126}$$

Engelund and Hansen (1967) operated with the stream power too and expanded Bagnold's equation, while Ackers and White (1973) argued that the additional dissipation is only an effect due to the fine sand component.

Looking at the sediment transport in the Colorado River, Vanoni (1960) showed how much the results calculated by the different formulas diverge, a fact found also in other river systems. For this reason, transport equations were also proposed based on regressions reduced from an immense dataset (Shen and Hung, 1972; or Karim and Kennedy, (1983). A newer overview of equations with better turbulent models can be found in Lyn (1986).

3.5 Critical Remarks

Before one can start with a new description it is always prudent to know the deficits of the old theories. We have already made several remarks of this kind, however a series of additional questions should help.

3.5.1 Form Drag

First of all we encounter a paradoxon. The flow develops by interaction with the erodible bed and through the sediment transport bed forms, which deviate from the flat bed. However, according to the equation of Bernoulli every elevation formed produces a higher velocity on its top, and therefore a higher capacity to erode. The flat bed should be the most stable one, which is not the case. In other words, a feedback must exist between the flow and the sediment transport, which allows the nonplanar bed forms. We will show later that this is a result of separation processes.

The bed forms preferentially have—in good agreement with the observations—a wavy character with a quasi-periodicity; they can therefore be approximated as a wavy bed, which is described by a frequency and an amplitude (often in a spectrum). The essential point however is that they should not be treated as waves because they do not propagate like waves but by a material transport at the surface. Therefore, they cannot be described by the wave equation.

The flow separation on these forms is of different size than the one on single grains, and an additional length scale has to be defined. This length scale should be equivalent to the one used when partitioning the flow field. Until now this fact was taken care of by introducing a wall shear stress containing three components

$$\tau_w = \tau_v + \tau_r + \tau_f \tag{3.127}$$

The part given by viscous forces is often neglected, and the remaining ones are due to the sand roughness and the bed forms, producing the so-called form drag. This splitting was introduced by Einstein and Barbarossa (1952), who also gave a function for the form drag of bed forms consisting of fine sand. Chuna (1967) extended this approach for gravel-like sediment. For the splitting into these three contributions, a series of statements had been introduced. The best known one are those of Engelund (1966a, 1966b) Lovera and Kennedy (1969), Alam and Kennedy (1969), Raudkivi (1976), and van Rjin (1984). At this moment, we will not go into detail further because some reservation with respect of a τ_f representation must be made. τ_f is the mean value of an inhomogeneous flow over a separation cell in which the flow is accelerated and decelerated. Therefore this splitting is rather problematic. Although it is used worldwide and helps to estimate the transport in case of bed forms, it does not say anything about the development of such forms.

3.5.2 Manifestation of Separations

Once more we stress that the whole momentum transport of the flow to a rough bed is controlled by the pressure distribution due to flow separations on the roughness elements, a fact that is not reflected in the classical transport equations. This is a deficiency one recognizes easily since most of the self-organizing processes are coupled with a feedback system in which the separations are essential.

In this respect, two categories of separation have to be distinguished. The separation on the single grain producing a local flow field is important for the stability of its neighbor, with respect to its erosion as well as its deposition. The second one is the separation on bed forms, which is essential for their life cycle.

In other words, flow separations must be incorporated into a description of the sediment transport, and we have to formulate some questions which we have to discuss further:

—*Separations produce flow structures, what is their importance?*

—*What are the feedback processes due to flow separations?*

—*What is the influence of flow separations on the bed stability?*

—*Can a roughness theory be formulated based on a statistic of separating flow elements?*

3.5.3 New Knowledge of the Turbulent Flow

The flow is described by the Navier–Stokes equation, and here it is not the place to go into its mathematical details. However, many essential properties of this equation have been neglected in the past, and this has to do with the rather confined use of turbulence as an important element of the sediment transport. Most modern descriptions make use of the velocity fluctuations, which is insufficient since the Navier–Stokes equation is an integro-differential equation. As, e.g., Tsinober (2003) pointed out the equation has a nonlocal part, which is represented by the pressure or by the vorticity, both are normally lacking in the description of transport. Therefore, we have to investigate the implication of such simplifications on the sediment transport since turbulence could not be described yet. The effect of the three big N, nonlinearity, nonintegrability, and non-locality, needs to be elucidated.

3.5.4 The Universality of a Sediment Transport Equation?

Sediment transport was thought of as a phenomenon, which can be described in the frame of classical mechanics, since it deals with forces and motions of particles. Therefore, it was commonly postulated that the sediment transport could be described by one universal equation, of course very complex and containing a series of parameters and many constraining equations. The transport therefore should be representable by modules, classical units for a numerical treatment of the problem. In the introduction we mentioned some thoughts of Einstein as a key for newer developments. His basic idea was that a universal equation does not exist since the problem is by far too complex. However if a description by a single equation was found his equation would be non-linear and therefore probably have chaotic solutions. We have to be more humble and admit that we are describing special cases for which we have studies and observations on similar systems. This is why Einstein insisted on case studies and their classification. In a discussion, Kennedy went even further by mentioning that the theoretical elements we use are no theories but are of more or less empirical character. The reason for this situation is not so much ignorance but has to do with the complexity of the coupling mechanisms, which remains an unsolved problem as long as we cannot describe turbulence by a closed theory. In a common lecture series at ETH Kennedy and Brooks pointed to the fact that all practical problems in sediment transport are instationary and depend on stochastic parameters. In other words, if a universal equation exists, it is not a deterministic one but a stochastic description with a series of dependent elements. Therefore what remains is a classification of the different states as proposed by Einstein. However, as Einstein taught it is very sensitive. He did so by showing pictures of riverbeds taken at rivers with practically identical parameters. They were still differing considerably in their geometry. It is therefore a pity that Einstein's picture library was lost for the scientific community.

With these remarks we close a first part with the question:

Up to which degree are the elements describing sediment transport of chaotic nature?

4 Saturation and Asymptotic States

Besides the attempts to describe the sediment transport by mechanistic processes, another approach was pursued by interpreting the sediment transport as a dynamical response to an extremum principle. In this case, the local mechanisms are not important; the system reacts and we need not know how it responds in detail. An important property of such a system theory is the saturation of the transport load, which is a quantity that was missing in the classical theories. Grass' representation was the exception, however neither there the saturation was incorporated with all its consequences. Therefore, part of the results found by the investigation of the dynamical processes can be incorporated into the older theories, and several authors have already done this in the past.

4.1 Sediment Transport as a Dynamical Process

An alluvial channel system responds continuously to changes of discharge, grain distribution, or other parameters. It does so by modifying the flow field and the channel geometry mainly by changing the bed form. The system tends to create a state of equilibrium, which is another way of saying that nature always tends to minimize an existing gradient. A change in the bed form must be due to a transport process. At one place sediment is eroded to be transported and deposited at another location in the channel. The transport needs the fluid as carrier, and the interaction is governed by a feedback process.

Any transport in a control section is determined by two quantities, namely the transport capacity and the available material, the supply. The classical transport equations are incomplete equations of capacity without restrictions for the saturation of the system. The difference between the influx and the outflow of material was neglected, which means the continuity equation for the sediment must be included by an additional equation. The material flux is composed of a convective input to the control volume, and a source term originating from the bed. Although, this kind of continuity equation is trivial, it is a good help for estimates. It can be formulated as

$$\frac{\partial(\rho_s c_V)}{\partial t} + \operatorname{div}(\rho_s c_V \underline{u}) = 0 \tag{4.1}$$

$$\wedge \quad \rho_s = \text{const.} \quad \therefore \quad \frac{\partial c_V}{\partial t} + \operatorname{div}(c_V \underline{u}) = \dot{c}_V = \dot{V}_s / V \tag{4.2}$$

and Eq. 4.2 can be integrated using Gauss' theorem

$$\int_V \frac{\partial \rho_s c_V}{\partial t} \mathrm{d}V + \int_A \left(\rho_s c_V u_x \frac{\partial x}{\partial n} + \rho_s c_V u_y \frac{\partial y}{\partial n} + \rho_s c_V u_z \frac{\partial z}{\partial n} \right) dA = 0 \tag{4.3}$$

$\partial x/\partial n$, ... are the direction cosines of the angle between the normal $\underline{n}$ to the surface element and the axes of the Cartesian coordinate system (x, y, z). For an incompressible medium the first integral is zero, and one gets the simplified equation,

$$\Delta_{2-1}\left(AU_x\right) + L_\mathrm{B} X_\mathrm{sal} \frac{\mathrm{d}h}{\mathrm{d}t} = 0 \tag{4.4}$$

with $\underline{u} = (U_x, 0, \mathrm{d}h/\mathrm{d}t)$, h the bed height, L_B the channel width and a mean saltation length X_sal.

In the stationary case $\mathrm{d}h/\mathrm{d}t = 0$ and we get

$$Q = A_1 U_{x1} = A_2 U_{x2} \tag{4.5}$$

The deposition is governed by the saturation of the transported sediment. The river is not capable to transport the whole amount of material that was eroded. In other words, a simplified form of a criterion for the saturation can be formulated

$$\Delta_{2-1}\left(AU_x\right) \leq 0 \quad \wedge \quad L_\mathrm{B} X_\mathrm{sal} \frac{\mathrm{d}h}{\mathrm{d}t} \geq 0 \tag{4.6}$$

where the control cross-section 2 lies downstream of 1. However, this is merely a rough description; a real criterion must contain the dynamical flow parameters.

In the case of steady-state conditions, which is presumed for most of the transport equations, the channel would remain unchanged and the transport could be described by an equation for the transport capacity (influx = outflux). However, the steady-state case is the exception. The sediment transported into a control section usually originates from further upstream, where it was eroded. In the general case of nonsteady-state conditions, even if the source of the sediment is of homogeneous particle size and the grain distribution is constant, the geometry of the channel has to change by erosion or deposition, e.g., by a change in slope. A nonstationary description must therefore be supplemented by additional arguments, such as the erosion of the side banks by slides, which would compensate for the bed-erosion.

The behavior over time can also be incorporated by supplementing the capacity equation with a continuity equation, like Eq. 4.1 or 4.3, to a system.

To achieve this, the calculation needs to be split into sections, where the resulting transport of an upward element defines the initial condition of the downward element. Another argument in favor of a sectional splitting of flow system is the large time scale needed to change the geometry of the bed. It is thought that rather instantaneous transport events can be treated as stationary. However, data for short time investigations are rare. With this splitting as a method of calculation, we sketch a possible solution of the problem. Without further analysis of the physics, the question remains, whether a superimposed universal law exists, which restricts the interaction processes, as has been found in the form of Gibbs' potential in thermodynamic.

In the past, this idea was proposed periodically in a variety of theories based on conservation laws and restrictions formulated as laws of extreme behavior, so-called minimizing or maximizing principles. The astonishing fact about these kinds of theories was that they came to contradictory results, and therefore further analysis should help in reaching a better understanding of the interaction mechanisms.

4.2 Hypotheses of Extremum Principle

A series of hypotheses on extremum principles exists in the literature and can be classified by the functionals through which they are defined. We distinguish categories based on the classical conservation laws for mass, momentum, and energy.

4.2.1 Continuity Arguments

- Minimization of the discharge Q. For a given slope S and a sediment concentration c, the characteristic parameters of a channel change as a whole, such that Q remains minimized.
- Maximization of the sediment concentration. For a given Q and S, the channel width L_B will develop such that the transported sediment rate q_s is a maximum (Singh, 1961; White et al., 1982).

In both criteria all other parameters were held fixed. Additionally the first criterion strictly is a momentum criterion, since a minimal Q is correlated with a maximum friction. In general, the investigated parameters depend too much on all other variables, and the strong restrictions of fixing all other parameters cannot be considered universal.

4.2.2 Momentum Arguments

- A minimal stream power: Eq. 3.121 or in the form

$$P = \rho_f QS \tag{4.7}$$

  Given the boundary restrictions, Chang (1980) postulated a necessary and sufficient condition for the occurrence of an equilibrium state in requiring that P per river length is minimal. Then, an alluvial channel with Q and Q_s as the independent variables changes its width and slope such that P is minimized. For a given Q that means that also the slope is a minimum.
- A minimal Froude number Eq. 1.33. For a given Q and Q_s, the width of the river establishes itself such that Fr becomes minimal.
- The total friction drag is minimal. For a given Q and Q_s the total friction drag of the river becomes a minimum.
- The friction factor λ (Eq. 2.39) is minimal. This is equivalent to the former statement; however, it is disputed in this formulation.
- The friction factor λ (Eq. 2.39) is maximal. Davis and Sutherland (1980) concluded this based on the following observation: beginning with a flow over an initially flat bed, the shape of latter is changed by sediment transport resulting in an increased drag, due to the form drag of the bed structures. This deformation ends, when the local friction factor has reached a maximum. They conclude that the equilibrium state of a nonplanar bed, which developed by a self-organization of the system, must be accompanied locally by a maximum of the friction coefficient.

In other words, different researchers found diametrically opposed extremum principles. The water in a river would continuously accelerate due to gravity if friction would not decelerate it. This means the friction must regulate the equilibrium state. The fluid friction is a result of momentum flux from the fluid into the bed. As we know, there are two sources of friction, the surface friction and the form drag resulting from flow separations. Now the above criteria can be formulated using only one or both parts of the friction. If both contributions are used it is best to shift to a description using the wall shear stress τ_w. In this representation, it is easily seen that all the above laws of extremum principles are not independent

$$\lambda = 8\left(\frac{u_\tau}{U}\right)^2 \quad \because \quad u_\tau = \sqrt{\frac{\tau_\mathrm{w}}{\rho_\mathrm{f}}} \quad \therefore \quad \lambda = 8\frac{\tau_\mathrm{w}}{\rho_\mathrm{f} U} \tag{4.8}$$

Now it is evident that λ and τ_w can be used to form the same laws of extremal principles. Similarly, also the Froude hypothesis can be formulated equivalently

$$\mathrm{Fr} = \frac{U}{\sqrt{gH}} \quad \because \quad \tau_\mathrm{w} = \rho_\mathrm{f} gHS \quad \therefore \quad \lambda = \frac{8gHS}{U^2} = \frac{8S}{\mathrm{Fr}^2} \tag{4.9}$$

Since Fr and λ are inversely proportional, a minimum in Fr corresponds to a maximum in λ. The stream power is a quantity, which describes the equilibrium state implicitly, and with Eq. 4.9, Eq. 4.7 becomes

$$P = \rho_\mathrm{f} QS = \frac{Q\tau_\mathrm{w}}{gH} \tag{4.10}$$

Here a minimum in P is equivalent to the statement that the friction factor is a minimum. This means we have contradictory statements and it seems of value to discuss it. The key to this contradiction lies in the form drag, which is poorly represented by the wall shear stress, and we will come back to this fact later. However to complete the representation of laws of extremum principles based on the momentum exchange, we have to mention some newer approaches incorporating the form drag. One of the very first steps will be to recognize the ambient shear conditions, as they notably control the local value of the energy slope. This requires that we are able to predict some form of a resistance coefficient, which is explicitly connected to the boundary roughness quantity, for a given river at any location. The pure skin friction introduced by an equivalent sand roughness can be used for the two plane bed conditions. The first is the so-called lower-stage plane bed (LSP), a flat bed at very low transport, and the second is called the upper-stage plane bed (USP) for the flat bed regime occurring in the two-phase condition. They differ considerably in their sand roughness (Nomicos, 1956; Julien and Raslan, 1988). For a bed deformed into morphological features, practitioners are advised to use the so-called form drag approach described in Klaasen (1980) and Karim (1995) which builts empirical relationships between geometrical properties of the bed forms, e.g., dune height and wavelength, and the roughness they create. We will discuss the physics of the separation process later in much more detail. However, it has to be mentioned here that Verbanck (2004) noticed a bed condition in the upper alluvial regime, passing from USP to the antidune standing wave (ASW), which actually corresponds to a decrease in the friction factor. Verbanck explains this drag reduction by saying that in the region of topographic forcing, streamline curvature is noticeable and inhibits turbulence (by rearrangement or even "relaminarization" of streamlines over the dune crest). This effect is eventually assisted by the very high levels of suspended bed-material load (Galland, 1996; Garg et al., 2000; Hsu et al., 2003). One of the consequences of this turbulence damping is that shear stress is locally reduced and allows the bed form to grow in height and maintain itself (Nelson et al., 1993). Verbanck (2004) speculated therefore that this condition represents the most effective way to evacuate extreme flow discharges as it minimizes the alluvial resistance.

4.2.3 Energy Arguments

- Minimal energy dissipation. Brebner and Wilson (1967) or Yang et al. (1981) postulated that the system is in equilibrium, if the energy dissipation becomes a minimum.

Potential energy is transformed into kinetic energy due to the slope of the topography, and in the equilibrium state this amount of energy has to be dissipated. The dissipated energy is therefore fixed and cannot be subject of an extremum condition. The transformation from one form of energy to another one always is in equilibrium too and therefore has to be formulated in a different way. However, a river system has to overcome a given difference in altitude and can do so by changing also its length, in other words, its slope. The above-mentioned criterion therefore has to be reformulated

- A river chooses a slope S, such that for a given Q and Q_s the energy dissipation ε becomes a minimum.

The dissipation rate ε of the flow energy is the sum of a viscous and a turbulent contribution, of which only the latter is of importance. ε therefore can be replaced by the dissipation rate found in turbulence theory. If one neglects rheological deviation from a Newtonian fluid one finds (Tennekes and Lumley, 1972), e.g.,

$$\varepsilon = 2\nu\overline{s'_{ij}s'_{ij}} \quad \because \quad \overline{s'_{ij}} = \overline{\frac{1}{2}\left(\frac{\partial u'_i}{\partial x_j} + \frac{\partial u'_j}{\partial x_j}\right)}$$

$$\vee \quad \varepsilon = 2\nu\int_0^\infty k^2 E(k)\mathrm{d}k = 15\nu\frac{u'^2}{\lambda_{\mathrm{Ta}}} \approx \overline{\omega'_i\omega'_j} \tag{4.11}$$

where $E(k)$ is the energy spectrum as function of the wave number k, and λ_{Ta} is the Taylor microscale. The energy dissipation rate is therefore proportional to the square of the deformation rate, which is approximately equal to the enstrophy, given by the fluctuations in the vorticity. To minimize the energy dissipation is therefore equivalent to an increase of the size of the turbulent structures or a decrease in the enstrophy production. In case the production and the dissipation are in equilibrium, we have

$$-\overline{u'_iu'_j}S_{ij} = 2\nu\overline{s'_{ij}s'_{ij}} \tag{4.12}$$

In rivers, the flow can be described in good approximation by a constant stress density and we get

$$U_z = 0 \quad \wedge \quad -\overline{u_x' u_z'} = u_\tau^2 \tag{4.13}$$

Equation 4.13 can be inserted into Eq. 4.12 and it will result into Eq. 4.11. Using Eq. 4.8 it follows that

$$\varepsilon = u_\tau^2 S_{ij} = u_\tau^2 \frac{\partial U_x}{\partial z} = \frac{\tau_\mathrm{w}}{\rho_\mathrm{f}} \frac{\partial U_x}{\partial z} = \lambda U_x^2 \frac{\partial U_x}{\partial z} \tag{4.14}$$

In all these equations ε was a local value, whereas for a global extreme-condition the mean dissipation rate has to be used. Therefore, we have to calculate the mean value over a cross section of the channel as it was done for the mean velocity, $\overline{U_x} = Q/A$ and we get

$$\bar{\varepsilon} = \frac{\iint_A \varepsilon \mathrm{d}A}{A} \tag{4.15}$$

In addition to the momentum arguments, we include the velocity profile in such a representation, and this means that the flow parameters are considered in a more complex form. However, just as for the momentum arguments, contradicting results can be found. For example, in a river where the roughness becomes larger ε increases in contradiction to the extremal criterion. On the other hand, in a meandering river the energy dissipation can decrease per unit length and the criterion would be correct.

Laws of extreme would be a wonderful tool for simulations, because when such limiting laws exist, the problem could be treated in a variational approach, since the equilibrium state would be an asymptotic state. Due to this promising approach, however with experimental data supporting contradicting hypotheses, we have to investigate, why we get contradictions. Can the state of a river switch from one to the other, and what would be the physical background of such an instability? As mentioned earlier, we find a possible explanation in the behavior of the bed resistance given by the sand roughness and the form drag of the bed forms. For the sand roughness we have

$$\frac{\partial \tau_\mathrm{w}}{\partial Q} > 0 \tag{4.16}$$

If dunes exist, the above value can become negative. A positive gradient is compatible with the assumption that a maximum dissipation rate exists; a negative value stands for a minimum dissipation. This result is, however, only correct for a straight channel and the additional condition that $Q_{\mathrm{s\,in}} = Q_{\mathrm{s\,out}}$. Yet, in its essence, the paradox is clarified. When the system switches from one to the other regime this will lead to the formation or the disappearance of bed forms, which we will discuss separately. In other words, a universal constrain based on conservation laws is not sufficient since a feedback of the system was not considered.

On the other hand, these laws can be useful too, as they show the importance of the sediment flow rate Q_s, which is equivalent to a generalization of the continuity equation. In Fig. 4.1, formulating a new universal law shall elucidate this

$$\nabla Q_\mathrm{s} \xrightarrow{t \to \infty} 0 \tag{4.17}$$

This formulation says that Q has to be maximal in case (a) and minimal in case (b) see (Fig. 4.1).

Fig. 4.1 The behavior of a river with a sudden change in slope *S*. (a) In the part with a large slope, the bed will be eroded and *S* is decreasing. (b) In the part with a smaller slope sediment is deposited and the slope is decreasing too. Moreover this adjustment will move upstream

The process itself is very complex and not part of the extremum approach, however, additional conclusions can be drawn from observations. For case (a) in Fig. 4.1, where $\nabla Q_s > 0$, the channel is straight. The reason for this fact is a stabilization of the flow by secondary vortices. For $\nabla Q_s < 0$ the bed is rising, but since the discharge must remain Q, the river searches for a new river bed if the slope becomes too small. This process is chaotic and results in a new topography, as it can be observed when a river forms a delta, or as partly chaotic producing a meander. This kind of channel is extremely interesting since it is longer than the straight one overcoming a given difference in altitude, and it is capable of a higher discharge. Asymptotic behaviors in nature are always suited for studying complex mechanisms and the meander is just a good example in the context of sediment transport. An overview on this topic can be found in Calander (1978). Here we will restrict ourselves to the attempt to describe meanders by an extremum process in the sense of thermodynamics, as postulated, e.g., by Scheidegger (1967). The basic idea was proposed by Langbein and Leopold (1966) (see also Leopold and Langbein, 1966) showing that there exists a finite probability $f(k)$ with

$$k = \frac{\mathrm{d}\varphi}{\mathrm{d}s} \tag{4.18}$$

which means a river can bend by an angle $\mathrm{d}\varphi$ against the straight line direction during its passage through a section of the length $\mathrm{d}s$.

The authors arbitrarily assumed that $f(k)$ could be given by a Gaussian distribution

$$f(k) = C\mathrm{e}^{\left[-(1/2)k^2/\sigma^2\right]} \tag{4.19}$$

with σ the standard deviation of φ. This is identical with the random walk problem as treated by von Schelling (1951, 1961, 1964). He found for x and s, the arc length

$$\begin{gathered} s = \frac{1}{\varphi}\int \frac{\mathrm{d}\varphi}{\sqrt{2(\gamma - \cos\varphi)}} \quad \wedge \quad x = \frac{1}{\varphi}\int \frac{\cos \mathrm{d}\alpha}{\sqrt{2(\gamma - \cos\varphi)}} \\ \because x: \text{abscissa}, y: \text{ordinate} \quad \wedge \quad \varphi = \operatorname{arc}tg\,\mathrm{d}y/\mathrm{d}x \quad \wedge \quad \gamma = 1 - \cos 2\omega \end{gathered} \tag{4.20}$$

Using this crude method, which does not contain the flow parameters, one finds the meander bending in rather good agreement with observations. Instead of random walk approach, Scheidegger (1967) suggested to use a variation principle for the change in direction encountered in meanders in analogy to the Boltzmann velocity distribution in gas dynamics. He concludes that a description could be formulated in complete analogy if we could postulate the Gaussian distribution. This would have to be ascertained experimentally, and in the stationary case $f(k)$ has the linear relation,

$$\overline{\left(\frac{\partial\varphi}{\partial s}\right)^2} = a + bs \tag{4.21}$$

Since then the Gaussian distribution was searched for (Thakur and Scheidegger, 1968; Peschke, 1973), but not always found. [Surkan and Van Kann (1969) who therefore rejected the idea.] The problem remains unsolved and a universal extremum condition failed once more. In this case, one probably has to introduce also a system feedback taking care of the changes in the parameters by the curvature of the channel. We will not go deeper into this problem but reproduce the results of Onishi et al. (1976). These authors investigated the drag and the sediment transport in channels with bends. They introduced a bend loss coefficient K_L by the difference in resistance between the head loss for one bend and the loss for an equal length of a straight channel as

$$K_L\left(\frac{U^2}{2g}\right) = \Delta\left(\frac{U^2}{2g}\right)_{\mathrm{B}\wedge\mathrm{L}} \tag{4.22}$$

The result was

$$\begin{aligned} &K_L \upuparrows \quad \wedge \quad \mathrm{Fr}; r_d, b_r \\ &\mathrm{Fr} = \frac{U}{\sqrt{g r_b}} \quad \wedge \quad r_d = r_b / d_{50} \quad \wedge \quad b_r = B / r_c \end{aligned} \tag{4.23}$$

which means that the bend loss coefficient increased with the Froude number. r_d is the hydraulic radius and measured relatively to the mean grain size, and b_r is a normalized span where B and r_c are the span and centerline radius. In certain cases, K_L was negative that means the head loss in the curved channel was less than that in the straight channel. Their result for sediment transport showed that a wide curved channel $b_r = 0.274$ performed better than the straight channel and a narrow curved channel $b_r = 0.137$ not so well as the straight channel. The explanation of Onishi et al. (1976) was that this drag reduction was due to the secondary flow and its interaction with riverbanks.

The search for a universal law of an extreme is probably an unrealistic task. For many investigations the approach was already a helpful tool to gain a better understanding. The only result, which will remain consistent with all observations, is the relation (Eq. 4.16) giving rise to the conditions in Fig. 4.1a and b and with it Eq. 4.17, the only universal condition.

4.3 The Expanded Description of Grass

Let us start with the bed-load transport. The sediment transport depends on the flow conditions usually characterized by the wall shear stress τ_w. If a critical shear stress τ_{wc} is reached, a given grain will move. This criterion is not fulfilled over the entire bed since it is caused by the instantaneous conditions given by the turbulent flow. On the other hand, everywhere the condition is met, all grains belonging to this class of transportable grains move. In the description of Grass, these conditions are reflected by probability functions as shown in Fig. 3.4. The empirical determination of q_s by a bed-load formula has to be a value given by the overlapping of the two probability functions $P(t)$ sketched by the gray area in Fig. 3.4. Such a description assumes that the probability distributions are constant, which means that the eroded material has to be replaced by material consisting of the same grain size and density distribution. The sediment source can be the bed itself or an upward section of the river. Such equilibrium rarely exists since a hydrological regime is in most cases nonstaedy state.

This description can, however, be rendered dynamic by introducing time dependent probability functions. The interaction with the flow asks for an iterative solution of the process. Therefore, the computation becomes very extensive, which shows how restricted even the classical statistical descriptions are.

In Fig. 3.4 a system was shown in which not all grain fractions are moving. In this case, we speak of a partial bed-load transport. When not enough sediment is supplied from the upward flow the bed starts to become armored, which results in the grain distribution of the covering layer of the bed becoming rougher and rougher until no more grains are transported. No genuine bed-load transport exists and the sediment discharge is due to feeding from upper sections. If this source is also drained, the bed-load transport stops, and we can speak of a static stable bed configuration. This is a first asymptotic state, and the complex details of this stabilization process will be discussed later.

Besides this special case some other interactions can be studied in the Grass representation (Fig. 4.2).

In Fig. 4.2a, the probability function of τ_w exists for all particles, which can be transported due to τ_{wc}. At first glance one would assume that all particles are in motion. But the Grass representation shows that the part of the grain distribution exceeding the transport capacity will remain at rest (the white area in the figure). On the other hand in Fig. 4.2b, the distribution shows that there exists a surplus in capacity capable to move all grains still at rest. This case has not been studied by experiments, but it is evident that the classical transport equations are problematic in this respect too. A description as given by Grass, cannot be used to calculate the transport capacity and is therefore still of empirical quality best for cases close to the one used for the calibration. The supply as well as the capacity can be limited resulting in a variation of the transport equations. The case where the eroded and deposited material is in equilibrium is called a dynamic equilibrium. Often this status is called the saturated case; however, this terminology should be used only for a flow not capable of transporting additional material, although the criterion τ_{wc} is fulfilled. We speak of a saturated system when the capacity of the transport is limited, although there is enough transportable sediment available. On the other hand, a status of limited supply causes a change in the roughness of the bed.

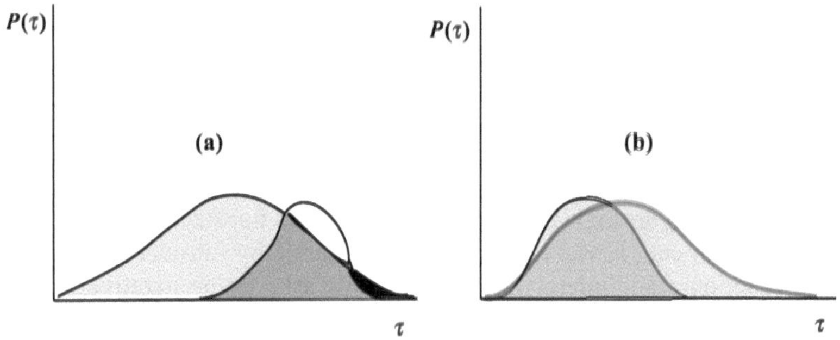

Fig. 4.2 The joint probability of supply and capacity. (a) Part of the supply (the white area) cannot be transported; on the other hand we have an excess capacity (the black area). (b) Capacity and supply do not overlap entirely, however, by the integral concept all material is transported

For a suspension we can argue very similarly, if the transport is defined by a critical value like $u_{\tau c}$ (Eq. 3.69 or 3.70). However if a suspension and a bed-load transport exist simultaneously, the critical value for the suspension is always fulfilled. The suspension can interact with the system only through its interaction with the flow structures or by the interaction of colliding particles. Such a model cannot explain the fact that material of the size of the suspended particles still exists within the cover of the bed. For this, a theory for the hiding parameter must be developed.

In case of a steady-state distribution of τ_w (see the comment Eq. 3.32) than an increase or decrease of the slope demand that $\underline{x}(t)_w$ must depend on both variables. The equations have to be solved iteratively. This shows clearly that a time- and space-dependent transport equation is missing. This deficit of the classical representations is known since they do not incorporate the previous history of the river so important for the actual transport.

This deficiency was passed over by two methods. One is by treating the system iteratively that means not only the transport is calculated iteratively but also the bed configuration, which in the classical theories is thought as a steady-state boundary condition. Another possibility is to treat the river system, as it would be exposed during limiting sequences in time to extreme flow conditions, since these events are responsible for most of the sediment transport.

4.4 Limitations

Alluvial systems are always limited in a sense. These limits are given not only by the precipitation or the amount of erodible material but also by the grain distribution if only part of it can be transported. The probabilistic representation of Grass pointed out the problems in a limited form, since this representation assumes universal distribution functions, which never exist in real life. Especially τ_{wc} depends very much on the grain distribution and the arrangement of the grains so that an empirical input is needed. The transport saturation is calculated by a chosen transport formula and the results therefore deviate extremely depending on the selected theory. When we use a calibrated formula for an empirical input, the quality of the result depends on how closely the investigated example is reflected by this calibration. The Grass representation is basically correct, although not universal. An attempt to overcome its deficits lies in the partitioning of the grain distribution and the assumption that every fraction can be treated as sediment of uniform grains. Here the τ_{wc} distribution would become very narrow for each fraction, and a collapse into a single line is defined by its value in the Shields diagram Fig. 3.1 for a single size. The transport calculated by this method is usually too large and it is always larger than the value calculated assuming a uniform grain size of d_{50}. An improvement cannot be expected as long as the influence of the turbulence is not introduced with a higher resolution in time. In other words, τ_w as a mean value remains the most problematic value.

This sounds rather pessimistic; however, the flows showing saturations or a limitation are extremely important. For example an armored bed defines a grain distribution producing a static stability although the cover of the bed contains grains, which should be moved according to Shields diagram. This is possible only because in an armored bed nonmovable ones protect some transportable grains. In other words, they can hide behind them. Since this configuration is static stable it can be used to test or to evaluate Einstein's hiding factor ξ (Eq. 3.24). This is the main reason why investigations of extreme cases, like system in a static or dynamic stability, are so important for theoretical purposes.

5 Problematic Issues

Beginning with this chapter, we change the presentation of the material somewhat, since in Chaps. 1–4 the classical material was sighted and compressed, so that the reader can calculate the sediment transport as this has been done until now, while also being aware of the deficits of the classical methods. From now on, we will concentrate on aspects that would help us widening our knowledge and improving the predictability for a given system. That means not the recipes will be the goal but the descriptions of the circumstances that should help to approach difficult questions more autonomously. Often this is not more than inspiration or a suggestion; however, it also sharpens our thoughts for future research.

The often-cited complexity of the sediment transport has its origin in the large variety of the sediments and their specific properties on one hand and in the turbulence which is the crucial phenomenon of the flow on the other hand. One problematic issue of the grain distribution has already been discussed for the bed load but not yet for suspensions. Even more problematic is the description of the interaction of the turbulent flow with the sediment. It is so complex that it is usually given by a superposition with some corrections because simplifications are needed to reduce the transport problem to one which can be investigated with the available tools.

By describing a flow phenomenon, one has to distinguish cases in which the instantaneous and local 3D description is essential and those where it is sufficient to use mean statistical quantities. Both representations have their advantages, and often a mix can improve the prediction quality.

For example, at the moment the pure fluid starts to transport sediment, the fluid becomes a two-phase mixture and its properties start to deviate from the original Newtonian behavior defined by Eqs. 1.14 and 1.15. The rheology of such systems is extremely complicated; however using a modified representation of the viscosity, one can obtain good results. Questions of this kind will be treated in the first part of this chapter. In the second part, the features of a turbulent flow and their influence on the transport will be investigated. Particularly we will discuss the consequences, which have to be drawn from the fact that a turbulent flow is nonlocal and nonlinear. Nonlinearity is not only a property of the flow itself but also of most interaction mechanisms which results in the occurrence of chaotic phenomena. In the third part of the chapter, we will investigate the influences of such processes.

5.1 Assumptions and Consequences of Rheological Nature

Rheology describes the state of stress of matter as a consequence of its deformation. Usually when we talk of sediment transport it is in a geophysical context, and we assume that the fluid is water or air, both of which are classical Newtonian fluids. The transported material will alter, e.g., the density of the fluid or even modify its bulk rheological properties. Therefore, we will first discuss the properties of a Newtonian fluid allowing a later description of the modifications introduced by the sediment.

The fluids are described by their constitutive equation; these are equations relating suitably defined stress and deformation variables. For a Newtonian fluid, the constitutive equation is defined by a linear relation between the stress tensor $\underline{\underline{T}}$ of the fluid and the deformation tensor $\underline{\underline{D}}$, (Serrin, 1959)

$$\underline{\underline{T}} = (-p + \lambda \operatorname{div}\underline{u})\underline{\underline{I}} + 2\mu \underline{\underline{D}}$$
$$\underline{\underline{T}} = -p\underline{\underline{I}} + 2\mu \underline{\underline{D}} \tag{5.1}$$

with

$$\underline{\underline{I}} = \delta_{ij}; \quad \underline{\underline{T}} = \tau_{ij}; \quad \underline{\underline{D}} = d_{ij} = \frac{1}{2}\left(u_{i,j} + u_{j,i}\right) \tag{5.2}$$

and p the thermodynamic pressure, and λ and μ scalar functions of the thermodynamic state. For incompressible fluids, p is a fundamental dynamical variable and μ a scalar function of temperature, whereas $\underline{\underline{T}}$ and $\underline{\underline{D}}$ are symmetric tensors. This is in agreement with a linearization of the Cauchy relation for a Stokes fluid

$$\underline{\underline{T}} = \alpha \underline{\underline{I}} + \beta \underline{\underline{D}} + \gamma \underline{\underline{D}}^2 \tag{5.3}$$

with α, β, γ being scalar functions of the principal invariants of $\underline{\underline{D}}$, that is

$$\alpha = \alpha\left(I, II, III\right) \quad \text{etc.} \tag{5.4}$$

The principal invariants may be defined as the coefficients of the characteristic polynomial from the expansion of the determinant given by Eq. 5.5

$$D(\lambda) = \left\|\left(\lambda \underline{\underline{I}} - \underline{\underline{D}}\right)\right\| = \lambda^3 - I\lambda^2 + II\lambda - III \tag{5.5}$$

It follows in particular that I = Trace $\underline{\underline{D}} = \operatorname{div}\underline{u} = \Theta$. The eigenvalues d_1, d_2, d_3 of $\underline{\underline{D}}$ are roots of the equation $D(\lambda) = 0$; they are all real since $\underline{\underline{D}}$ is symmetric. Clearly the principal values of $\underline{\underline{D}}$ are functions of the principal invariants.

A definition of the functions α, β and γ gives rise to a definite type of viscous response of the fluid. The definition of the viscosity in a Newtonian fluid is given by the Cauchy–Poisson law of viscosity Eq. 5.1 for a Stokes fluid, which is defined by the following conditions:

1. $\underline{\underline{T}}$ is a continuous function of the deformation tensor $\underline{\underline{D}}$, and is independent of all other kinematic quantities.
2. $\underline{\underline{T}}$ does not depend explicitly on the position $\underline{x}$.
3. There is no preferred direction in space.
4. When $\underline{\underline{D}} = 0$, $\underline{\underline{T}}$ reduces to $-p\underline{\underline{I}}$.

These criteria allow to assess whether a Newtonian formulation is appropriate or not.

The pertinent flow equation is the Navier–Stokes equation (NSE) (1.20). Let us assume that we incrementally add sediment to such a fluid. The increasing sediment content will change the constitutive equation and with it the rheological behavior. The sediment load will be given by the volume-concentration c_V respectively the volume occupied by the sediment V_s. The main forces acting on the particles are surface forces, and it is therefore appropriate to use the volume description instead of the mass description. Three groups of forces are most relevant: the colloidal forces, the forces resulting from Brownian motion, and the viscous forces, which are of hydromechanical origin even for turbulent flow conditions. Older literature on small particle interactions with the fluid can be found in Happel and Brenner (1965).

5.1.1 Influence of Colloidal Forces

The colloidal forces, which are described by the laws of van der Waals, are of electrostatic origin and can therefore be of attractive or repulsive nature. The surface charge distribution on a particle depends not only on its form but also can be strongly influenced from outside, mainly by charges deposited on the surface of the particles due to active molecules such as surfactants. Especially particles capable of forming electrical double layers have to be mentioned such as clay or bentonite, both consisting of microscopic plates. When such material gets into contact with salty water, the rheology can change drastically because these plates self-assemble similarly to a house of cards where the voids are filled with water. Events like this can be relevant in certain cases, e.g., when a river carrying clay sediment discharges into the sea. We will not treat these special cases but refer to the literature. A good overview, especially on the implications of the various forces, can be found in Israelachvili (1994) or for colloidal forces in Hunter (1987).

When the colloidal forces are dominant, flocks are created by attractive forces, whereas in case of repulsive forces pseudo-lattices occur. In case of very high-repulsive forces, even pseudo-crystals can be detected by light scattering. It is therefore necessary to have a criterion determining the importance of the colloidal forces. A meaningful approach necessitates that the distance between neighboring particles is known in the mean, $\langle r_{a1} \rangle$. This property can only be evaluated with special measuring methods. WIth this input, a criterion for the importance of the colloidal forces can be formulated (Woodcock, 1985) as

$$\frac{\langle r_{a1} \rangle}{d_s} \geq \left[\left(\frac{1}{3\pi V_s} + \frac{5}{6} \right)^{1/2} - 1 \right] \tag{5.6}$$

For spherical particles of 1-μm diameter at a distance $\langle r_{a1} \rangle$ of 10 nm, the colloidal forces for particles suspended in a volume-concentration of 0.575 in water become dominant. For the same distance a volume-concentration of 0.268 already gets critical for particles of the size 0.5 nm. The changes in the rheology resulting from such colloidal forces are rather numerous and cannot be discussed here in detail. Usually colloidal forces can be neglected; however, the influence of high concentrations of fine material in mixtures has not been investigated until today, although in such cases some deviations can be expected. For simplicity therefore the goal is to reduce the complexity by formulating a functional for the viscosity instead of constitutive equations.

5.1.2 The Influence of Brownian Motions

Particles are stochastically moving due to impacts of fluid molecules according to the thermodynamic state of the mixture. For larger particles, the Brownian forces can be neglected since the averaging of the forces tends to zero and the inertia of the particle is high. For small particles, this is different and the Brownian motion is relevant for particles of d_s<1 μm. This is an empirical value; however, it can be confirmed by the theory of Brownian motion

$$E_k = \frac{3}{2} kT \quad \wedge \quad d_s \approx 1 \mu\text{m} \rightarrow m = 10^{-15}\,\text{kg} \quad \therefore$$
$$u_{\text{th}} = \sqrt{3kT/m} \approx 3\text{mm}/\text{s} \tag{5.7}$$

with the Boltzmann-constant $k = 1.38 \times 10^{-23}$ J/K and T the temperature in K. Not every change of direction can be observed, however, the mean shift-value x in time t can. For the quadratic shift distance it was found

$$\overline{x^2} = 3\text{D}t = \frac{kT}{2\pi \mu r_s} t \quad (\text{Einstein and Smoluchowski}) \tag{5.8}$$

The Brownian forces are homogeneous and isotropic and therefore responsible for a homogenization of the particle distribution. In this case, the particles have a distance to each other close to the statistical mean and the mean distance is the third root of the number of particles in a unit volume.

Viscous forces are acting on the particles. The viscous forces are proportional to the local velocity difference between the particles and the surrounding fluid. Hence, the way these affect the suspension viscosity enters via the viscosity of the continuous phase which then scales all such interactions. Suspension viscosity is usually considered as the viscosity relative to that of the continuous phase. The macroscopic viscosity is strongly dependent on the microstructural details. Isolated particles imply that the flow lines deviate from the undisturbed flow and hence produce a suspension viscosity, which increases with the concentration of the particles.

In a two-phase flow, which is homogeneous in the previous sense and exposed to a weak shear rate $\dot{\gamma}$ Eq. 1.13, the particles would have to flow around each other or collide with each other. This is equivalent with an increase in drag or in other words, it would correspond to an increase in viscosity. However, the distribution of particles remains essentially constant because of Brownian motion, which restores the randomness. In case the shear rate is increased, the imposed velocity gradient induces an orientation of the particle structure, which is not restored by the Brownian motion. The particles start to form chains, which can move in the fluid with a smaller drag, and therefore the viscosity decreases, a so-called shear thinning can be observed. The existence of chains and sheets can be observed by light scattering. However, at a certain critical value $\dot{\gamma}_c$, these packages break and the viscosity increases with increasing shear rate.

5.1.3 The Influence of the Viscosity

The viscous forces acting on the particles are proportional to the relative velocity u_r between the grains and the fluid (Chap. 3, Sect. 3.3.1). In other words, the reaction of the flow to the suspended particles is equivalent to a viscosity so that the influence of the suspensions can be described as an alteration of the viscosity (Chap. 5, Sect. 5.1.2). The "suspension viscosity" is in this case formulated as a relative viscosity compared with the one of the pure fluid

$$\mu_s = \mu_f f(V_s) \quad \rightarrow \quad \mu_s = \mu_f f(V_s, d_s, ...) \tag{5.9}$$

For $V_s < 10\%\ V$ for Eq. 5.9 the formula derived by Einstein (1906, 1911) can be used,

$$\mu_s = \mu_f (1 + 2.5V_s) \tag{5.10}$$

A derivation of this formula using the creeping condition can be found in Faber (1995).

Equation 5.10 is independent of the grain size and the positions of the spherical particles. If the interactions between the particles are to be included, higher orders of V_s have to be considered. Such a higher order equation for the elongational viscosity was derived by Batchelor (1977).

$$\mu_s = \mu_f (1 + 2.5V_s + 6.2V_s^2) \tag{5.11}$$

For the shear viscosity, a variety of coefficients for the V_s^2-term have been published varying between 5 and 15 (Barnes, 1981). For higher concentrations of suspensions, the theoretical description becomes extremely difficult and needs averaging processes over neighbor influences. Ball and Richmond (1980) did this by using the sum of continuously added new particles, that means Einstein's equation (Eq. 5.10) is transformed into an infinitesimal equation, which can be integrated,

$$\mathrm{d}\mu_s = \left(\frac{5\mu_s}{2}\right)\mathrm{d}V_s \quad \therefore \quad \int_{\mu_f}^{\mu_s} \frac{\mathrm{d}\mu_s}{\mu_s} = \int_0^{V_s} \frac{5}{2}\mathrm{d}V_s \quad \therefore \quad \mu_s = \mu_f e^{(5V_s/2)} \tag{5.12}$$

The volume of the sediment appears to be larger, due to the enclosed fluid, than its real value and can be corrected by a so-called accumulation factor K by the definition

$$\mathrm{d}V_s \rightarrow \mathrm{d}V_s / (1 - KV_s) \tag{5.13}$$

and Eq. 5.12 can be written as

$$\mu_s = \mu_f (1 - KV_s)^{-5/2K} \tag{5.14}$$

For $V_s = 1/K$, the viscosity tends to infinity, which corresponds to the maximum packing of the suspension. This quantity V_{sm} can be evaluated experimentally (Wakeman, 1975) and with this value Eq. 5.14 is identical to the finding of Krieger and Dougherty (1959) who used the intrinsic viscosity $[\mu]$, which also is an empirically evaluated quantity.

$$\mu_s = \mu_f (1 - V_s / V_{sm})^{-[\mu]V_{sm}} \tag{5.15}$$

Herein $[\mu]$ is the extrapolation of μ_s for $c_V \rightarrow 0$ and $\dot{\gamma} \rightarrow 0$. Considering this, the ratio V_s/dV_s seems to be a good value for normalizing the concentration of suspensions. It is important to consider that the viscosity depends also on the grain distribution, since V_{sm} depends on it. When one varies the grain mixture which is composed of, e.g., two species where the bigger ones takes 0.6 of the volume fraction, the viscosity has a minimum; one speaks of the Farris-effect (1968). For a binary mixture, the following equation can be used

$$\mu_s = \mu_f \left(1 - V_{s1} / V_{sm1}\right)^{-[\mu_1]V_{sm1}} \left(1 - V_{s2} / V_{sm2}\right)^{-[\mu_2]V_{sm2}} \tag{5.16}$$

For a mixture of grains of three different sizes, an equivalent equation can be formulated. In case the particles deviate from the spherical shape, the viscosity increases even more (Barnes, 1981).

Many, but not all, concentrated suspensions are shear thinning and not all are forming 2D substructures, which is important for zones near walls. In addition, Krieger (1972) found that V_{sm} as well as $[\mu_s]$ are stress dependent but independent of the grain diameter. He suggested that a modified Péclet number should replace the shear rate,

$$Pe = \frac{uL}{D} \quad \rightarrow \quad \tilde{P}e = \frac{6\sigma r_s^3}{kT} \quad \text{(Krieger, 1972)} \tag{5.17}$$

Pe stands for the advection–diffusion ratio, and D stands for the diffusion respectively dispersion factor and $\tilde{P}e$ stands for the ratio of the viscous forces to the forces of the Browian motion acting on the particle. The Ellis model can represent the viscous–Péclet relation

$$\frac{\mu_s - \mu_\infty}{\mu_f - \mu_\infty} = \frac{1}{1 + b\left(\tilde{P}e\right)^p} \tag{5.18}$$

with b and p dimensionless parameters.

These derivations show that the viscosity not only depends on temperature as was assumed in the classical description of a suspension flow (Eqs. 1.12 and 3.115) but also depends on the concentration. The concentration of the suspension, however, is a function of the height above the bottom Eq. 3.100. Therefore in the suspension transport equations, a viscosity must be used, which is a function of the concentration. The implicit description can only be treated realistically in an iterative way. This is a first expansion of the classical theories in case of higher concentrations.

Until now it was assumed that the main rheological problem is related to fine sediment and can be ignored if the grain distribution is coarse enough. But for larger grains and mixtures of higher concentration, the main problems arise because the grains start to influence the transport process, e.g., larger particles remain in a suspended state for a shorter time. For those cases, a real rheological description is necessary in the sense that the Newtonian fluid has to be replaced by a new type of fluid. Cases of that kind are rare; however, they can become very dramatic. When one encounters such a regime, it is best to consult the special literature on two-phase flows because this would go beyond the limits of this work, e.g., one would have to explain the stability criteria for the fluid at which the two phases separate due to a change in the flow conditions etc. On the other hand, one would like to estimate the sediment transport close to such extremes, and therefore we continue by introducing simplifications.

5.1.4. Rheological Aspects of Two-Phase Flows with a Large Grain Size Distribution

We will discuss flows with a fairly dilute as well as fairly high-sediment concentration for which a separate treatment of the suspension and the bed-load part is not meaningful, because the superposition in Eq. 3.117 is not valid anymore. The fluid containing a too high concentration of solid particles looses its Newtonian character, and we have to introduce new constitutive equations. Davis (1996) and Denn (1996) pointed out the significance of the non-Newtonian behavior for such mixtures but went more for a description of the two-phases than for a change in the constitutive equations. The reason for this approach was that by a rheological description all the physics are contained in the constitutive equation, and one looses the possibility to introduce mechanical models. On the other hand, the rheological approach has the advantage to be a continuum-mechanical description. Rheological models based on physical considerations were proposed by Brenner (1970, 1974) for small c_V. For higher densities where particle–particle interactions must be taken into consideration, we refer for spherical particles to Milliken et al. (1989) and for cylindrical ones to Shaqfeh and Koch (1990) and if migration effects should be included, see Phillip et al. (1992).

The change in state can most easily be understood when one looks at a rheogramm. In that representation the shear stress is plotted against the shear rate (Fig. 5.1).

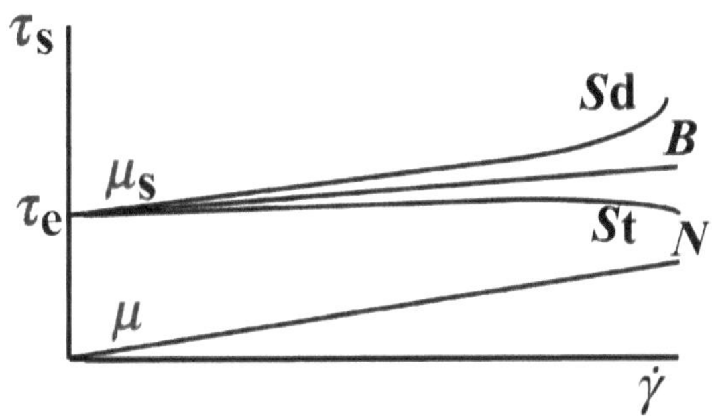

Fig. 5.1 Rheogramm of a Bingham-fluid (B) in relation to a Newtonian fluid (N), where the Bingham-fluid can be shear thinning (St) or shear thickening (Sd). See also eq. (5.19)

By definition of the Newtonian fluid (Eqs. 5.1 and 5.2 respectively 1.13) $\tau(\dot{\gamma})$ is in the plot Fig. 5.1 a straight line through the origin having the slope μ. Concentrated mixtures can be shear thinning or shear thickening if the shear stress decreases or increases with increasing shear rate. They usually also exhibit a yield stress like a Bingham fluid would do. A Bingham fluid is one which behaves plastic like and is defined by the constitutive equation

$$\tau_s = \tau_e + \mu_s \dot{\gamma} \tag{5.19}$$

τ_e is the extrapolation of the shear stress for $\dot{\gamma} \to 0$, a definition in analogy to the definition of the intrinsic viscosity Eq. 5.15 usually called yield stress. The problem involved with the extrapolation can be realized by the fact that τ_e depends on the decades used in the extrapolation. The smaller the shear rate, the smaller τ_e is and it is tending asymptotically toward a Newtonian fluid, however in a flow of very slow velocity. τ_e is therefore a fictive value, and one has to declare how it was evaluated whenever it is used (Barnes and Walters, 1985).

Equation 5.19 postulates that μ_s again represents a slope, however only in the special case of a straight line, normally with a negative curvature for the shear thinning case St and positive curvature for the shear thickening case Sd. Often the slope for μ_s and μ are compared.

The Bingham model is quite good for fine material (Qian and Wan, 1986), however for coarser material we observe deviations. These alterations are explained by scattering mechanisms, and following Bagnold (1954) represented by a dispersive shear stress, τ_d, a contribution Bagnold evaluated experimentally in the laboratory. The following constitutive equation takes care of this effect is

$$\tau_d = c_{Bd}\rho_s\left[\left(\frac{0.615}{c_V}\right)^{1/3}-1\right]^{-2} d_s^2\left(\frac{du_x}{dz}\right)^2 \quad \wedge \quad \tau_d = f\left(c_V^{2/3}, d_s^2, \dot{\gamma}^2\right) \tag{5.20}$$

with the empirical parameter, named after Bagnold, $c_{Bd} \approx 0.01$ in good approximation.

The dispersion is not only the result of a scattering of small particles by the bigger ones but also has originates in the turbulent flow too. A model, which takes care of this additional source of dispersion, was proposed by O'Brien and Julien (1985). In that model, a turbulent dissipation parameter was used as a coefficient for the quadratic term in the shear rate

$$\begin{aligned}
&\tau_s = \tau_e + \mu_s\frac{du_x}{dz} + \zeta\left(\frac{du_x}{dz}\right)^2 \quad \wedge \\
&\zeta = \rho_m l_m^2 + c_{Bd}\rho_s\left[\left(\frac{0.615}{c_V}\right)^{1/3}-1\right]^{-2} d_s^2 \quad \wedge \\
&\rho_m = \frac{\rho_s V_s + \rho_f V_f}{V_s + V_f} \quad \vee \quad l_m \rightarrow (\text{Eqs. 3.95 and 3.97})
\end{aligned} \tag{5.21}$$

where l_m is the mixing length of the mixture. All these quantities are very difficult to obtain and have to be measured experimentally. For example, l_m is a characteristic value of the turbulent flow and is damped by the increasing concentration, and since κ is not a universal constant anymore, one needs several tricks to evaluate this quantity. The problem becomes even more pronounced when one tries to evaluate the think velocity u_V in such a mixture. The results found for dilute concentration therefore are extrapolated and used for cases of higher concentrations, although it is known to be incorrect. But because of a better understanding, they shall be mentioned for completeness. Julien and Lan (1991) introduced

$$u_{\mathrm{v}}^{2}=\frac{4}{3}\frac{gd_{\mathrm{s}}}{c_{\mathrm{D}}}\frac{\rho_{\mathrm{s}}V-\rho_{\mathrm{m}}V_{\mathrm{s}}}{\rho_{\mathrm{m}}V} \quad \wedge \quad c_{\mathrm{D}}=\frac{24}{Re_{\mathrm{B}}}+\frac{2\pi \mathrm{He}}{Re_{\mathrm{B}}^{2}}+1.5$$
$$\wedge \quad Re_{\mathrm{B}}=\frac{\rho_{\mathrm{m}}d_{\mathrm{s}}u_{\mathrm{v}}}{\mu_{\mathrm{m}}} \quad \vee \quad \mathrm{He}=\frac{\rho_{\mathrm{m}}d_{\mathrm{s}}^{2}\tau_{\mathrm{e}}}{\mu_{\mathrm{m}}^{2}} \tag{5.22}$$

by using the Hedstrom number He. Equation 5.22 is a system of equation since Re_{B} is formed by using u_{v}.

With this remark we close that discussion because we have shown the limits of the theories available today. Fortunately, sediment transport at high concentration of particles in the fluid is rather rare, however of utmost importance in some extreme cases, like the Yellow River in China. The problem formulation for these asymptotic cases will need a representation by a constitutive equation, probably calibrated by measurements. The so-found shear stress relation has to be inserted into the general flow equation as formulated by Cauchy (1828)

$$\rho\frac{\mathrm{d}\underline{v}}{\mathrm{d}t}=\rho\underline{f}+\mathrm{div}\underline{\underline{T}} \tag{5.23}$$

This is a very tedious undertaking because it needs a new theory for the turbulence of such a fluid, which is the reason why no big progress was made in the past in that direction.

5.1.5 Two-Phase Flows

Instead of a rheological description of the mixture, which means that the mixture can be treated as a uniform fluid, one can treat the mixture as a continuous interactive one in which particles move within a Newtonian fluid. This has some numerical advantages as long as turbulence can be described by the velocity fluctuation of a Newtonian fluid. This is generally not the case, but for dilute concentrations this simplification can be exploited.

Greimann (2003) summarized first results of such a description using mainly the publications Greimann et al. (1999) and Greimann and Holly (2001). These investigations became possible because new measurement techniques became available (Nezu and Rodi, 1986; Kiger and Pan, 2000). All these investigations were, however, made with uniform grains and only a very dilute suspension of $c_{\mathrm{Vs}} << 0.1$. The results were in good agreement with those of Francis (1973) or Hu and Hui (1996), but this is not enough to propagate this method for real rivers with a higher amount of suspension. In addition, the method to collect empirical data will not be available in the near future since the sophisticated measurement techniques are applicable in laboratory conditions only.

5.2 Nonlocal Properties of the Flow Field

It is no secret that in the past the description of the sediment transport was based on a rather primitive model of turbulence. Turbulence was thought as a state of the fluid in which velocity fluctuations occur statistically distributed in time and space and described by their moments and correlations. Usually those fluctuations are represented by the Reynolds stresses Eq. 2.9, which appear in the NSE after averaging the equation using the Reynolds decomposed velocities (Eq. 2.3). This and the energy dissipation are the main terms for describing the sediment transport. This reflects the fact that neither in time nor in space the status will be constant, however in the mean a stationary flow like in a river can exist. Often such a representation of turbulence is sufficient, but there are cases where we need a better formulation of turbulence, and we will now investigate where the usual model has to be supplemented. As a reference how turbulence is represented in our days one should consult Tennekes and Lumley (1972, updated editions). For the specific issue investigated in this chapter we would like to refer to Tsinober (2003).

5.2.1 Non-local Properties of the Flow Field Without Sediment

At several occasions, it was pointed out that the NSE (1.23) is an integro-differential equation and therefore mathematically nonlocal, then as a local equation all terms would have to be differentials. Most prominent one can see the nonlocality at the pressure term, the pressure is local, but its value in the incompressible case has to be evaluated by integration over the velocity term of the entire flow field

$$\overline{p}+p'=\frac{1}{4\pi}\int_{V(\underline{x})}\frac{\partial^2}{\partial x_i\partial x_j}\left\{\left(\overline{u}_i+u_i'\right)\left(\overline{uj}+u_j'\right)\right\}\frac{\mathrm{d}V(\underline{x})}{\left|\underline{x}'-\underline{x}\right|} \tag{5.24}$$

$$p'\left(\underline{x}'\right)=\frac{1}{4\pi}\int_{V(\underline{x})}\left\{2\frac{\partial\overline{u}_i}{\partial x_j}\frac{\partial u_j'}{\partial x_i}+\frac{\partial^2}{\partial x_i\partial x_j}\left(u_i'u_j'-\overline{u_i'u_j'}\right)\right\}\frac{dV(\underline{x})}{\left|\underline{x}'-\underline{x}\right|} \tag{5.25}$$

with $\underline{x}'$ the position at which we have to evaluate the pressure. In Eq. 5.25, the first term in the parenthesis stands for the interactions of the mean and the turbulent velocity gradients. The second term stands for the effect of the Reynolds shear stress fluctuations upon the mean Reynolds stresses. In shear flows, this term is usually much smaller than the first one. This shows that the NSE has to be solved together with the integral equation for the pressure.

Forces on the grain surface are responsible for the transport, and as we know these are the pressure (dominant for larger grains) and the shear forces (dominant for smaller grains), and they particularly are very difficult to measure with the needed accuracy. Due to the progress in the optical methods, large advances were made by measuring the Lagrangian velocities of small seeding particles, which allow to evaluate u_i the 3D components of $\underline{u}$ for a rather large volume in space (Maas, 1996; Rösgen and Totaro, 2003). With such a resolution of the velocity field, the local pressure distribution can be calculated in principal by Eq. 5.25. This is easier said than done, and the reason is that we have to integrate over the complete space participating on the flow, what would need, even if the total field information would be available, an extremely high computer capacity. The Eq. 5.25 is therefore only of help if the pressure can be evaluated from an integration of much smaller dimension, however, the integrand decreases by $1/r$, which requires that the integration-volume must be rather large. The truncation of the integration-volume is therefore a first measure for the nonlocality of the flow field. It is evident that this truncation can be stronger for a homogeneous field than for an inhomogeneous one. Nevertheless, for certain flow fields the integration of Eq. 5.25 can be executed, and we have a new tool to estimate the forces excerpted by the flow field.

To overcome the difficulties encountered with the pressure term, it was proposed to apply the curl-operator on the NSE and doing so, eliminate by this procedure the pressure as variable, and have instead an equation for the vorticity $\underline{\omega}$ of the flow field with

$$\underline{\omega} = rot\underline{u} = \left[\nabla \times \underline{u}\right] \quad \Leftrightarrow \quad \omega_i = \varepsilon_{ijk}\frac{\partial u_k}{\partial x_j} \tag{5.26}$$

and we can write Eq. (1.23),

$$\begin{aligned} &\frac{\partial \omega_i}{\partial t} + u_j\frac{\partial \omega_i}{\partial x_j} = \omega_j\frac{\partial u_i}{\partial x_j} + \nu\frac{\partial^2 \omega_i}{\partial x_j \partial x_j} \\ \rightarrow \quad &\frac{\partial \omega_i}{\partial t} + u_j\frac{\partial \omega_i}{\partial x_j} = \omega_j s_{ij} + \nu\frac{\partial^2 \omega_i}{\partial x_j \partial x_j} \\ \wedge \quad &s_{ij} \equiv e_{ij} = \frac{1}{2}\left(\frac{\partial u_i}{\partial x_j} + \frac{\partial u_j}{\partial x_i}\right), \ \text{(Eq.1.14)} \end{aligned} \tag{5.27}$$

in other words, increasing the order of the differential equation by one can eliminate p. It seems that the integro-differential equation became a differential one, and we have to ask us if the nonlocal character of the flow equation has disappeared, too. As Tsinober (2003) showed this is not the case since the relation between ω and s_{ij}, the deformation tensor is a nonlocal one (kinematically nonlocal). The old and the new nonlocality are related with each other but are not the same.

Nonlocal means that interaction occurs in the flow field over long distances. Is a fluid element, e.g., part of a vortex than the flow of this element is linked to the flow over a distance of the size of the vortex. Therefore, the velocity and its fluctuations are in a strict sense nonlocal. The real nonlocal elements are the velocity derivatives and therefore also the vorticity, and one has to raise the question how the elements of the divers classes are interacting with each other. Tsinober (2001) showed that one cannot assume that a coupling does not exist if the correlation disappears. For example, in a homogeneous turbulent flow the correlation between the velocity and the deformation disappears:

$$u_i; s_{ij} \quad \Rightarrow \quad \left\langle u_i s_{ij} \right\rangle = 0 \tag{5.28}$$

However, vanishing correlation does not necessarily mean absence of dynamically important relations. This can be seen when we investigate the terms most essential for the turbulent flow

$$\begin{aligned} &(\underline{u} \bullet \underline{\nabla})\underline{u} \equiv \underline{\omega} \times \underline{u} + \underline{\nabla}\left(\frac{u^2}{2}\right) \\ \wedge \quad &\underline{\omega} \times \underline{u} \end{aligned} \tag{5.29}$$

Both contain with the given definition of scales, large scales and small scales, and it is irrevocable that a coupling between them exists. This directly points to one of the main unsolved questions in turbulence, how are small and large scales interacting, and how the self-amplification of small scales (ω_i and s_{ij}) works. Since turbulence is a dissipative phenomenon, this kind of amplification has the character of a paradox. In this context Tsinober (2003) draws our attention to the terms

$$\omega_i \omega_j s_{ij} > 0 \quad \vee \quad -s_{ij} s_{jk} s_{ki} > 0 \tag{5.30}$$

as well as to the appertaining terms in the equations for ω_i and s_{ij}, which contribute more than the shear or the outer forcing.

The self-amplification is also the reason why the enstrophy ω^2 or the total deformation s^2 are not invariants for the non-viscous case, as this holds for the kinetic energy u^2. The conclusion is that small objects cannot be treated as passive elements. This result should considerably influence the numerical treatment of turbulent flows because this wrong assumption is often used in the computation of simulation models.

Due to the nonlocal relation, very often small-scale elements, like the vorticity or the deformation, are responsible for the development of elements like the velocity. With this remark, we encounter one of the most central questions describing the sediment transport, the cascade of the large elements in physical space. Therefore the opinion, that there exists a cascade, and the viscosity affects only the small-scale elements is wrong. These are good news since it is known for long that in rivers an inverse cascade must exist since all the vorticity is created at the bed. First, therefore, out of this rather fine structures large eddies have to form before they can decay in the usual cascade process. A lot of speculations can be found in the literature how this inverse cascade develops, e.g., by vortex pairing, etc. To know the exact mechanisms would be identical to the understanding of the turbulence, but the only thing we know is that all scales coexist in a statistical relation to each other. Only when the energy input stops, the system starts to decay in a form that can be described in form of a cascade. By the way, historically seen, the cascade theory also originated from the representation of the decaying turbulence in the wake of a grid in a wind tunnel experiment.

We have to conclude that the viscosity as a material property remains important also at very high Reynolds numbers, although statistical parameters and properties may become independent of it in a turbulent flow. In addition, Tsinober showed what it means to attach a certain diffusivity to the material streamlines as a result of the viscosity. If it exists, the streamline becomes aligned in direction with the largest eigenvalue of the deformation tensor, whereas passive vectors without diffusivity align with the intermediate eigenvalue as this is known for the vorticity vectors. The significance of this understanding can be seen best in its importance for a numerical calculation, but it is also physically important since it allows developing new models.

It was one of the main pillars in turbulence theories that small and large scales are uncoupled and that there exists a local isotropy in the small-scale range at high Reynolds numbers. With the new knowledge of the nonlocal interactions and the mutual interference between small-and large-scale objects, it has to be investigated what this means with respect to the structure of turbulence.

A first manifestation is the observable anisotropy of the small structures, which implies the small scales only partially forget the anisotropy of the large structures. This becomes even more important for the transport because large structures developed from smaller ones, and therefore in an anti-Richardson–Kolmogorov-cascade, the anisotropic memory is even stronger (Cimbala et al., 1988).

The turbulence production depends strongly on the wall configuration. Λ-vortices are created at smooth walls (Chap. 3, Sect. 3.2.4). and they are completely different from the vortices produced by separation at roughness elements. Both are large structures but with a different anisotropy, which will show up in the small scales. Since these large structures depend on the flow geometry, there must exist an interaction that cannot be generalized. This is extremely important for the development of numerical simulation programs, and it puts some doubts on the large eddy simulation (LES) method in particular.

This is of importance in a physical sense when the production mechanisms of the large structures are modified by additives as they can even cause drag reduction (Gyr and Bewersdorff, 1995). The additives must not be of molecular size since also sand can also act as a drag reducer by interacting with the generation of the large-flow structures. Some aspects of such a mechanism will be discussed later.

In his reflections on nonlocal influences, Tsinober (2003) goes even a step further and explains instability processes as nonlocal appearances in time.

5.2.2 Non-local Properties of a Sediment Laden Flow Field

The interaction between particles and the flow field occurs on the scale of the grains, the feedback scale, however, can exceed these by far. For sedimenting suspensions see Bernard-Michel et al. (2002), or Maxey et al. (1996) for streaming particles and bubbles, especially for pipe flows see Ljus et al. (2002). Gyr and Kinzelbach (2004) discussed the scaling problem of the feedback processes creating bed forms. Feedback processes can be the reason for clustering in suspension flows, producing inhomogeneous concentration distributions and flow volumes develop with high-particle concentration surrounded by fluid of rather dilute particle concentration.

Particle forcing origins mainly from its inertia either through gravitational forces or in form of centrifugal acceleration forces when it happens to be embedded in a vortex. The velocity deficit behind the particle has its direct feedback on the small scales of the flow. Therefore, also the shape of the particles is of importance and the displaced volume by the solid grain can have a feedback effect as already mentioned in the discussion of the Magnus effect. This effect is even in place when the particles are neutrally buoyant with the same density as the fluid where the developments of clusters can still be observed (Cartwright et al., 2002 and the literature mentioned therein).

Formation of clusters can be observed especially during sedimentation, a particular case of sediment transport, since it corresponds to a free fall through a fluid at rest. The motion of the particles initiate fluid motion, the more, the denser the local concentration. The explanation for the cluster formation is given in analogy to a double diffusion process. The so-called intrinsic convection—the upward flow of displaced fluid volume of the descending particles due to continuity reasons—gives rise to a large-scale inhomogeneity of the concentration distribution. Zones of high density are formed, which have a higher than average fallout velocity. The micromechanical process is obviously more complicated and a first try was undertaken by Batchelor (1972) and Batchelor and Green (1972) who showed how neighboring particles start to group. Koch and Shaqfeh (1989) expanded this method, whereupon Druzhinin (1997) formulated a stability theory. It is an instationary development of the large-scale waves of the fluid flow interacting with c_V (Asmolov, 2003). The particle velocity has a correlation length λ, which is large compared to the diameter of the grains (Segre et al., 1997). A typical scale is $\lambda \approx 20\ d_s$, which can grow up to L/2 if L is the length scale of the formed cell. The velocity of the cell (cluster) u_{cl} is at constant $\langle c_V \rangle$ in pure fluid, ρ_f.

$$u_{cl} \propto u_v c_V L^2 \tag{5.31}$$

This result has some implications for particle swarms as mentioned in Chap. 3 (Sect. 3.3.2), however for the sediment transport it is more of basic interest than for practical use since sedimentation is a very particular case of transport. The circumstance that small-scale interaction can form large structures should also be observable for a turbulent flow condition, and in fact Druzhinin (2001) showed that for an isotropic turbulent state *clusters* could form.

It has to be investigated how the particles interact with the vorticity and deformation fields, respectively. Since the inertial effects are the most important ones governing that process, the particles should be characterized by Stokes-number Eq. 3.50. However, to strengthen the inertial character, the *Stokes-number* is now seen as the ratio of the response time of the particle t_p with respect to the smallest fluid time scale t_f.

$$\begin{aligned} &\mathrm{St} = \frac{t_p}{t_f} \quad \wedge \quad t_p = \frac{d_s^2}{18\nu}\frac{\rho_f}{\rho_s} \\ &\vee \quad (\mathrm{Eq.3.50}) \quad \mathrm{St} = \frac{u_v \nu}{d_s^2 \rho' g} \quad \wedge \quad (\mathrm{Eq.3.54}) \ u_v = \sqrt{\frac{4\rho' d_s g}{3\zeta}} \end{aligned} \tag{5.32}$$

where the Stokes drag law Eq. 3.55 was applied. The representative flow time of these scales is the Kolmogorov time scale t_K, which is given as function of the energy dissipation rate ε and the kinematical viscosity ν. The Kolmogorov scaling parameters are:

$$l_K \equiv \left(\nu^3/\varepsilon\right)^{1/4} \quad \text{Kolmogorov} \quad \text{length-scale} \tag{5.33}$$

$$t_K \equiv \left(\nu/\varepsilon\right)^{1/2} \quad \text{Kolmogorov} \quad \text{time-scale} \tag{5.34}$$

$$u_K \equiv \left(\nu\varepsilon\right)^{1/4} \quad \text{Kolmogorov} \quad \text{velocit-scale} \tag{5.35}$$

with

$$\varepsilon \approx u_x'^3 l \quad \vee \quad l \approx \kappa z \tag{5.36}$$

where u_x' is the velocity fluctuation measured in flow direction and l the integral length-scale of the turbulent flow.

Experimentally St = 1 was found as a criterion defining two main states (Eq. 3.101), and in the new formulation this criterion becomes

$$t_f = t_K \quad \therefore \quad \mathrm{St} \ll 1 \ \ (a) \quad \vee \quad \mathrm{St} \approx 1 \ \ (b) \quad \vee \quad \mathrm{St} \gg 1 \ \ (c) \tag{5.37}$$

(a) The particles follow the path lines.
(b) A cluster formation can be observed (Maxey, 1987), and the settling velocity is influenced (Aliseda, et al., 2002).
(c) The particles respond so slowly to changes of the flow that they can be treated as they would fall out undisturbed.

It was shown numerically by Squires and Eaton (1991), Ferrante and Elgobashi (2003), and others that particles of St = 0.1–1 form clusters, which was experimentally confirmed by Fallon and Rogers (2002). Since it is difficult to distinguish between inertial and gravitational influences, a number S_g related to the Stokes-number was introduced:

$$S_g = \frac{t_p}{t_g} = \frac{U_d}{u_f^{'L}} \quad \wedge \quad U_d = g t_p \quad \vee \quad (\ )^L \rightarrow L: \text{Lagrange} \qquad (5.38)$$

Here, defining S_g, the drift velocity U_d has been introduced together with the mean turbulent fluctuation velocity along the path line. If gravitational effects are of importance, S_g becomes large and the particles escape the trajectories of the fluid flowing around them. Through this process, the concentration homogenizes and therefore for St = 1 the gravity counteracts the cluster formation. Also centrifugal forces can give rise to separation in the grain size distribution, which is of utmost importance for the interaction of particles with the so-called coherent structures where they can influence the structural formation process.

The astonishing result is the intrinsic significance of St spanning a very large range of flow timescales, starting from the Kolmogorov scale and ending with the scale characterizing the settling of rather large particles. The physical significance for the cluster formation lies in coupling the smallest dynamical scales of the flow and the scale of the particle-flow interaction, when their timescales are comparable. We have therefore to rely for clustering on the small timescales, which implicates that only feedback processes can describe the effects of sediment transport on large-scale structures as, e.g., coherent structures.

The turbulent structures altered by the interaction with the particles are, with the exception of some rare cases, unimportant for the sediment transport because the cluster formation is not essential for the transport. Once the sediment suspended, it is transported more or less independent of its inhomogeneity in flow direction. The vertical distribution is much more important because here the large-scale structure is connected to the deposition and resuspension at the bed in direct interaction with the flow. This becomes important when large amounts of sediment get suspended and we find a high-concentration gradient $\partial c_V / \partial z$. It is evident that such high concentration must influence the turbulent structures, and the laminarization at high concentrations became a popular field of interest (Einstein and Chien, 1955; Vanoni, 1960; Coleman, 1986). With the occurrence of laminarization, a drag reduction can be observed and the velocity profile changes drastically. An energy budget argument as discussed in Chap. 3 (Sect 3.3.3.3) was brought forward as cause for the laminarization. The energy needed to hold the particle in suspension must be supplied by the turbulence and the turbulence therefore must be damped. This formulation still sheds no light on the mechanistic process, but the main idea was confirmed by several experiments. Hopfinger and Linden (1982) found by studying the decay of the turbulence produced by oscillating grids that the turbulent velocity fluctuations were damped. They inserted a buoyancy term into the general dynamical description of the turbulent decay as formulated by Batchelor (1953) representing the sink of the energy. The grid experiment was repeated by Huppert et al. (1995) and the results were confirmed. McLean and Smith (1979) proposed a correction in the eddy viscosity, which is defined by the Richardson number and therefore based on the volume-concentration c_V.

The laminarization concept is physically problematic since in an asymptotic case this kind of flow is hardly capable of suspending particles. The state of the flow of high-concentrated suspensions close to the bed is therefore still an open question. For a smooth bed, the drag reduction could be explained in analogy to Lumley's theory (1969, 1977) for the drag reduction of dilute polymer solutions, in which he showed that a thickening of the viscous sublayer is the cause of the drag reduction. There are also contradictory results with respect to the laminarization concept. To mention here are the results found by Best et al. (1997) and Cellino and Graf (1999) who found in open channel flows with suspensions a stimulation of the turbulence close to the wall. This reopened the debate on how a suspension is created and how it interacts with the flow in the zone of the turbulence production. In practice, however, it is still recommended to use the Rouse concentration profile with its calibration value Eq. 3.100, until no better theory is found. This classical representation does not need a laminarization process for its reasoning, in contrary the particle–particle interaction as described by Leighton and Acrivos (1987a) supports the diffusion theory as described in Chap. 3 (Sect. 3.3.3.2). In analogy to the diffusion, this process was called hydromechanical diffusion, and Davis and Hassen (1988) verified it experimentally. By this theory, the micromechanical description of the particle migration is incomplete because only the large structures are considered and often only monodisperse spherical particles were treated as suspended load (Leighton and Acrivos, 1987b; Ham and Homsy 1988). Binding (1988) proposed a variational principle which however could never be proven, although it yields quite good results.

The question, whether turbulence is stimulated or damped if suspensions are present and how this behavior can be defined by objective criteria is now revived because of the enormous progress made in the optical measuring techniques during the last decade, an overview can be found in Adrian (1991). Of special interest for the measurements in suspensions are methods using a matching technique of the refractive index *(Müller and Wiggert, 1989). An even more important progress is the analysis of a whole turbulent flow field by particle tracking methods particle tracking velocimetry (PTV) (Lüthi et al., 2005).

In the spirit of our time, it was postulated that there must exist a relation between the suspending of particles and the so-called coherent structures (Sechet and Le Guennec, 1999). This problematic issue will be discussed later in more detail, as the moment it is only the feedback mechanism that is not understood. This would be necessary if the suspension transport should be understood in physical terms.

This deficit can also be seen when one evaluates the different numerical simulation methods. We have no exact equation and the given equations often have not even well-defined boundary conditions. A rare exception is, e.g., given by Brady and Bossis (1988) by a two-phase flow description of spherical particles at high-density concentration, which is named Stoke's dynamic.

5.3 Nonlinear Processes

Since we lack a closed turbulence theory, it is mentioned that we have to take care at least of its nonlinear behavior, which is an intrinsic property of the NSE, whenever turbulence is involved in a flow process. The main crux is, the description of the of the particle-flow interaction, when we are not able to integrate the nonlinearity of the pure fluid flow. The nonlocal phenomena show that small objects interact with large ones, which is a rather phenomenological description of the essential mechanism. It is however nonlinear, and it is exactly for this unsolved mathematical problem why a theory of turbulence has failed. The ways to crack this nut are manifold, and every investigation contributes a puzzle piece to the understanding.

One property of nonlinear systems is the chaotic behavior of its solutions, and the application of dynamical systems theory to turbulence was therefore obvious. What is a dynamical system? It consists of a finite number of time-dependent variables, which develop as described by a system of ordinary differential equations. When the system has three or more degrees of freedom, it can develop chaotic behavior. This understanding was first introduced by Pointcaré (1892) but not used in fluid dynamics until Arnold and Hénon (Arnold, 1963, 1978). Arnold and Avez (1968) and Hénon (1976) pointed out that the complexities of the kinematics of certain 3D stationary flows are in fact chaotic solutions, exhibiting strange attractors which have bifurcation properties. This means the solution is complex, nonperiodic and probabilistic, however, actually deterministic, and extremely sensitive to the initial conditions. Smallest deviation in these initial conditions ends up in an exponential separation of the solutions, which is also known as the butterfly effect. Furthering the problems, the NSE is a partial integro-differential equation and as such conceptually an infinite manifold of ordinary differential equations. Herewith it contradicts the restriction that only dynamical systems of low order can realistically be treated. We can expect that the chaotic solutions in fluid mechanics belong to the class of flows between the laminar and turbulent state, so the best we can expect is to get solutions for the transition (Aref, 1996). Not one of the ideas found for dynamical systems of low degree could be transferred to a continuum field, in particular to a flow field covering the whole domain (Lumley, 1996). The only thing we can say is that the turbulent flow measured in fixed points behaves chaoticly as a function of time, deterministic in the mean quantities but shows a butterfly effect. More than this cannot be said that there exists a strange attractor, but this representation is rather semantic. Lumley therefore proposed that such attractors should be treated statistically.

One can ask: Why we did discuss this rather problematic issue here in the framework of a representation for the sediment transport. There are two reasons for this. The first one is to elucidate the connection between the nonlinear interactions and the coherent structures. The second one is to present some analogies of the sediment transport equations with the flow equation. In a locally linearized form, the latter can fulfill the conditions of a dynamical system of low degree of freedom. However, this idea was never used mainly because we cannot expect a simplification of the computing methods. On the other hand, such considerations can be helpful in planning projects in the field of sediment transport.

As mentioned in Eq. 1.34, the sediment transport depends on a rather large number of variables, some of which are time dependent. The interaction of these variables is described by nonlinear equations; evidently it follows that the sediment transport is of chaotic nature too.

A system of so many parameters possesses a solution path in the phase-space. The number of parameters defines a space in which every parameter represents its own dimension. If the value of one parameter changes, the solutions shifts in this space so that the changes in all other parameters occur due to the variation of the selected one.

Assuming the sediment transport is given by a universal law, the solution should follow a single path. However, a chaotic system has bifurcations and one would like to know which parameter contributes most to the chaotic behavior. This is of essential significance, when single parameters of the system are held fixed. For example, if we assume the wall shear stress fixed instead of using the time-dependent Reynolds shear stress. We are trying to answer, whether in the entire phase-space, there exist subspaces in which certain parameters are nonchaotic and the transport can be described in a deterministic form. It is the assumption that the entire phase-space behaves nonchaotic, which is taken for granted in the classical sediment transport description.

The main purpose of a theory is to have a predictive model. In case of such a complex problem as sediment transport, this is mainly done by simulation methods, which are useful if they are robust. However, it is just the robustness of the simulation which depends on the chaotic behavior of the equations. This leaves us with the problem of testing the chaotic properties of a system.

In the fictitious dialog with Einstein (Chap. 1, Sect. 1.1), we mentioned that he was convinced that sediment transport was so dependent on the initial and boundary conditions that no universal representation could be expected. Translated into the new language, sediment transport is affected by a butterfly effect, but a very special one insofar as it does not affect the asymptotic behavior.

The main property of a chaotic system is its sensitivity with respect to variations in the initial conditions. Neighboring trajectories diverge exponentially, however, more in the sense of a folding and stretching distance than an Eulerian distance since the phase-space and initial conditions of, e.g., a strange attractor may limit the solution globally.

Picking up this line of thought, the drift rate can be quantized by the Lyapunov exponent, and therefore it should be one of the goals in the sediment transport research to evaluate the Lyapunov exponents for the process or for parts thereof.

Let us start by investigating the criteria for local chaos. The sediment transport is given by a function of type (Eq. 1.34) so, for generality, we will consider a system governed by the differential equations,

$$\frac{\mathrm{d}x_i}{\mathrm{d}t} = F_i\left(x_1, x_2, \ldots\ldots, x_n\right) \quad \wedge \quad i = 1, \ldots\ldots, n \tag{5.39}$$

the x-values correspond to the variables, which span the phase-space. To evaluate the stability of the system at a certain location, a linearization of the equations of motion about that point is needed. We now linearize the equations about any reference orbit

$$x = \hat{x}(t) = \left(\hat{x}_1, \ldots\ldots, \hat{x}_n\right) \tag{5.40}$$

to yield the tangent map

$$\frac{\mathrm{d}\delta x_i}{\mathrm{d}t} = \sum_{j=1}^{n} \delta x_j \left(\frac{\partial F_i}{\partial x_j}\right)_{x=\hat{x}(t)} \tag{5.41}$$

With the norm

$$\mathrm{d}(t) = \sqrt{\sum_{i=1}^{n} \delta x_i^2(t)} \tag{5.42}$$

we have a measure for the divergence of two neighboring trajectories, that is the reference trajectory $\hat{x}$ and its neighbor with their initial conditions

$$\hat{x}(0)+\delta x(0) \tag{5.43}$$

with

$$\delta \underline{x}=\underline{x}-\hat{\underline{x}}(t) \tag{5.44}$$

During the motion along a trajectory, an originally spherical volume of radius d(0) deforms into an ellipsoid, where the half-axes deform as $\delta x_i(t)$, assuming we stay in its main axis system. The mean rate of exponential divergence is defined as

$$\sigma_i = \lim\begin{pmatrix} t\rightarrow\infty \\ \mathrm{d}(0)\rightarrow 0\end{pmatrix}\frac{1}{t}\ln\frac{\delta x_i(t)}{\mathrm{d}(0)} \quad \wedge \quad \mathrm{d}(0)=\sqrt{\sum_{i=1}^{n}\delta x_i^2(0)} \tag{5.45}$$

This is a set of n such quantities

$$\sigma_i, i=1,\ldots\ldots,n \tag{5.46}$$

The σ_i are called the Lyapunov characteristic exponents which can be ordered by size, such that

$$\sigma_1 \geq \sigma_2 \geq \ldots\ldots \geq \sigma_n \tag{5.47}$$

The so-defined quantities are real numbers, where the characteristic one in direction of the trajectory is zero. The sum of the Lyapunov exponents defines the mean variation of the volume element of the phase-space during the motion along the trajectory. The local relative change in volume is at any location of the trajectory given by the divergence

$$\mathrm{div}\underline{x}=\mathrm{div}\delta(\underline{x}) \quad (\mathrm{Eq.}4.44) \quad =\left(\frac{\partial F_i}{\partial x_i}\right)_{x=\hat{x}(t)} \quad \wedge \quad \mathrm{d}(t)=\sqrt{\sum_{i=1}^{n}\delta x_i^2(t)} \tag{5.48}$$

and hereby, the mean divergence rate is defined as

$$\sigma=\sum_{i=1}^{n}\sigma_i = \lim_{t\rightarrow 0}\frac{1}{t}\int_0^t \mathrm{div}\delta(\underline{x})\,\mathrm{d}t \tag{5.49}$$

For regular motions the exponents are zero since d(t) increases linearly.

If the function F in Eq. 5.39 stands for q_s, then the local characteristic Lyapunov exponent is σ_{qs}, a function of the parameter expansion in the phase-space. The function Eq. 1.34 is unknown and therefore the Lyapunov exponent has to be evaluated experimentally.

What does that mean? We investigate the Lyapunov exponents locally since any location in the phase-space has its own value constituted by a set of the form Eq. 5.47. For evaluating the exponents at any place in the phase-space, we would have to execute an enormous amount of experiments. In terms of Einstein, we have to consider any existing river as a realization given as one point in the phase-space. This is helpful since many case studies are published and can now be reevaluated with the goal to find the exponents. Here we need to find areas in the phase-space with a large Lyapunov exponent since this would be an indication of high nonlinear interactions of the parameters in this range. These are areas for which we cannot expect an unequivocal transport law.

The ideal result of such an investigation is a lean data set that ascertains areas in the phase-space in which the system behaves regularly, that is $\sigma = 0$, because d(t) grows linearly with time. In these areas, the sediment transport can be described by unequivocal equations. A schematic representation of this idea is found in Fig. 5.2.

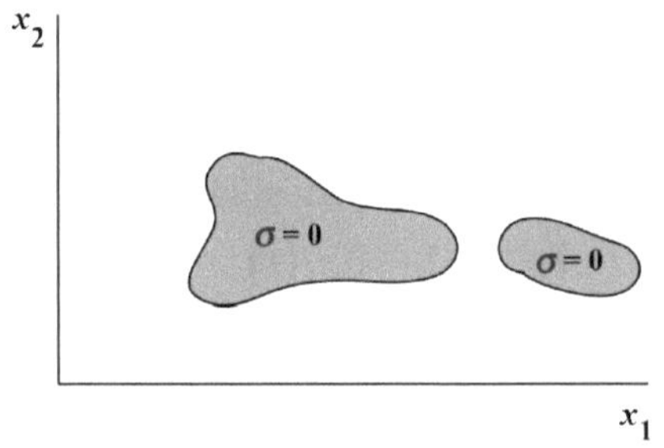

Fig. 5.2 A schematic of unequivocal islands in a phase-space, shown for two-dimensions. In two areas, the Lyapunov exponent σ is ≈ 0, herein we have an unequivocal solutions whereas outside the nonlinear behavior of the system is so strong that such solutions are not probable

The extraordinary idea of this representation is that we investigate the time dependence of the system. This can end up in some misunderstanding as we want to show. Let us take as an example the grain size or the grain density, they do not change in time and therefore do not contribute to a Lyapunov exponent directly. From Eqs. (5.39 and 5.41) it follows

$$\frac{dd_s}{dt} = 0 = F_{d_s} \quad \therefore \quad \frac{d\delta d_s}{dt} = 0 \tag{5.50}$$

What can we expect from a variation of these two quantities? A variation in these parameters has the consequence that we investigate the transport at another location in the phase-space. Since the exponent is a local property, at that location the behavior of the system could have changed completely because the Lyapunov exponent has changed, and therefore also the sensitivity did this too. With this excursion we show that variations of even time-independent parameters can result in a transport that differs in its nonlinear behavior. How can this be explained using the above example? Typical time-dependent quantities are those given by the turbulent fluctuations, e.g., a change in the pressure field can result in a transport of heavier grains, which cause a new perhaps stronger nonlinear interaction.

Einstein observed alluvial systems which were quasi-identical in their parameter set and behaved quite differently. Such systems should be discussed in detail since they show a nonlinear behavior, and we could learn which of the parameters is the most critical one for this apparent difference. Investigations of this kind are new and we cannot know yet what we will benefit. One benefit can be seen directly since the above method is similar to the stability investigations, which are very relevant for the sediment transport, e.g., that stability depends on the location of the system in the phase-space.

It is common to investigate whether a certain state is stable or not, by testing whether in a test section sediment is deposited or eroded by linear stability theories. One distinguishes only between infinitesimal and finite disturbance of the system. The latter would be more appropriate but is practically never applied. This problematic issue will be discussed in greater detail when we discuss the formation of bed forms.

In the past, the development and stability of an alluvial flow was tested in laboratory experiments. This necessitates very long test sections, usually much longer than the channels used in laboratories since sediment can be transported in very long "bed waves". The inappropriate length of the channel is often a function of the mismatch between the timescale of the flow and the transport in suspension. In laboratories this problem is addressed by upstream sediment feeders.

A new form of investigation of the stability would be to test the development of a forced artificial disturbance by observing whether the disturbance grows or decays. However, this method is very rarely used although, it would be a nice tool to check not only the stability but also the sensitivity of the system. For example, Schmidt and Gyr (1998) used this technique to check the stability of ripple propagation on bed of fine sands.

6 Scales

An alluvial system is characterized by a variety of scales, where one has scales of the flow structures as well as scales of the transported material. These scales are essential for the description of the diverse sediment transport mechanisms at work. We will have to distinguish between matching scales that do and nonmatching scales that do not interfere with each other. Whereas the grain sizes are usually known, the flow scales are free to rearrange themselves due to the presence of other flow structures, the transported material, and also in interaction with the developing bed forms.

It is essential to group the scales so that one can compare the right length scales with the appropriate time and velocity scales. We classified them in following tables, being well aware that it is not possible to give sharp criteria for their separation. We therefore supplemented the classification with a discussion on the variety of concepts based on scaling ideas.

6.1 The River as a System and Its Hydrological Scales

The largest scales are of geological size, which are however not relevant for sediment transport since the latter cannot be related in a physical way to these scales. With respect to the relevant scales, we presume that the alluvial system is known. The length–time–velocity (LTV) set is given by the river itself and its temporal and velocity behavior (Table 6.1).

Table 6.1 LTV scales for the system as a whole

L: in km	T: in y, month, d, h	U: in m/s	System
10–100	y, month, d, h	1–10	The river and its branches, during the whole observation time. Annually, precipitation period, daily
1	H	1–10	Channels and test area
0.001–1		≈ 0	The width of the river $L = B$
0.0001–0.01		≈ 0	The depth of the river $L = H$

All length scales are designated L, the timescales T, and the velocity scales U. The largest length scale is the length of the main channel of the river, the next smaller one is the length of side branches, followed by the width and the depth of the river. Two classes of timescales exist, one for the observation and one for the prediction time, where the latter is usually larger than the first. Depending on the site, there exist hydrographs dating back a century and in other cases one is lucky to have a sequence of 3 years.

Flood situations have a high discharge of sediment, and it is most important to know about these events in an appropriate resolution, therefore daily and hourly measurements are of benefit. In these situations, the average velocity U is used. U therefore depends above all on the slope S, which usually varies from 0% to 0.5%, and the water depth of the river. An estimate for a typical Reynolds number is given by

$$\mathrm{Re} = \frac{UH}{\nu} \approx \frac{(5\cdot 10)(1)}{10^{-6}}\left[\left(\frac{m}{s}\right)(m)\left(\frac{s}{m^2}\right)\right] = 5\cdot 10^7 \tag{6.1}$$

Such a large Reynolds number confirms that for the outer scales, the turbulent structures are not important. However, on these scales the transport cannot be evaluated either.

The goal is to make a prediction for the sediment transport that should be valid for a rather long period so that the results can be compared within a hydrological investigation. To do this, we have to consider the hydrograph of the river for a long time period since the discharge rate of the sediment q_s varies depending on the discharge rate q of the river. Regulated rivers will have a rather constant transport rate, whereas rivers through arid areas will experience a sediment transport concentrated in a few events.

In the framework of these large scales, it is obvious to represent q_s as a function of Q. When the slope S is constant, Q can be represented by H, and q_s diverges from the computations using a capacity equation, since the supply is not considered in the computation and the capacity is not well defined. Therefore, field measurements are indispensable for choosing the right transport equation of the classical type. Williams and Julien (1989) have defined an applicability index for calculating the transport of single fractions with the various equations and compared the results with the total measured discharge. However, it seems better to check the applicability of the used equations by taking random samples and comparing them with the evaluated results. To caution the reader we mention that, when the supply of sediment has seasonal and regional variations, rivers with a limited supply possess high fluctuations in the transport quantity. This relationship is described in several papers and the authors introduced approximations. We do not enter into these representations since they are of hydrological and not physical nature.

6.2 The Scaling of the Turbulent Flow

Besides the evident large physical scales of a river, the flow is characterized by the turbulent state, which means that we are confronted with a large variety of flow scales. An upper limit for their size is given by the physical space available; however other large-scale structures like separations do not belong to the category of turbulent flows.

It is common to describe the turbulent flow by statistical means, mainly based on the velocity fluctuations. It is known and discussed in Chap. 5 (Sect. 5.2) that this representation is by far too simple. A statistical representation of turbulence, which goes beyond the one based on the first four moments of the velocity fluctuations, would belong to a new theory of the sediment transport using the newest results in turbulence research. Some remarks will be made but not more.

In the last two decades, the theory of the coherent structures of turbulent flows was discussed. This new interpretation was based on a much more deterministic view (Chap. 3, Sect. 3.2.4). These structures are somewhat large because they possess a very high correlation over a large distance when scaled with inner variables. They are nonlocal elements, although their physical sizes are rather small. Their presence is restricted by the properties of the bed, which must be more or less flat and smooth.

These coherent structures have a lot in common with the so-called large structures, which have been known for a long time and which do not belong to the turbulent state in a strict sense. They are the results of restrictions by the boundary condition mainly of geometrical type.

A special class of large structure is the one that originates from separations at bed forms or obstacles. It is this type of structure which will be discussed at length in Chaps. 7–11 as separations will be the main elements, which we will suggest as improvements for the classical sediment transport theories.

Especially for the suspension flow, the interaction of the tiny particles with the smallest flow structures is relevant, and we also have to characterize the smallest scales in a turbulent flow.

Based on these ideas, we characterize the different scales by some physical criteria. However, as mentioned in Chap. 5 (Sect. 5.3) turbulence is the result of the nonlinear property of the Navier–Stokes equation (NSE), and therefore a classification should always be seen as an approach to a situation we cannot describe at the moment. However, models are essential as schemes when constructing transport equations for a certain type of sediment under a given flow condition.

6.2.1 The Statistical Scales of the Turbulent Flow

The turbulent flow contains structures of many sizes; they are very complex in their geometrical form and in continuous dynamical interaction with each other. They constantly change their form, momentum, and energy content. An essential property of the turbulent flow is its vorticity distribution. Vorticity arises from the local fluctuations of the velocity components, which concur with spatial velocity gradients. Flow regions of high vorticity usually have a short lifetime. They are the nuclei of small-scale structures. Larger structures generally have a lower vorticity density. Untypical turbulent structures often depend on the mechanism of their creation, especially when they are more or less 2D, their dissipation rate is low and they therefore have a long lifetime.

Nevertheless, it is common to define the turbulent flow as an equilibrium mixture of vortices, which merge and decay. This simplistic concept has several reasons. Vortices are fairly simple features which can be treated mathematically and which are appropriate for thought experiments. This concept is also related to the linearized representation of the flow state as it is expressed, e.g., by a Fourier spectrum. We know the deficits, but often we have no better model to describe the interaction of the flow with the bed.

We assume that the flow structure capable to move a grain must be at least of the size of the grain,

$$l \approx d_s \tag{6.2}$$

If the grain is larger, then the influences of the smaller flow scales average out, and if it is smaller, a multitude of scales can bring the grain into motion. However, the length scale alone is not sufficient to initiate transport, also the force and the angular momentum must be strong enough (Eq. 3.10). The structure element must possess a sufficient momentum- and energy density within the size l. For larger elements, this is often given in the literature only in 2-D form, but what is needed is the volume or mass density, respectively. For example, for the energy density,

$$e = \frac{E_{\text{kin}}}{V_{\text{st}}} \quad \vee \quad \left(e = \frac{E_{\text{kin}}}{A_{\text{st}}} \right) \tag{6.3}$$

where V_{st} represents the volume and A_{st} the surface of the structure.

The energy per structural element can be calculated from the Fourier spectrum, which corresponds to the energy appointed to single oscillators, physically interpreted: the flow field consists of circular vortices. To get rid of the directional dependence, the spectrum has to be integrated over spherical shells around the origin of the wave number space, and thereby we usually sum up the spectra u'_i.

Using the correlation tensor R_{ij}:

$$R_{ij} = \frac{\overline{u'_i u'_j}}{\sqrt{\overline{u'^2_i}\,\overline{u'^2_j}}} \tag{6.4}$$

the spatial spectrum can be evaluated from this correlation,

$$\begin{aligned}
&R_{ij}(\underline{r}) = \overline{u'_i(\underline{x},t)\,u'_j(\underline{x}+\underline{r},t)} \quad \wedge \quad \Phi_{ij}(\underline{k}) \bullet\!-\!\circ\, R_{ij}(\underline{r}) \\
&\therefore \quad \Phi_{ij}(\underline{k}) = \frac{1}{(2\pi)^3} \int_{-\infty}^{+\infty}\!\!\iint e^{(-i\underline{k}\bullet r)} R_{ij}(\underline{r})\,\mathrm{d}\underline{r} \quad \wedge \\
&R_{ij}(\underline{r}) = \frac{1}{(2\pi)^3} \int_{-\infty}^{+\infty}\!\!\iint e^{(i\underline{k}\bullet r)} \Phi_{ij}(\underline{k})\,\mathrm{d}\underline{k} \\
&R_{ii}(0) = \overline{u'_i u'_i} = 3u'^2 = \int_{-\infty}^{+\infty}\!\!\iint \Phi_{ii}\,\mathrm{d}\underline{k}
\end{aligned} \tag{6.5}$$

With • -° the Fourier-transform-operator

The summation of the diagonal elements of the tensor Φ_{ij} represents the kinetic energy belonging to the given wave number vector $\underline{k}$. This has now to be integrated over the spherical shells of radius k, and we get

$$\begin{aligned}
&E(k) = \frac{1}{2} \oiint \Phi_{ii}(\underline{k})\,\mathrm{d}\sigma \ : \ k^2 = \underline{k} \bullet \underline{k} = k_i k_i \ \therefore \\
&\int_0^{\infty} E(k)\,\mathrm{d}k = \frac{1}{2}\int_0^{\infty}\left[\oiint \Phi_{ii}(\underline{k})\,\mathrm{d}\sigma\right]\mathrm{d}k = \frac{1}{2}\int_{-\infty}^{+\infty}\!\!\iint \Phi_{ii}(\underline{k})\,\mathrm{d}\underline{k} = \frac{1}{2}\overline{u'_i u'_i} = \frac{3}{2}u'^2
\end{aligned} \tag{6.6}$$

The factor 1/2 was chosen such that the content of the 3D spectrum is equal to the kinetic energy per unit mass.

Since l is the scale of the circular vortex

$$l = 2\pi \frac{1}{k} \tag{6.7}$$

Due to Eq. 6.7, the vortex has a kinetic energy of

$$E_{kin}\big|_k = kE(k) \tag{6.8}$$

and with it a characteristic velocity

$$u'(k) = \left[kE(k)\right]^{1/2} \tag{6.9}$$

and the strain rate exhibited to the neighborhood becomes,

$$s(k) = \frac{1}{l}\left[kE(k)\right]^{1/2} = \frac{1}{2\pi}\left[k^3E(k)\right] \tag{6.10}$$

When varying k, this quantity increases by a factor of $k^{2/3}$ and reflects the fact that the the influence by neighboring vortices is highest when of similar scale. The energy is also time dependent, which is shown schematically in Fig. 6.1.

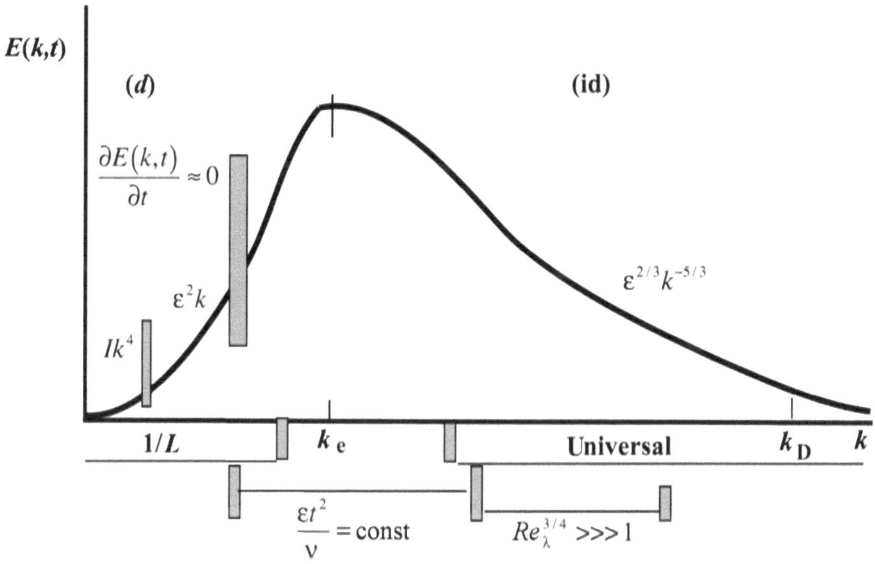

Fig. 6.1 Schematic representation of a 3D energy spectrum with the ranges (*d*) and (id). In the first range, the spectrum depends on the origin of the vortices in question, whereas in the second range it is independent thereof. The first range contains large vortices with low energy dissipation, which therefore exist permanently. The range of vortices containing most of the energy follows around k_e. The so-called universal equilibrium range with the inertial range as subunit follows for wave numbers larger than k_e

The wave-number k_e belongs to the vortices containing the highest amount of energy. By inserting this value into Eq. 6.9, we define the turbulent scale. The upper wave number limit is given by k_D, which is the scale where the energy is "totally" dissipated. It is given by the maximum dissipation

$$k_D = k^2E(k,t)\big|_{max} \quad \wedge \quad \overline{\left(\frac{\partial u'}{\partial t}\right)^2} \sim k^2E(k,t) \tag{6.11}$$

One would expect that $k_D \approx k_K = 2\pi/l_K$, however, the measured values are

$$0.09 \le k_K / k_D \le 0.5 \tag{6.12}$$

In the concept of a cascade, a nonlinear transport process carries energy from large to small vortices. This mechanism is also responsible for the increasing isotropy of smaller vortex classes at larger wave numbers, which is called an approach to local isotropy. This state appears when the Reynolds numbers defined by the turbulent length scale are of the order 100.

$$\underline{k}(\text{isotropic}) \quad \forall \quad \text{Re}_l \ge 100 \quad \wedge \quad \text{Re}_l = \frac{u'l}{\nu} \tag{6.13}$$

In Fig. 6.1, this range is called the universal range, and it overlaps with the range where dissipation occurs. *E(k)* becomes

$$E(k,t) = \nu^{5/4}\varepsilon^{1/4}E^*(kl_K) \tag{6.14}$$

where E^* is a value evaluated experimentally. In the zone of the vortices with the highest energy, the system is in equilibrium too and by a dimensional analysis *E(k)* is in this case given by:

$$E(k,t) = \nu^{5/4}\varepsilon^{1/4}E^*\left(kk_K\varepsilon t^2/\nu\right) \tag{6.15}$$

The range of the very large vortices contributes about 20% to the total kinetic energy budget. For these large vortices, respectively, the small vortices compared with a mean length scale L_m it was found that

$$\begin{aligned} E(k,t) &= \varepsilon_\nu^{8/3} I^{-1/3} E^*\left(k\left(I/\varepsilon_\nu^2\right)^{1/3}\right) \quad \wedge \quad L > L_m \\ E(k,t) &= \varepsilon_\nu^{5/4}\varepsilon^{1/4}E^*\left(k\left(\varepsilon_\nu^3/\varepsilon\right)^{1/4}\right) \quad \wedge \quad L < L_m \end{aligned} \tag{6.16}$$

with ε_v the eddy viscosity and *I* the Loitsianski integral (Landau and Lifschitz, 1991).

Kolmogorov postulated that there exists a wave-number range for small vortices in which the viscous dissipation must be negligible compared to the inertial redistribution of energy. In this subrange only the energy dissipation $\varepsilon(t)$ is of importance, and with this assumption the often-cited $k^{-5/3}$-law can be derived

$$E(k,t) = ¢\,\varepsilon(t)^{2/3}\,k^{-5/3} \tag{6.17}$$

This assumption holds if the Reynolds number as defined by Eq. 6.13 is high enough, $Re > 10^5$ or $Re_l > 10^3$. This is an important assumption, which is fulfilled by most of the investigated alluvial systems, however, not for most laboratory experiments (Eq. 6.1). It is therefore very intriguing that in the literature one finds spectra with that subrange, although it is known that the law in Eq. 6.17 cannot exist in the published experiments. It shall be a warning that the turbulent structures differ from the ones found in most of the laboratory experiments. Also for the numerical computation, the above results are of importance. It makes sense to consult the literature treating the computation of turbulent flows in these cases (Reynolds and Cebeci, 1976, Lumley, 1996; Moin, 1996; Germano, 1999; Leonard, 1999).

In this representation, the turbulent fluctuations are explained by a statistical vortex distribution. The advantage of such a representation is a virtual gain of information since a model now defines the fluctuations. With such a model, one can describe the interaction of the sediment with single vortices and by using the vortex distribution to extrapolate the total sediment transport or evaluate the local influence of the vortices by averaging over the distribution. This way one can evaluate the local mean strain and shear stress. This requires, however, that the flow field needs to be expanded in vortices instead of waves (Lumley, 1970). He proposed to declare a vortex of wave number k as a disturbance, which has its energy in the range of $0.62k$ and $1.62k$. By this definition, the energy is centered around k in logarithmic scaling because $\ln(1.62) = \ln(1/0.62) \approx 1/2$.

Whether one calculates the strain rate or the shear stress by Eq. 6.10, one finds the result described in Chap. 2 (Fig. 2.2), and we are confronted with the problem that $u'\, v'$ has a wide probability distribution (Eqs. 3.1 and 3.15), the standard deviation is.

$$\sigma(u'v') \rightarrow O(10) \tag{6.18}$$

This is a hint that the phase relation between u' and v' has a stronger correlation than expected. In other words, several vortices act in a correlated way at a given measuring location, and this thought model corresponds to a first step in the direction of coherent structures.

6.2.1.1 Dimensional Analysis

Another approach to evaluate the scaling relations of the turbulent flow is by dimensional analysis. By assuming that the smallest energy dissipating vortices do not interfere with the large scales, the motion of these vortices is limited by the energy flux and the dissipation rate. This assumption is inherent with most turbulence theories. These are the basic assumptions of Kolmogorov's equilibrium theory, as discussed in Chap. 5 (Sect. 5.2.2), Eqs. 5.33–5.36. These small scales are represented by ε and ν, and with these scales one can define a Reynolds number

$$Re = \frac{l_k u_k}{\nu} = 1 \tag{6.19}$$

A flow on these scales is dominated by viscous forces, and their order of magnitude can be evaluated only if the energy dissipation is known. By assuming an energy-transport-equilibrium, the dissipation rate of the small scales has to be evaluated from the energy feeding rate from the larger scales. The amount of kinetic energy per unit mass in the large scale turbulence is proportional to u'^2; the rate of transfer of energy is assumed to be proportional to u'/l, where $l = L_{\text{int}}$ is the size of the largest eddies. The length scale l relates to the integral scale L_{int} of turbulence, measured by statistical methods.

$$E_{\text{kin}} \propto u'^2 \quad \wedge \quad E_{\text{trans}} \propto u'/L_{\text{int}} \tag{6.20}$$

The rate of energy supply to the small-scale eddies is thus of the order

$$u'^2 \cdot \frac{u'}{l} = \frac{u'^3}{l} = \frac{u'^3}{L_{\text{int}}} \tag{6.21}$$

This energy is dissipated at the rate of ε, which should be equal to the supply rate

$$\begin{aligned} &\therefore \quad \varepsilon \propto u'^3 / L_{\text{int}} \\ &\wedge \quad L_{\text{int}} = \int_0^\infty R(x_i, x_i + l)\,\mathrm{d}l \quad \wedge \quad R_{ij} = \frac{\overline{u'_i u'_j}}{\sqrt{\overline{u'^2_i}\,\overline{u'^2_j}}} \end{aligned} \tag{6.22}$$

Here, the spatial correlation in x-direction was used to estimate L_{int}. Here, dissipation is clearly seen as a passive process in the sense that it proceeds at a rate dictated by the inviscid inertial behavior of the large eddies. This can be assumed for a scaling theory, although in a strict sense it is incorrect as shown in Chap. 5 (Sect. 5.2.1). Nevertheless, the resulting Eq. 6.22 is of fundamental importance as it tells us that the large eddies lose a significant fraction of their kinetic energy within one turnover time. Tennekes and Lumley (1972) found that the nonlinear mechanism that generates small eddies from of large ones is as dissipative as its characteristic time permits. In other words, turbulence is at large scales a strongly damped nonlinear stochastic system.

The direct dissipation of the large eddies is given by the timescale of their decay,

$$l^2/\nu \tag{6.23}$$

The viscous energy loss proceeds at a rate

$$\nu u'^2/l^2 \tag{6.24}$$

which is small compared to ε in Eq. 6.22 if the Reynolds number

$$Re_{\text{int}} = \frac{u'L_{\text{int}}}{\nu} \tag{6.25}$$

is large. In this Reynolds number representation, we can formulate the ratio between the Kolmogorov scales and the new Re as

$$\begin{aligned} l_{\text{K}}/L_{\text{int}} &\propto \left(u'L_{\text{int}}/\nu\right)^{-3/4} = Re_{\text{int}}^{-3/4} \\ t_{\text{K}}u'/L_{\text{int}} &\propto t_{\text{K}}/t = \left(u'L_{\text{int}}/\nu\right)^{-1/2} = Re_{\text{int}}^{-1/2} \\ u_{\text{K}}/u' &\propto \left(u'L_{\text{int}}/\nu\right)^{-1/4} = Re_{\text{int}}^{-1/4} \end{aligned} \tag{6.26}$$

These relations indicate that the length, time, and velocity scales of the smallest eddies are very much smaller than those of the largest eddies. For example,

$$\begin{aligned} &l_{\text{K}}/L_{\text{int}} \propto \left(u'L_{\text{int}}/\nu\right)^{-3/4} = Re^{-3/4} \\ &\wedge\ \ u' \approx 5\%U, \quad L_{\text{int}} \approx H \ \vee\ l_{\text{K}} \approx HRe^{-3/4} \rightarrow \ \sim 0.2\text{mm} \end{aligned} \tag{6.27}$$

In addition to the integral length scale, it is common to define a Taylor microscale λ_{Ta}, it is not a physical length but a statistically evaluated mean defined by Eq. 6.28 (Fig 6.2).

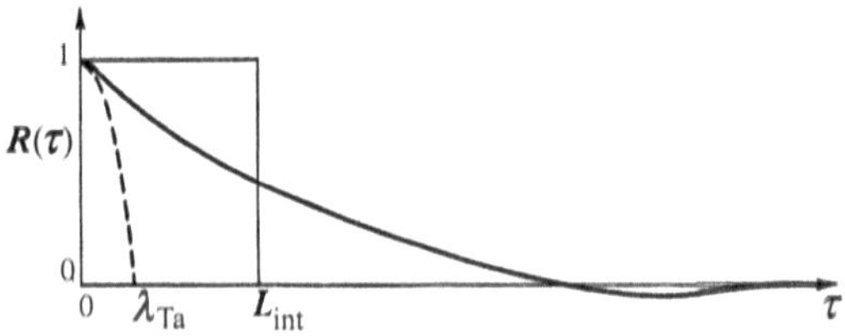

Fig. 6.2 Definition-sketch of the Taylor microscale using the autocorrelation of the velocity fluctuations in flow direction (Eq. 6.34)

$$\varepsilon \propto \nu \frac{u'^2}{\lambda_{\mathrm{Ta}}^2} \quad \therefore \quad \lambda_{\mathrm{Ta}} \sim u' \sqrt{\frac{\nu}{\varepsilon}}; \quad \overline{\left(\frac{\mathrm{d}u'}{\mathrm{d}t}\right)^2} = \frac{\overline{2u'^2}}{\lambda_{\mathrm{Ta}}^2} \quad \therefore \quad s_{ij} \sim \frac{u'}{\lambda_{\mathrm{Ta}}}$$
$$\lambda_{\mathrm{Ta}}^2 = \int_0^\infty E(k)\mathrm{d}k \Big/ \int_0^\infty k^2 E(k)\mathrm{d}k \qquad (6.28)$$
$$\lambda_{\mathrm{Ta}}^2 = -\frac{R_{\mathrm{uu}}(0)}{8R''_{\mathrm{uu}}(0)}$$

R'' is the curvature of R. In the original paper by Taylor (1935) for an isotropic turbulence, the relation was given by

$$\varepsilon \sim 15\nu \frac{\langle u_x'^2 \rangle}{\lambda_{\mathrm{Ta}}^2} \qquad (6.29)$$

(Eq. 4.11).

In analogy to Eq. 6.26, one can now also introduce the ratio of the Taylor microscale to the integral scale,

$$\frac{\lambda_{\mathrm{Ta}}}{L_{\mathrm{int}}} \sim Re^{-1/2} \qquad (6.30)$$

In a viscous fluid, we have an exchange of momentum due to molecular motion, which is much smaller than the turbulent exchange. It is a diffusion process, and the length scale depends on the time the diffusion has been acting

$$l_{\mathrm{Dif}} \sim \sqrt{\nu t} \qquad (6.31)$$

This quantity is important for the estimation of the vorticity diffusion, which will be always encountered when one works with vortices.

In analogy to Eq. 3.82, instead of ε_{i} one can introduce for the momentum diffusion an eddy viscosity or a turbulent exchange coefficient ν_{t}, which can replace the main term of the Reynolds-stress tensor

$$-\overline{u_x' u_z'} \equiv \nu_{\mathrm{t}} \frac{\partial U}{\partial z} \quad \wedge \quad \nu_{\mathrm{t}} \sim u' l_{\mathrm{int}} \qquad (6.32)$$

A Reynolds number formed with this integral scale of Eq. 6.25 is equal to the ratio of the eddy viscosity to the kinematic viscosity,

$$\frac{\nu_{\mathrm{t}}}{\nu} \sim \frac{u' L_{\mathrm{int}}}{\nu} = Re_{\mathrm{int}} \qquad (6.33)$$

This shows quite nicely that for high Reynolds numbers the molecular diffusion can be neglected, and the ratio of these two viscosities is a good tool for estimates.

Most of the scales have been derived as length scales; however, often it is easier to use time or velocity scales. A typical example is the Taylor microscale evaluated from the autocorrelation in time, using a signal as measured by most instruments.

$$R_{ij}(\tau)=\frac{\overline{u'_i(t)u'_j(t+\tau)}}{\sqrt{\overline{u'^2_i(t)}}\sqrt{\overline{u'^2_j(t+\tau)}}}\rightarrow R(\tau)\approx 1-\left[\frac{\tau^2}{-\frac{1}{2}\frac{\partial^2 R}{\partial \tau^2}}\right]_{\tau=0} \tag{6.34}$$

$$\therefore\quad \lambda=\lambda_\tau\equiv-\sqrt{\frac{1}{2}\frac{\partial^2 R}{\partial\tau^2}}$$

The direct dissipation of large scales by the viscosity is small as can be seen by a dimensional analysis of those timescales. The largest elements of size H decay with a viscous timescale of:

$$t_{\mathrm{D}}\propto\frac{L_{\mathrm{int}}^2}{\nu}\quad\rightarrow\quad t_{\mathrm{D}}\sim 10^6\, s \tag{6.35}$$

the direct energy dissipation rate of the large scale is therefore,

$$\varepsilon_{\mathrm{D}}\propto\frac{\nu u'^2}{L_{\mathrm{int}}^2}\quad\therefore\quad(6.22)\rightarrow\frac{\varepsilon_{\mathrm{D}}}{\varepsilon}=\frac{\nu}{u' L_{\mathrm{int}}}\sim\frac{1}{Re_{\mathrm{int}}} \tag{6.36}$$

For large Reynolds numbers, the energy dissipated directly by the large scales is negligible compared to the total dissipation.

An estimate for the molecular diffusion time needed to equalize the momentum (or temperature) in a cell of size H^2 is,

$$T_{\mathrm{m}}\propto\frac{H^2}{\nu} \tag{6.37}$$

This time is much larger than the turbulent mixing time defined by replacing the kinematic viscosity with the eddy viscosity ν_{t}, Eq. 6.32

$$T_{\mathrm{t}}\propto\frac{H^2}{\nu_{\mathrm{t}}}\quad\therefore(6.33)\rightarrow T_{\mathrm{t}}\propto\frac{L^2}{\nu Re_{\mathrm{int}}} \tag{6.38}$$

All these scales will be found in Table 6.2 and 6.3 and estimated for a river with H=1m and a mean velocity U= 2 m/s

Table 6.2 LTV turbulent scales

Length scale	Time scale	Velocity scale	Comments
Integral scales			
$L_{\mathrm{int}}\approx H$	$T\propto H/U$	U	For the largest elements
$\approx$ 1 m	$\approx$ 0.5 s	$\approx$ 2 m/s	
L_{int} Eq. 6.22	$t=L_{\mathrm{int}}/u'$	$u'\approx 5\%U$	For mean elements
	$\approx$ 10 s	$\approx$ 0.1m/s	
Taylor micro scales			
λ_{Ta} Eq. 6.30	λ_τ Eq. 3.34		
$\approx$3mm			
Kolmogorov scales, the smallest turbulent scales Eq. 6.26			
l_{K}, Eq. 6.26 $\approx$ 0.2 mm	t_{K}, Eq. 6.26 0.03s	u_{K}, Eq. 6.269 $\approx$ 5.6 mm/s	It is better to use the estimate of the magnitudes
O(0.1mm)	O(0.1s)	O(1 mm/s)	

6.2.2 Elements of Large Scales

The large flow scales belong to two classes of structures, both of which do not belong to the turbulent equilibrium system. These large scales are given by the geometrical restrictions of the channels and by separations on obstacles of all sizes, from blocks up to bed-forms like dunes. The geometrical restrictions appear as truncation of the low wave number spectrum, whereas the separated flow structures are vortices which do feed into the turbulent system and are contributing to the largest energy containing scales as shown in Fig. 6.1.

6.2.2.1 Geometrical Restrictions

In z-direction, flow structures scale with the distance to the bed and are therefore limited by the water surface. Therefore, H determines the largest scales, which are equivalent to the largest integral scales.

The limitation of the structural sizes by the distance to the bed was already noted in the derivation of the velocity profile (Chap. 2, Sect. 2.1.3.1). With this concept in mind, one may argue which sizes of vortices interact with the bed. For a proper mechanistic description of the interaction process, one would need a size distribution of the vortices. It is this distribution one has to strive for if one would like to introduce a more deterministic representation based on large-scale elements. Another unknown quantity is the relative velocity or the convection velocity of the vortices of different scales compared to the mean flow velocity U.

To elucidate this question, Favre et al. (1967) introduced the space-time correlation by measuring simultaneously at points a short distance downstream of each other. Applying a cross correlation to the signals, one found the maximum value shifted in time and with a certain direction. This can be thought of as a sort of a convection velocity of the random signals; Favre speaks of it as celerity of the eddying motion. Figure 6.2 shows this celerity plotted for eddies of various frequencies and hence presumably of various sizes L. As might be expected, the convection velocity for the small eddies is simply the mean flow velocity; they are just swept along with the stream. The larger eddies, however, seem to keep better in step across the layer and move as a whole with an average eddy convection velocity v_c of about 0.8–0.9 units of U at a height of about 0.8δ. Favre mentioned a lower value of about 0.7–0.8 which is in better agreement with Landahl's (1967) calculated values (Fig. 6.3.)

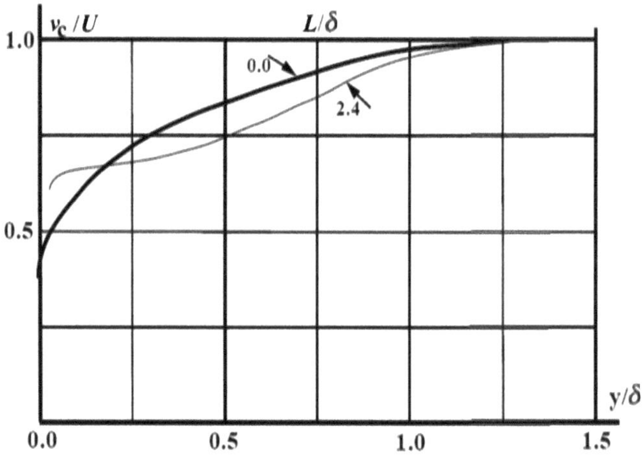

Fig. 6.3 Celerity divided by external flow velocity for eddies of various size L, plotted against distance y from the wall divided by boundary-thickness δ. ($y = z$ for the bed distance)

6.2.2.2 Scales of the Separation Vortices

The scale of a separating flow structure depends primarily on the obstacle size at which the flow separates. On sediment elements of all sizes, this scale is proportional to the diameter d_s for rocks as well as for sand grains. The scales of the separations on bed forms are different although; they depend on the height of the bed form as well as the size of the separation bubble, which differs from the bed-form wavelength. Further details will be discussed in Chaps. 7–11, where flow separations and their feedback are treated as a new and important element of the sediment transport. These two sorts of scales are strongly related, but often one would just like to have an idea about the sizes, which are given later for the example used in Tables 6.1–6.3.

Table 6.3 LTV Diffusion scales

Length scale	Time scale	Velocity scale	Comments
Viscous scales			
l_{Dif}, Eq. 6.31	t_D Eq. 6.35	$u_D = l_{\mathrm{Dif}}/t_D$	
≈ 3 mm	$\approx 10^6$ s	$\approx 3 \cdot 10^{-9}$ m/s	
Turbulent scales using eq. (6.33) $\nu_t \tilde{}\ \frac{1}{2} Re_{int} = 0.1\ m^2/s$			
L_t	T_t Eq. 6.38	U_t	
≈ 0.95 m	≈ 10s	≈ 0.1 m/s = u′	
≈ O(1 m)			

The timescales are defined by the shedding frequency of the obstacles at a given Reynolds number. This frequency is measured by the Strouhal number Str (since the usual abbreviation St as found in the literature was used in this text for the Stokes number).

$$
\begin{aligned}
&\mathrm{Str} = \frac{fk_\mathrm{s}}{u_0} = \frac{k_\mathrm{s}}{t_\mathrm{a} u_0} \\
&\wedge \quad \overline{u_0} = \frac{u_\tau}{k_\mathrm{s}} \int_{\approx 0}^{k_\mathrm{s}} \left[\frac{1}{\kappa} \ln \frac{z}{k_\mathrm{s}} + 8.5 \right] \mathrm{d}z = 8.5 u_\tau + u_\tau \int_{\approx 0}^{1} \ln x \mathrm{d}x \\
&u_0 \approx u_\tau \left(8.5 - \frac{1}{\kappa} \right) = 6 u_\tau \\
&\therefore \quad f = \frac{1.2 u_\tau}{k_\mathrm{s}} \quad \wedge \quad t_\mathrm{a} = \frac{k_\mathrm{s}}{1.2 u_\tau} \rightarrow t_\mathrm{a} \approx 0.1\mathrm{s}
\end{aligned}
\tag{6.39}
$$

In the range of the discussed example, Str is in good approximation a constant value

$$
Str = \frac{f k_\mathrm{s}}{u_0} \rightarrow \approx 0.2 \tag{6.40}
$$

The scales for the sediment bodies are listed in Table 6.4, whereas the separation is too dynamical and will therefore be discussed later. However, also the sediment bodies, like dunes, propagate and for their motion a good approximate value can be given, which is valid over a rather large range

$$
\frac{u_\mathrm{D}}{U} \approx \frac{h}{H} \quad \because \quad H \sim 5h \quad \therefore \quad u_\mathrm{D} \approx 0.2U \tag{6.41}
$$

which allows estimating the transport of sediment by the propagation of the dunes as explained later.

Table 6.4 Scales of sediment bodies and the sedimentation

Length scales	Timescales	Velocity scales	Comments
≈ 10 m	≈ 25 s	≈ 0.4 m/s	For a dune
≈ 0.5 m			Small boulder
≈ 0.1 m			Small cobble, ripple length
≈ 0.01 m			Medium gravel
≈ 1 mm		1 mm/s	Course sand, sink velocity
≈ 0.01 mm		>0.01 mm/s	Fine silt, sink velocity

6.2.3 Higher Moments of the Fluctuations and Related Scales

With the Reynolds decomposition Eq. 2.3, a scaling of the flow structures was introduced. In several cases, the fluctuations are relevant for the sediment transport, and for those cases the scales of the fluctuations and their distributions have to be discussed. This analysis is of importance when results found in laboratory experiments are compared to alluvial flows, especially when results found by small-scale experiments are scaled up. A mismatch of the turbulent scales is unimportant as long as turbulence as such is irrelevant or can be described by a simple wall shear stress. However, when a higher resolution of the flow field is needed, this upscaling can become problematic. In an alluvial channel the flow is fully developed, which means all fluctuations and their higher moments are in equilibrium. In a laboratory channel, this is only achieved at a certain distance downstream from the inlet. Most of the channels have no adequate length with this respect, and much of the misunderstanding in the literature is rooted therein.

As a criterion one can use:

$$\begin{aligned} L &\approx 80B \quad \rightarrow \text{for } (u') \\ L &\approx 160B \rightarrow \text{for } (u')^m \end{aligned} \tag{6.42}$$

The length of the channel must be at least 80 times its width for the fluctuations to be fully developed and again twice as long until the higher moments are in the equilibrium state also. For a laboratory channel of 0.3-m width, the measurements for the simple fluctuations have to be made 24 m downward of the inlet. For a fully developed turbulent flow, the channel would have to be longer than 48 m, until the equilibrium state starts. A short look into the literature shows that only very few experiments fulfill this criterion.

The timescale involved in this problem is easier to treat and can be evaluated empirically by taking longer and longer measuring sequences until one finds that the mean values do not change anymore.

We know that in the past the lack of our knowledge of turbulent flows hindered the development of sediment transport descriptions, however, it would go beyond the limits of this text to cover the entire complexity of this flow state. We have to refer the reader to the literature on turbulence; however, some basic discussion for the velocity and its fluctuations will be given briefly.

We apply a gating circuit on $u(t)$, which turns on when the signal is between two adjacent levels. If we average the output of the gating circuit, we obtain the percentage of time $u(t)$ spent between the two levels. In the limit of $T \Rightarrow \infty$, we define a quantity $B(u)$, probability density given by

$$B(u)\Delta u = \lim(T \rightarrow \infty)\frac{1}{T}\sum(\Delta t) \tag{6.43}$$

The probability that $u < u(t) < u + \Delta u$ is equal to the proportion of the time spent there.

$$B(u) \geq 0; \quad \int_{-\infty}^{\infty} B(u)\mathrm{d}u = 1 \tag{6.44}$$

The time average Eq. 2.2 can now be written as

$$\overline{u} = \lim(T \rightarrow \infty)\frac{1}{T}\int_{t_0}^{t_0+T} f(u)\mathrm{d}t = \int_{-\infty}^{\infty} f(u)B(u)\mathrm{d}u \tag{6.45}$$

The mean values of the various powers of u are called moments.

The first moment

$$\overline{u} = \int_{-\infty}^{\infty} f(u)B(u)\mathrm{d}u \tag{6.46}$$

is the familiar mean velocity.

Subtracting the mean velocity from the signal one obtains the fluctuating part

$$u' = u - \overline{u} \quad \wedge \quad \overline{u'} = 0 \quad \therefore \quad B(u) = B(\overline{u} + u') \tag{6.47}$$

The moments formed with the fluctuating part u'^m and $B(u')$ are called the central moments. The first moment is zero.

The second moment σ^2 is called the variance or second central moment,

$$\sigma^2 \equiv \overline{u'^2} = \int_{-\infty}^{\infty} u'^2 B(u)\,\mathrm{d}u = \int u'^2 B(u')\,\mathrm{d}u'$$
$$\sqrt{\sigma^2} = \sigma \quad \text{standard deviation} \tag{6.48}$$

This value is not affected by any lack of symmetry in $B(u')$ about the mean value.

The third moment

$$\overline{u'^3} = \int_{-\infty}^{\infty} u'^3\, B(u')\,\mathrm{d}u' \tag{6.49}$$

depends on the lack of symmetry in $B(u')$ and

$$\overline{u'^3} = 0 \tag{6.50}$$

when is $B(u')$ symmetric about the mean value.

The skewness

$$S \equiv \frac{\overline{u'^3}}{\sigma^3} \tag{6.51}$$

is a dimensionless measure for the asymmetry of the fluctuations. It characterizes the anisotropy of the flow field, which is most dominant near the bed. The skewness also helps defining the boundaries of diverse flow layers for fluctuations in analogy to Reynolds stresses in Eq. 2.34.

The fourth moment made dimensionless with σ^4 is called kurtosis or flatness factor K

$$K \equiv \frac{\overline{u'^4}}{\sigma^4} = \frac{1}{\sigma^4}\int_{-\infty}^{\infty} u'^4 B(u')\,\mathrm{d}u' \tag{6.52}$$

The value of K is large, when the values $B(u')$ in the tails of the probability density are relatively large. That is to say, a function that frequently has values far away from the mean will show a large kurtosis. This is an important value since it is known that the extreme values are the most relevant for the transport (Eq. 3.1).

In addition to the characterization by fluctuations we also have to consider the correlations and the joint moments, which are part of joint statistics. For statistically independent probabilities, we get the simple relation

$$B(u_x' u_z') = B_{u_x'}(u_x')\,B_{u_z'}(u_z') \tag{6.53}$$

If the independence is given, a correlation function has to be used. An especially relevant relation shall be pointed out here

$$-\overline{u_x' u_z'} \,/ \left(\overline{u_x'^2}\,\overline{u_z'^2}\right)^{1/2} \approx 0.4 \tag{6.54}$$

The "magic number" of 0.4 appears in many different types of shear flows.

The higher moments over flat, smooth walls are reproduced in many textbooks, and they will not be reproduced here. The usual situation is quite different and not universal because of the rough walls, and therefore the moments have to be evaluated for the special case or they have to be looked up as the relevant literature.

6.2.4 Structure Functions

A special diagnostic tool in turbulence are the so-called structure functions. They allow determining whether the fluid behaves Newtonian, and therefore they are relevant for the sediment transport occurring in form of suspensions.

The structure functions are more physical entities than the function mentioned in Chap. 6 (Sect. 6.2.3) as they do not use the Reynolds decomposition; therefore their scales are not biased. They are defined as

$$F_i(r,t) = \left\langle \left[\underline{u}(\underline{x}+\underline{r},t) - \underline{u}(\underline{x},t) \right]^i \right\rangle; \quad i = 2,3,... \tag{6.55}$$

Based on Kolmogorovs' theory it is known (Lesieur, 1987; Frisch 1995; Tsinober, 2001) that the second structural function obeys the following scaling law,

$$F_2(r) \propto (\varepsilon r)^{2/3} \quad ; \quad l_K < r < L_{\text{int}} \tag{6.56}$$

Using the so-called surrogate dissipation Eqs. 6.29 and 5.33, which is estimated via the relation between the Taylor microscale and the Kolmogorov length scale the energy dissipation can be given as:

$$\varepsilon = \frac{15\nu\sigma_{(u)}^2}{\lambda_{\text{Ta}}^2} \tag{6.57}$$

Even more important is the third structure function, which—for Newtonian fluids—is given by the relation

$$F_3(r) \propto \varepsilon r \tag{6.58}$$

or in the form

$$S_3(r) = \left\langle \left\{ \left[\underline{u}(\underline{x}+\underline{r},t) - \underline{u}(\underline{x},t) \right] \frac{\underline{r}}{r} \right\}^3 \right\rangle \tag{6.59}$$

where the value

$$S_3(r) = -\frac{4}{5} \langle \varepsilon \rangle r \tag{6.60}$$

which yields the proportionality coefficient equal to –4/5 in Eq. 6.58. This value is related to the –4/5 Kolmogorov-law, (Kolmogorov, 1941), and is valid in the inertial range. While investigating the non-Newtonian character of drag reducing fluids, Gyr and Tsinober (1996) have pointed out that it is even more appropriate to investigate the skewness of the velocity derivatives

$$s = \left\langle \left(\frac{\partial u}{\partial x} \right)^3 \right\rangle \Big/ \left\langle \left(\frac{\partial u}{\partial x} \right)^2 \right\rangle^{2/3} \tag{6.61}$$

where s > 0 is the result of the vortex stretching mechanism. In addition one can compare

$$\left\langle \left(\frac{\partial u}{\partial u} \right)^3 \right\rangle_{\text{New}} \Big/ \left\langle \left(\frac{\partial u}{\partial x} \right)^3 \right\rangle_{n-\text{New}} \tag{6.62}$$

and it is this quantity, which is, at least approximately, proportional to the enstrophy generation (Batchelor and Townsend, 1949)

$$\langle \omega_i \omega_k s_{ik} \rangle . \tag{6.63}$$

6.2.5 The Scales of the Coherent Structures

Coherent structures can be observed in flows over flat, smooth beds, where the smoothness is somewhat relative, because coherent structures occur also on beds with a roughness above $z^+ = 5$, and has to be defined. The coherent structures could only be verified for a roughness up to $k^+ < 10$. This is about twice the height of the viscous sublayer, which was postulated for a hydraulically smooth bed. In other words, for the sediment transport coherent structures are relevant in laboratory experiments but rarely in alluvial channel flows.

$$5 \geq z^+ \geq k_s^+ \quad \wedge \quad k_s^+ = \frac{k_s u_\tau}{\nu} \quad \therefore \quad k_s = \frac{5\nu}{u_\tau} \rightarrow O(5 \cdot 10^{-5}\,\mathrm{m}), \text{alluvial} \qquad \text{lab} \rightarrow O\left(5 \cdot 10^{-4}\,\mathrm{m}\right) \tag{6.64}$$

Nevertheless, we will discuss them here with certain persistence. The reason is that it it became popular to speak of coherent structures even for cases which do not belong to this category of structures. See for example the proceedings of a conference about the possible impact of coherent structures on the sediment transport, Ashworth et al. (1996) and especially the article of Smith (1996) therein. The root of this misinterpretation is the fact that the rather tiny structures have an analogous counterpart in the large structures.

The main length scales for the coherent structures are given by the distance from the bed and the length scale related to their quasi-periodicity. As mentioned in Chap. 3 (Sect. 3.2.4), these structures are governed by the instability processes responsible for the momentum transport in the viscous dominated boundary layer close to the bed. The relevant scales must therefore be the viscous scales $()^+$. Since the scaled variables are dimensionless, it is essential to also know the physical dimensions

$$z^+ = \frac{z u_{\tau_W}}{\nu} \quad \vee \quad z = \frac{z^+ \nu}{u_\tau} \quad \wedge \quad u_\tau = \sqrt{\frac{\tau_W}{\rho_f}}; \quad z^+ = 1 \rightarrow z \approx 10^{-2}\,\mathrm{mm} \wedge u_\tau = 0.1\mathrm{m} \tag{6.65}$$

Using a measured nondimensional time period of 60 (Fig. 6.3),

$$t^+ = \frac{t u_\tau^2}{\nu} \quad \rightarrow \quad t^+ = 1 \rightarrow t = \frac{\nu}{u_\tau^2} \quad t^+ = 60 \rightarrow t \approx 0.4s \tag{6.66}$$

$$u^+ = \frac{u}{u_\tau} \rightarrow O\left(\frac{u}{0.1U}\right) \rightarrow u^+ \approx 5u \tag{6.67}$$

The main dimensions are shown in Figs. 3.5 and 3.6. Another "magic number" is the separation of the longitudinal vortices in cross-stream direction, whereas the spacing in flow direction is much wider and therefore has to be interpreted as a mean value,

$$\lambda_y^+ \approx 100 \quad \therefore \quad \lambda_y \approx 1mm; \quad \lambda_x^+ \approx 1000 \quad \therefore \quad \lambda_x \approx 10mm \tag{6.68}$$

When one understands these structures as the result of an instability process by which the structures develop until their abrupt decay, the corresponding time is declared as the burst period T_P. A thorough explanation of the matter is worked out in Holmes et al. (1996), and their result is shown in Eq. 3.46. Here in Fig. 6.4, we reproduce the results of Schmid (1985), showing the width of the burst period distribution.

This is a rather unexpected result for a periodic instability process, and one wants to know the reasons, one of which is the possible triggering of the instability process by large-scale structures of the outer flow regime. This has implications insofar as the instability structures triggered by those outer events should also scale with outer flow parameters. Since the trigger mechanisms also come from the outer flow, a mixed scaling of the coherent structures based on set of outer and inner scaling parameters is often found in the literature.

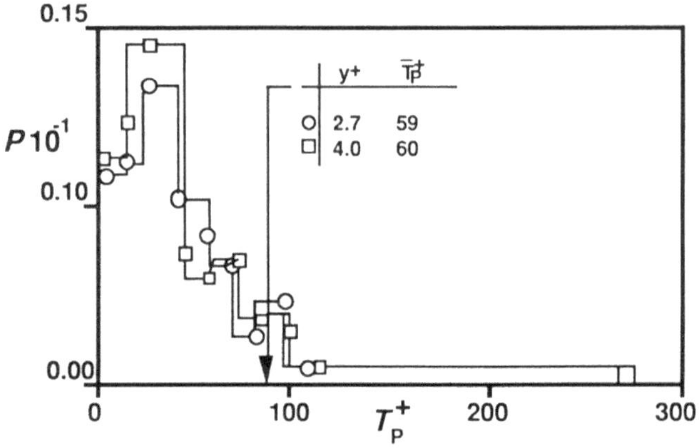

Fig. 6.4 Histogram of the burst period defined as the time between two equivalent consecutive events (Schmid, 1985). The shown measurements are taken at $u_\tau \approx 0.014$m/s and k^+_s=2.7

This interpretation started with the publication of Morrison et al. (1971), who found that the extreme load of the structures on the bed propagated like waves, which the authors called shear-waves. The convective velocity of these waves is

$$u_c \approx 8u_\tau \tag{6.69}$$

and allows estimating at which height the triggering structures must be located. This kind of wave model was compatible with the theoretical work by Benney and Lin (1960) and Benney (1961). Rao et al. (1971) pointed out this problem and declared that only the interaction between the inner and outer region of the flow could explain the formation of structures, and therefore both scales have to participate in their description. The authors proposed describing the length scales in viscous units and the burst frequency in outer parameters. In fact, when measuring the burst rate per unit length in viscous units it appears to be independent of Re. To understand this result better, a simplified picture of the outer trigger mechanism is shown in a 2D sketch in Fig. 6.5. This sketch is supported by the visualizations of Head and Bandyopadhyay (1981). The idea is that large-scale structures are cotransported with the outer flow and thus produce two typical secondary flows, the so-called ejections, and as counterpart the so-called sweeps. The ejection transports slower near wall fluid to the outer flow, whereas the sweep transports fast outer fluid toward the wall.

The topology of the interaction is sketched in Fig. 6.5 too. Two saddle points (*S*) and a (spiraling) nodal point (*N*) characterize the cell containing the large structure, whereas two half-saddle points (S/2) represent the attachment and separation of the flow close to the wall. The 3D instability structure as postulated by Perry et al. (1981) is shown in Fig. 6.6.

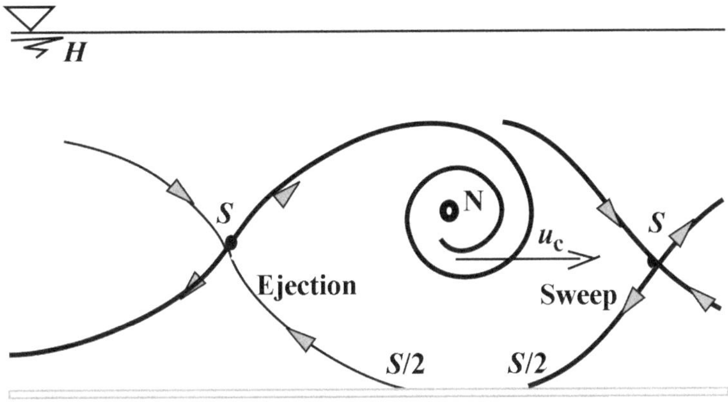

Fig. 6.5 A 2D sketch of a convected large scale outer flow structure propagating with a velocity u_c and interacting with the wall

Raupach (1981) confirmed that, at least for the ejection-period, the mixed scale-set to best represent T_{PE} was given by,

$$T_{PE} u_\tau / H = 0.13 \tag{6.70}$$

Evaluating a similar relation for the sweeps is difficult since those structures are very hard to detect in the outer flow (Alfredsson and Johansson, 1984). The main problem with these mixed sets of scaling parameters is that we do not know which and under what circumstances large structures trigger the burst cycle. The instability can be associated with structures characterized by different bed distances. Hence, it is recommended to measure these scales, or to use a set of values like the values given by Eq. 6.71

$$\begin{aligned} &L_{int} \propto U T_B \\ &\langle T_B \rangle \approx 5 \div 6\ H/U \qquad (z = 0.8H) \\ &\langle T_B \rangle \approx 2.5 \div 3\ H/U \quad (z = 0.2H) \end{aligned} \tag{6.71}$$

Chen and Blackwelder (1978) combined the velocity and temperature (passive tracer) measurements and found that a temperature front of size 0.03–0.63 z/H is transported together with a burst. In other words, the burst is relevant over a large part of the water depth. As we saw in Fig. 6.4 (Schmid, 1985), the burst period can be evaluated in the mean, and nevertheless scaled using viscous units. This results in a nondimensional burst time of about 60 at two distances close to the bed. It is thought that a Λ-vortex (also called horseshoe-vortex) can represent these bursts, which determine a large extent of the flow in the near wall zone.

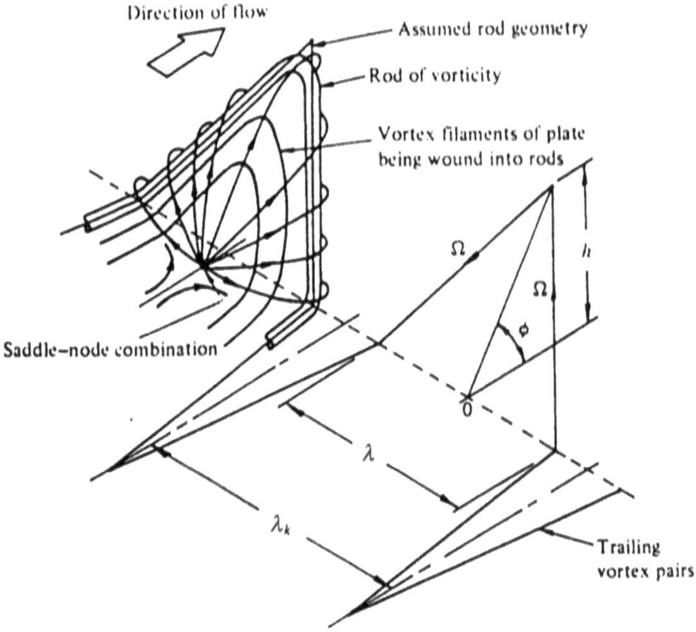

Fig. 6.6 A 3D representation of a Λ-vortex. After Perry et al., 1981

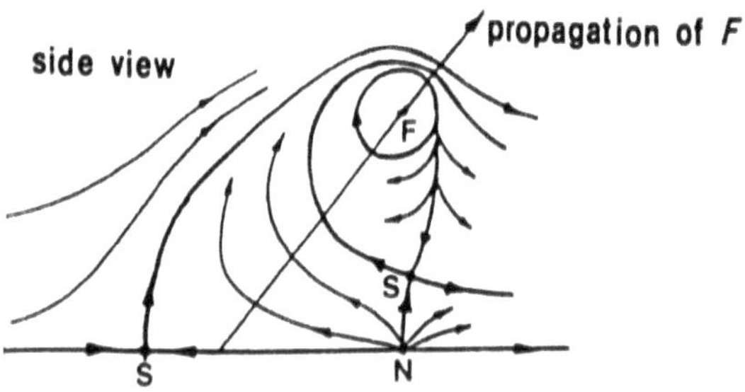

Fig. 6.7 Topology of a Λ-vortex in its central plane given by a side-view

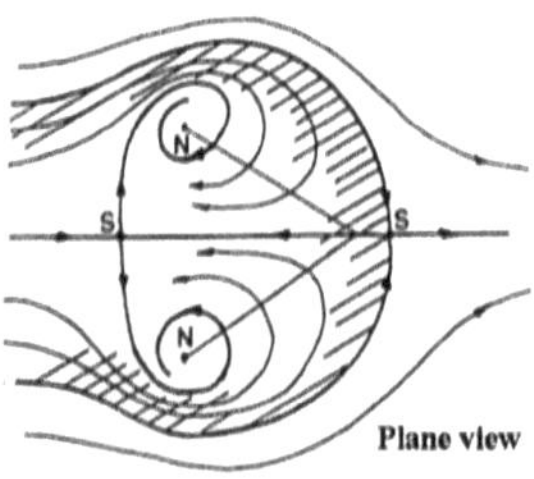

Fig. 6.8 Topology of a Λ-vortex close to the wall shown as a plane view

These structures whose origin was described in Chap. 3 (Sect.3.2.4), and sketches of which were given in Figs. 6.6–6.8, cause 3D ejections. We will discuss the structure of the sweeps as counterparts of the ejections later.

The quasi-periodic wavelength in cross-stream direction of $y^+ \approx 100 \pm 10$ was often called the magic number of the coherent structures. This value was explained by the so-called minimal flow unit for low-dimensional models (Lumley et al., 1999) and is based on the results found by Jimenez and Moin (1991).

These authors investigated the smallest range of a flow which is still considered turbulent and found that below $\lambda_y^+ \approx 100$ this is not anymore the case. According to these authors, a turbulent statistic needs a larger area, thus providing a physical interpretation of the magic number 100. The instability process, together with the convection velocity of the outer large-scale triggering structure, determines the length scale in flow direction. Therefore, it has a much higher variation but can be well approximated by $\lambda_x^+ \approx 500$ in the mean. The area of the bed influenced by a burst has a certain length, however is rather narrow in width, and can interact efficiently only with grains smaller than the magic number found. This again is important when comparing alluvial and laboratory conditions,

$$\begin{aligned} &d_s^+ \le y^+ = 100 \quad \therefore \quad d_s \le \frac{100\nu}{u_\tau} \quad \rightarrow \text{Eq.6.64} \\ &\text{Alu}: d_s \approx 1\text{mm} \quad \text{Lab}: d_s \approx 10 \text{ mm} \end{aligned} \tag{6.72}$$

An event where a sweep and ejection follow in close spatial and temporal proximity is called a burst. These events have a short lifetime and it is therefore important to know how they behave statistically, which necessitates criteria defining them. This problem was discussed in Chap. 3 (Sect. 3.2.4). The method used evaluating such structures is pattern recognition supported by conditional sampling techniques. This is a very dangerous attempt because we do not know what the structures look like. Since we only have a vague idea, we have to agree with Lumley (1981): "… one can find in statistical data irrelevant structures with high probability….". Therefore in Chap. 3 (Sect. 3.2.4), the quadrant method was used in spite of its known deficiencies. Results found by this method close to the bed can be interpreted as signatures of coherent structures. In Fig. 6.9, a typical joint probability density function (JPDF) of the velocity fluctuations is shown in a quadrant representation (Table 3.2).

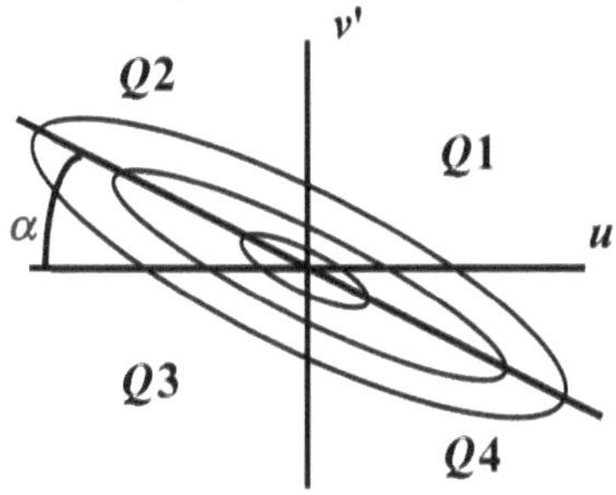

Fig. 6.9 A typical quadrant representation of the JPDF $(u'_x u'_z)$ with α being a mean incidence angle of the sweep and ejection events

The JPDF close to the bed is characterized by its elliptical shape of iso-probability lines centered at the origin of the coordinate net. The larger half axis cuts the origin at an incidence angle α. This angle corresponds to the mean incidence angle in $Q2$ for the ejections and in $Q4$ for the sweeps. The strong events, which are called the coherent structures created by the burst process, are responsible for the highest momentum transport to and from the wall. Comparing the JPDFs at different distances from the bed, one can conclude how fluid is transported toward the wall and away from it, which is a very relevant information for the sediment transport. A typical set of such angles is

$$\begin{aligned} &\alpha(Q2) \approx 166°(z^+ = 10) \wedge 149°(z^+ = 100) \quad \wedge \\ &\alpha(Q4) \approx 351°(z^+ = 10) \wedge 329°(z^+ = 100) \end{aligned} \tag{6.73}$$

In other words, close to the wall the events are only slightly inclined toward the bed, whereas the incidence angle increases with an increasing wall distance, compatible with their production mechanism (Fig. 6.5).

Another result found in the JPDF representation is that ejection and sweeps are antiparallel at an angle of 180°, which can be interpreted as a reflection on the wall. In the earlier-mentioned model, they belong to the same large structure, which can be approximated by a circular vortex.

6.2.6 The Quadrant Representation

It is known that the JPDF as shown in Fig. 6.9 is valid also when no coherent structures are proven to exist as, e.g., in flows over fairly rough beds. This has significant consequences, as mentioned already in Chap. 3 (Sect. 3.2.4), because the quadrant representation is thought to contain information on coherent structures. This is not at all the case since a simple mixing length theory would give the same results. Therefore, the notions of ejection and sweep are often used, although the structures do not exist at all. Therefore, it is strongly recommended to use the notions ejection and sweep only for the coherent structures as produced by the instability process in a smooth and flat boundary layer flow.

The quadrant method has much more in common with the statistical representation in Chap. 6 (Sect. 6.2.3); to say, whenever a JPDF is evaluated from the velocity fluctuation, the results should not be noted as structures but named contributions by the quadrants Qi (I = 1,..,4).

The similarity between the quadrant decomposition and the model of coherent structures has a deeper physical reason: the structures have similarities, but they are of different scale. It is still an unsolved question why the flow develops similar structures of all scales. This similarity was the origin of theoretical speculations that the flow contains structures given by fractal laws (Sreenivasan, 1996 and literature mentioned therein).

Also the influence of the wall (Chap. 6, Sect. 6.2.2.1) must be seen in the quadrant representation even if no coherent structure existed. In fact, approaching the wall the ellipse becomes narrower increasing the anisotropy, which can be derived by the higher moments as described in Chap. 6 (Sect 6.2.3). Typical values of these moments can be found in Johansson and Alfredsson (1982), e.g.,

$$\begin{aligned}
&S>0 \quad z^+\le 12 \quad \max \quad S=0.9 \quad z^+=3\div 4\\
&S\approx 0 \quad 12\le z^+\le 100\\
&S<0 \quad z^+\ge 100 \quad \max \quad S=-0.4 \quad z^+\approx 100\\
&K\left(z^+=3\right)=4 \quad K\left(z^+=10\right)=2.5\\
&K\left(z^+=100\right)=2.4 \quad K\left(z^+=1000\right)=3.25
\end{aligned} \tag{6.74}$$

When one uses statistics as Grass (1970) both K as well as S are of importance for the sediment transport.

6.2.7 Some Remarks on the Energy Equation and the Pressure Field

In the classical sediment transport equations, the wall shear stress was used as a measure for the forces involved. For the motion of a grain, we saw that a force and an angular momentum condition must be fulfilled (Eq. 3.10), where the dynamical buoyancy is more important than the tangential components of the flow force on the grain. Cotransported low-pressure volumes, as they are produced by vortices, can explain this buoyancy. It is therefore rather amazing that the relation between flow structures and pressure distribution was used only sporadically within a theory of sediment transport. Here we will discuss the problem for a turbulent, hydraulically smooth flow. The investigation will make use of the energy balance of the flow for which the energy equation for the mean flow is given by:

$$U_j\frac{1}{2}\frac{\partial U_iU_i}{\partial x_j}=\frac{\partial}{\partial x_j}\left(-\frac{P}{\rho}U_j+2\nu U_iS_{ij}-\overline{u'_iu'_j}U_i\right)-2\nu S_{ij}S_{ij}+\overline{u'_iu'_j}S_{ij} \tag{6.75}$$

For incompressible flow we find

$$-P\delta_{ij}S_{ij}=-PS_{ii}=-\rho\frac{\partial U_i}{\partial x_i}=0 \tag{6.76}$$

Therefore the mean pressure does not contribute to the deformation work.

Applying the Reynolds decomposition allows to determine the energy in the fluctuating turbulent flow field as the total energy minus the energy of the mean flow

$$\begin{aligned}
&U_j\frac{1}{2}\frac{\partial\left(u_i'u_i'\right)}{\partial x_j}=-\frac{\partial}{\partial x_j}\left(\frac{1}{\rho}\overline{u_j'p'}+\frac{1}{2}\overline{u_i'u_i'u_j'}-2\nu\overline{u_i's_{ij}}\right)-\overline{u_i'u_j'}S_{ij}-2\nu\overline{s_{ij}s_{ij}}\\
&\wedge \quad S_{ij}\equiv\frac{1}{2}\left(\frac{\partial U_i}{\partial x_j}+\frac{\partial U_j}{\partial x_i}\right);\ s_{ij}\equiv\frac{1}{2}\left(\frac{\partial u_i'}{\partial x_j}+\frac{\partial u_j'}{\partial x_i}\right)
\end{aligned} \tag{6.77}$$

Let us evaluate the order of magnitude of the terms in Eq. 6.77 with special attention to the pressure term

We know that

$$s_{ij}\sim u'/\lambda_{\mathrm{Ta}}\ \text{(Eq.6.28)} \quad \wedge \quad \lambda_{\mathrm{Ta}}/L_{\mathrm{int}}\sim Re_{L_{\mathrm{int}}}^{-1/2} \tag{6.30}$$

Applying this value, we have

$$-\frac{\partial}{\partial x_j}\left(\frac{1}{\rho}\overline{u_j' p'}\right) \sim \frac{\sim u'^3}{L_{\text{int}}} \quad \wedge \quad p' \sim \rho u'^2 \tag{6.78}$$

and

$$-\frac{\partial}{\partial x_j}\left(\frac{1}{2}\overline{u_i' u_i' u_j'}\right) \sim \frac{u'^3}{L_{\text{int}}} \tag{6.79}$$

$$2\nu\frac{\partial}{\partial x_j}\left(\overline{u_i' s_{ij}}\right) \sim \nu\frac{u'^2}{L_{\text{int}}^2} \sim \frac{u'^3}{L_{\text{int}}}\text{Re}_{L_{\text{int}}}^{-1} \tag{6.80}$$

For incompressible fluids, the continuity equation also yields

$$\overline{p'\frac{\partial u_1'}{\partial x_1}} + \overline{p'\frac{\partial u_2'}{\partial x_2}} + \overline{p'\frac{\partial u_3'}{\partial x_3}} = \overline{p'\frac{\partial u_i'}{\partial x_i}} = 0 \tag{6.81}$$

where the pressure redistributes the energy between the different velocity components without changing the total energy of the flow. In addition, the divergence of gradP satisfies a Poisson equation, which can be shown by applying the divergence operator on the NSE.

$$\nabla^2 P = -\rho\frac{\partial^2 u_i u_j}{\partial x_i \partial x_j}$$

$$\therefore \nabla^2\left(\langle P\rangle + p'\right) = -\rho\frac{\partial^2}{\partial x_i \partial x_j}\left(U_i U_j + U_i u_j + U_j u_i' + \langle u_i u_j'\rangle\right)$$

$$\therefore \quad \nabla^2\langle P\rangle = -\rho\frac{\partial^2}{\partial x_i \partial x_j}\left(U_i U_j + \langle u_i' u_j'\rangle\right) \tag{6.82}$$

$$\vee \quad \therefore \quad \nabla^2 p' = \frac{\partial^2}{\partial x_i \partial x_j}\left(+U_i u_j' + U_j u_i' + u_i u_j'\right)$$

$$\wedge \quad \frac{\partial p'}{\partial x_i}\Big|_{\text{w}} = \mu\frac{\partial^2 u_i'}{\partial x_j \partial x_i}\Big|_{\text{w}}$$

We showed in Chap. 5 (Sect. 5.2) that the pressure is a quantity of nonlocal nature and gave Eqs. 5.24 and 5.25 and Eq. 6.82 which allows to evaluate the pressure and its fluctuations uniquely from the velocity field. In Chap. 5 (Sect. 5.2), we pointed out that a truncation criterion is needed to calculate the pressure field from the velocity field. This, however, is especially problematic in the vicinity of the wall. Panchev (1971) showed that the pressure and the velocity components do not correlate much in homogeneous isotropic turbulence. This correlation is much higher in non-homogeneous turbulence close to the wall (Lilley and Hodgson, 1960). This result is important insofar as the sediment transport was defined by $u_x' u_z'$ and the integral in Fig. 3.8 needs to be evaluated. The volumes of low pressure are given by the location of the vortices which can be calculated using Eq. 5.24 if one measures the velocity field in the whole volume defining the integration space. To do this, we have to overcome two difficulties—the truncation of the integration space and the required high resolution of the velocity field within this space at any moment. With these remarks we caution those who derive the pressure distribution by measuring $u'_x u'_z$ locally above the bed.

An elegant way to overcome these problems is the numerical simulation of the turbulent flow field. Today it is possible to evaluate the vorticity field to a satisfactory resolution, which is not yet realizable experimentally, although huge progress was made in the last years, see also the comment of Tsinober (2001). But the vorticity again is only a local quantity that has to be evaluated when we are searching for the pressure field. Moreover, the vorticity must belong to a vortical system to create a significant pressure head, contributing to the pressure field in its vicinity. What is needed, therefore, is a definition of a vortex. We will use the term vortex for the following feature embedded in the flow

- A vortex exists if the instantaneous streamlines projected on a plane perpendicular to the vortex axis are quasi-circles or spirals, if they are observed from a coordinate system, which moves with the center of the vortex core.

With this definition, the low-pressure volume can be calculated by using the NSE in cylindrical coordinates (u_r, u_φ, u_z)

$$\frac{\partial u_r}{\partial t}+\left(\underline{u}\cdot\nabla\right)u_r-\frac{1}{r}u_\varphi^2=-\frac{1}{\rho}\frac{\partial p}{\partial r}+\nu\left(\nabla^2 u_r-\frac{u_r}{r^2}-\frac{2}{r^2}\frac{\partial u_\varphi}{\partial\varphi}\right) \tag{6.83}$$

If we assume that the streamlines are circular lines, Eq. 6.83 reduces to

$$\frac{1}{\rho}\frac{\mathrm{d}p}{\mathrm{d}r}=\frac{u_\varphi^2}{\mathrm{R}} \tag{6.84}$$

where R is the radius of curvature, u_φ the azimuth velocity along the streamline and the agreement that the pressure decreases toward the vortex axis. Since this flow is 2D and has only one vorticity component,
we have

$$\frac{\partial\omega}{\partial t}=\nu\left(\frac{\partial^2\omega}{\partial r^2}+\frac{1}{r}\frac{\partial\omega}{\partial r}\right)\quad\wedge\quad\omega=\frac{1}{r}\frac{\partial}{\partial r}\left(ru_\varphi\right) \tag{6.85}$$

Given the initial condition Γ_0, the temporal behavior (decay) of the vortex can be described by the Oseen-solution, which is given in viscous units in case we need to use this equation close to the wall, too.

$$u_\varphi=\frac{\Gamma_0}{2\pi r}\left(1-\mathrm{e}^{-r^2/4\nu t}\right)\quad\therefore\quad u_\varphi^+=\frac{\Gamma_0^+}{r^+}\left(1-\mathrm{e}^{-r^{+2}/4t^+}\right)$$
$$\wedge\quad u_\varphi^+=\frac{u_\varphi}{u_\tau};\Gamma_0^+=\frac{\Gamma_0}{2\pi\nu};r^+=\frac{ru_\tau}{\nu};t^+=\frac{tu_\tau^2}{\nu} \tag{6.86}$$

An estimation of the pressure distribution can be found by integration of Eq. 8.84

$$\frac{1}{\rho}\int_r^\infty \mathrm{d}p = \int_r^\infty \frac{u_\varphi^2}{r}\mathrm{d}r \quad \therefore \quad \frac{1}{\rho}\big(p(\infty) - p(r)\big) = \int_r^\infty \frac{u_\varphi^2}{r}\mathrm{d}r$$

$$\frac{1}{\rho}\big(p(\infty) - p(r)\big) = \int_r^\infty \frac{\Gamma_0^2}{4\pi^2 r^3}\left(1 - \mathrm{e}^{-r^2/4\nu t}\right)^2 \mathrm{d}r \tag{6.87}$$

$$\vee \quad p^+(r) = \int_{r^+}^\infty \frac{\Gamma_0^{+2}}{r^{+3}}\left(1 - \mathrm{e}^{-r^{+2}/4t^+}\right)^2 \mathrm{d}r^+$$

$$\wedge \quad p(\infty) = 0; \quad p^+ = \frac{p}{\rho u_\tau^2}$$

This is the result of a very simplified vortex model. Nevertheless, we will find it of high importance for flow fields dominated by vortices belonging to a vortex-skeleton system. Those can be found by evaluating the topology of the flow field.

7 Roughness and Roughness Elements

One of the main problems describing an alluvial flow system lies in the variety of existing roughness elements forming the bed. Their classification is therefore necessary to allow a systematic use.

As we know, roughness elements transfer momentum from the flow to the bed in a nonviscous form. The flow separation on the roughness element causes a pressure difference over the obstacles, which can be single grains, rocks of various sizes, or bed-forms. This form drag will be discussed separately in Chap. 8. Here we discuss the diverse types of grain roughness. The word grain is used here intentionally to mark the difference from the usual notion of the sand roughness as encountered for the first time in Chap. 2 (Sect. 2.1.3.2), which is the concept used in the classical theories.

7.1 Similarity Consideration in the Range of Constant Wall Shear Stress

Besides the momentum transport, the conversion of energy is also an important result of the flow–wall interaction. The results for the mean flow kinetic energy, Eq. 6.75, and the turbulent kinetic energy, Eq. 6.77 were derived in Chap. 6 (Sect. 6.2.7). Here we supplement the relation for the steady-state energy balance "turbulence production" = "viscous dissipation."

$$-\overline{u_i' u_j'} S_{ij} = 2\nu \overline{s_{ij} s_{ij}} \equiv \varepsilon \tag{7.1}$$

In a so-called 1D mean flow with $U = (U, 0, 0)$, these energy equations for a channel flow reduce to

$$\begin{aligned} &-\overline{u_x' u_z'} \frac{\partial U}{\partial z} - \frac{\partial}{\partial z}\left(\overline{p' u_z'} + \frac{1}{2} \overline{q^2 u_z'} \right) = \varepsilon \\ &\wedge \quad q^2 = \overline{u_i' u_i'} \quad \wedge \quad -\overline{u_x' u_z'} \frac{\partial U}{\partial z} = \tau \frac{\partial U}{\partial z} \end{aligned} \tag{7.2}$$

The first term is the production term and shows how much turbulent kinetic energy is drawn from the mean flow gradient. The second term, the so-called convective turbulent diffusion, gives the energy contribution fed to the turbulent flow by the vertical fluctuation in pressure and velocity. The sum of the first and second term is equal to the viscous dissipation. This kind of energy flux is shown in Fig. 7.1.

A fully developed turbulent flow over a smooth wall exhibits a thin fluid layer, the so-called viscous sublayer. Beginning with the no-slip condition at the wall this layer has a large velocity gradient in the wall normal direction. In the same layer, also the wall normal velocity fluctuations are much smaller than the fluctuations parallel to the wall. The turbulent energy production reaches a maximum above this viscous layer. In the near wall zone of thickness z_1, which is small compared to H, however large enough to extend into the zone of the fully developed turbulent flow, the turbulent energy production as well as the dissipation is larger than in all other parts of the flow. We assume therefore that the flow behavior depends little on the weak fluctuations in the inner (the wall-near) flow zone. With the said assumption, the boundary conditions at $z = 0$ and $z = z_1$ determine this important part of the flow. Since the turbulent fluctuations are related to the mean flow properties through Eq. 7.2, the boundary conditions at the outer boundary are given by the turbulent shear stress. Its value is about the same as at the bed, where the value defines the wall shear stress. Now, the energy production and dissipation, apart from a lateral energy exchange, depend only on the parameters τ_w, z_1, and ν. Using dimensional analysis it was shown that the solution is self-similar and becomes independent of z_1.

$$\frac{\partial U}{\partial z} = \frac{u_\tau}{\kappa z} f\left(\frac{u_\tau z}{\nu}\right); \quad -\overline{u_x' u_z'} = u_\tau^2 g\left(\frac{u_\tau z}{\nu}\right); \quad \varepsilon = \frac{u_\tau^{3/2}}{\kappa z} h\left(\frac{u_\tau z}{\nu}\right) \tag{7.3}$$

These equations are known as the law of the wall for smooth walls.

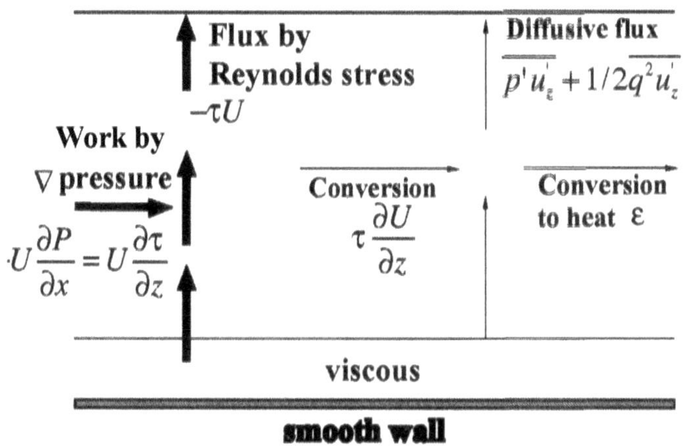

Fig. 7.1 Energy diagram of a channel flow over a smooth bed as given by Eq. 7.2

For large values of $\tau_w^{1/2} z/\nu$, the equations become independent of viscosity.

$$-\overline{u_x' u_z'} = u_\tau^2; \quad \varepsilon = \frac{u_\tau^{3/2}}{\kappa z}$$
$$\therefore \quad \frac{\partial U}{\partial z} = \frac{u_\tau}{\kappa z} \quad \therefore \quad U = \frac{u_\tau}{\kappa}\left[\ln\frac{u_\tau z}{\nu} + A\right] \quad A \approx 2.3 \tag{7.4}$$

This logarithmic profile applies to a flow with an equilibrium layer:

- In which the downstream pressure distribution is adjusted in such a way that their velocity profiles, if nondimensionalized with an appropriate velocity-defect law, are independent of the Reynolds number and of the downstream distance x
- Whose thickness is small compared to H
- In which the advection terms of the energy equation (Eq. 6.75) are small compared to the production terms of Eq. 7.2 and for which the variation in the shear stress over the equilibrium layer is small

These assumptions are not fulfilled in a flow over a rough bed.

7.2 Sand Roughness

Flow separation develops when the roughness elements are higher than the thickness of the viscous sublayer. It is therefore assumed that sand grains start to produce tiny separations as sketched in Fig. 7.2 as soon as they fulfill criterion Eq. 6.64. Then a layer of small-scale vortices covers the bed. These act similarly to the vortices, which are produced by the instability process on smooth beds as described in Chap. 3 (Sect. 3.2.4). Therefore, they are nothing else than a new boundary condition for the turbulent flow. After a short period and at a very small distance from the bed, the flow does not remember the origin of the turbulent fluctuations since the nonlinear turbulent redistribution mechanisms are so efficient. Therefore, the whole manifold of structures characteristic for a turbulent flow is present. The nonlocal properties (pressure field) of the flow equation act against this redistribution. However, they are of smaller order of magnitude, so we can neglect them in this context. Yet, as shown in Chap. 5 (Sect. 5.2), the theoretical consequences are very important since the physics of the interaction is different.

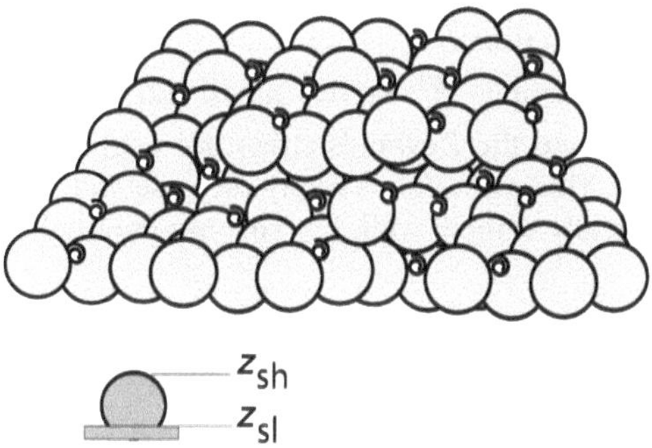

Fig. 7.2 A bed consisting of uniform spherical grains forming a typical sand roughness. The process is dynamic; some of the vortices are shed whereas others start growing

The main difference between a velocity profile for rough and smooth walls manifests itself only in a downward shift of the logarithmic part of the profile, shown in Eq. 2.35. This shift is a function of the effective roughness ε, where for sand roughness the value for the sand grains (Eq. 2.36), or for the equivalent sand roughness k_s, (Eqs. 2.40 and 2.41), has to be used instead of ε.

With this representation, a new problem emerged concerning the proper definition of bed. This cannot be answered as in the smooth case, where the solid surface is plane and given by a single elevation. In other words, a new parameter has to be introduced.

One possibility to overcome this problem is to fix $z = 0$ through a universal criterion. The new origin would have to be given by a fictitious plane on a level given beyond the lowest elevations and below the highest elevations of the grains. This could be the mean between the highest and lowest elevation with respect to the wall or another optional criterion. A series of such assumptions can be found in the literature. The most common ones are

$$z = 0 := \langle z_s \rangle \quad z_s \in \left(z_{sl} < z_s < z_{sh} \right)$$
$$z = 0 := z_{sh} - (0.2/0.3)(z_{sh} - z_{sl}) \tag{7.5}$$

Here z_{sl} and z_{sh} are the lowest and the highest elevation of the top bed layer, respectively.

This choice is well known and has been treated in a series of investigations; see especially the papers of Perry et al. (1969) and Perry and Abell (1977), where this type of roughness is called *k*-roughness. The most practical method for defining the origin is by extrapolation of the logarithmic velocity profile. This can be done since the velocity profile over the rough wall remains logarithmic down to the bed. Therefore, measurements taken from the outer flow determine the location of the wall due to the behavior of the shear–stress distribution shown in Fig. 2.2
The shear stress distribution is given by

$$\tau_{xz}(z) \approx \rho \overline{u'_x u'_z}\big|_z \quad z \in z^+ \geq 30 \quad \therefore \quad \tau = \tau_0 \left(1 - \frac{z}{H} \right) \tag{7.6}$$

To become independent of the type of momentum transfer at the wall, z must be much larger than k_s, and larger than $u_\tau/\kappa z$.

If the grain-Reynolds number k_s^+ is very small, Eq. 7.4 holds for values of z for which the mean velocity and the Reynolds stress are negligible, and the smooth-wall distribution is recovered. However, this is rarely the case and the criterion is a physical definition for a hydraulically smooth wall. For such regimes coherent structures have to be expected.

Ordinarily k_s^+ is not small and in such cases the integration yields an integration constant that depends on the flow conditions around the grains,

$$U = \frac{u_\tau}{\kappa} \ln \frac{z}{z_0} \quad z \in z >> k_s \quad \wedge \quad k_s^+ = \frac{k_s u_\tau}{\nu} \approx O(>>1) \tag{7.7}$$

with z_0 being a so-called roughness length. Together with u_τ, this length defines a "slip velocity" between the bed and a standard elevation in the layer of constant stress, whereby both quantities depend on the separation behavior. We define the sand roughness through the inner equilibrium layer, which reacts very quickly to changes of velocity, so that the separation vortices are shedding in an uncorrelated way. For similarity reasons the flow depends only on the size of the roughness elements, the friction velocity, and the viscosity, so that

$$z_0 = k_s f_s \left(k_s^+ \right) \tag{7.8}$$

For larger roughness the function becomes unity and we have $z_0 \approx k_s$, which is the most commonly used value for the roughness length.

For small roughness Reynolds numbers, the flow is hydraulically smooth and the roughness takes the value as in the case of the smooth wall, i.e.,

$$z_0 e^{-A} \nu / u_\tau \quad \wedge \quad A \to \text{Eq.7.4} \tag{7.9}$$

The ratio z_0/k_s is also responsible for the validity of the logarithmic profile and therefore the approximation that entered the relation (Eq. 7.6) can be checked. Using Eq. 7.9 for hydraulically smooth conditions, we find

$$z_0 \approx 0.1 \nu / u_\tau \tag{7.10}$$

Here the logarithmic profile starts at $k_s^+ \approx 30$ or in other terms at $z/z_0 \approx 300$. For really rough beds $z_0 \approx 0.1\ k_s$ and $z/z_0 > 50$ for the logarithmic range.

The viscous dominated flow zone near the wall becomes more important with smaller sand grains. This is also the reason why coherent structures can be observed even though the pure viscous sublayer criterion is violated. For a fully developed flow over sand roughness, one assumes that the main mechanism of momentum transport is achieved by the separation of the flow on all roughness elements. The shedding of the separation vortices occurs after the vortices have grown until they are so large that they are transported away. This is a probabilistic process, however they can also be shed in an organized form through the influence of a convected large outer structure. In the latter case, this would have direct consequences for the sand roughness theory since it was assumed to be independent of H. This independence is obviously not the case for the large structures and therefore a coupling between inner and outer structures occurs.

We have a last remark regarding Fig. 7.2, which shows spherical roughness elements. This is somewhat misleading because the grain form will influence the local separation and a sharp-edged grain can produce very different separation vortices. Historically, sand paper was thought to be the ideal model for sand roughness.

7.3 *d*-Roughness

The *d*-roughness is irrelevant for alluvial flows, however it shall be discussed here, since it allows introducing ideas that may become important when used for new types of roughness, especially, in view of the collective shedding process just mentioned. The notion of *d*-roughness was used for a roughness produced as a result of mechanic tooling of a surface as shown in Fig. 7.3.

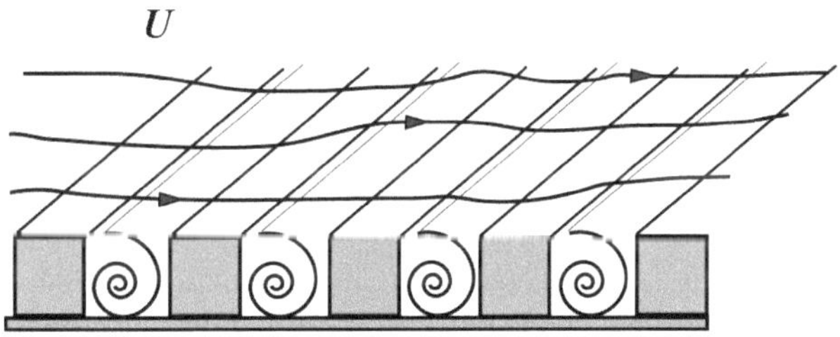

Fig. 7.3 The *d*-roughness has very long roughness elements more or less in spanwise direction. Separation vortices develop in the trough, which remain fairly stable and are stable against perturbations from small-scale fluctuations

The revolving tooling process produces a roughness of the wall consisting of long ripples in spanwise direction. If the distance between peaks and valleys is short, separation vortices occur in their troughs, which can remain stable against weak fluctuations of the flow.

However, large structures of the size H will raise the vortices, resulting in their collective shedding as sketched in Fig 7.4.

Separating the obstacles further decreases the system stability. The vortices start to shed individually and we have again a special kind of *k*-roughness.

The sketch is very similar to the one illustrating the stability processes producing coherent structures. The mechanism is also very similar, see Fig. 3.5, instead of a vorticity sheet we have discrete vortex tubes that are subject to a stability limit. Once the vortices have been shed, even the renewing process is analogous.

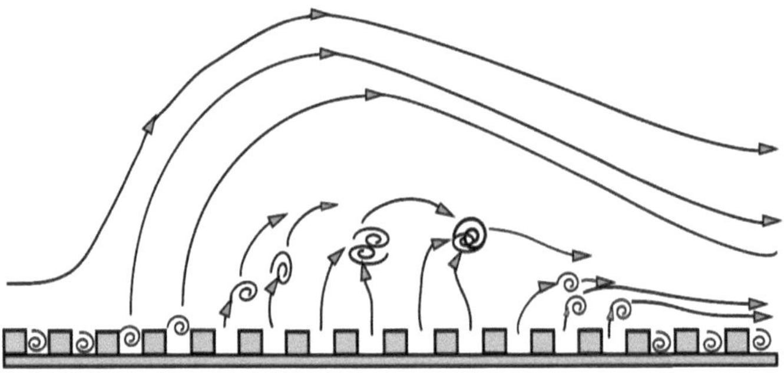

Fig. 7.4 A large flow structure raises the separated vortices in a coordinated fashion with the effect that some of the raised vortices are even pairing

Figure 7.4 shows a plane side-view of the situation but we need to be aware that the large structures are in fact 3D and are therefore most probably identical with the so-called "turbulent spots". They are described in detail by Cantwell et al. (1978), Cantwell (1981), Perry et al. (1981), or many other authors like Coles and Barker (1975), Wygnanski et al. (1982), Zilberman et al. (1977), Haritonidis et al. (1978), Coles and Savas (1980), Wygnanski (1981), and Wygnanski et al. (1979). The structure of a turbulent spot is probably the closest model for a large-scale structure, which is the reason for giving all the citations. It is therefore also very helpful for introducing topological concepts for the large structures.

Wygnanski and his coworkers stressed that the turbulent spots interact with the surrounding fluid. They are time varying even in a moving coordinate system, that is, they have a finite lifetime. A turbulent spot, as it appears at a laminar turbulent transition may coexist with the turbulent flow for long time. However, Coles thinks that the spots are part of an asymptotic state reducing the laminar boundary shear layer to a vortex layer on the bed. These views are different, nevertheless, not mutually exclusive. Coles and Barker went as far as declaring the large structures based on the known Λ-vortices themselves spots of a much smaller scale having two substructures analogous to the one shown in Fig. 6.4, but generated by another mechanism. Perry et al. (1981) criticized the concept of Coles and Barker, accepting it only as a phase mean. The essential truth is that until today we do not have a sufficiently detailed model. Perry et al. investigated the fine structures and concluded that the large structures are conglomerates of organized small Λ-vortices, which are pairing. This view holds, to a certain extent, for the structures observed over smooth, flat beds.

When investigating the different types of rough walls as described, the only common point is that the shedding of the small separation vortices is triggered by a rather weak low-pressure field originating from a spotlike large structure.

All the raised small vortices themselves exhibit a low-pressure core due to the concentrated vorticity. Vortices of similar sense of rotation are pairing due to their attractive induction. This dynamical process will be discussed in the next chapter. A compilation of the dynamical behavior of single and interacting vortices and the instability processes involved can be found in Lamb (1945) or in Saffman (1992).

7.4 Real Roughness

There exist rivers obeying simple rules, such as rivers with only fine sand as sediment or with velocity profiles as predicted by a sand roughness, however, the usual river is far from such an idealized configuration (Fig. 1.2). Separating vortices of the size of the sediment elements are present, and their shedding frequencies, given by the different Strouhal numbers in Eq. 6.39, also differ with size. The proper description of the roughness grows more complex, although parts of a river or areas of the bed may fit to the simpler models discussed in Chap. 7 (Sects.7.2 and 7.3). We briefly postpone the description of the real roughness, which needs detailed knowledge of the flow topology. Here we will discuss only the mean flow behavior.

In case of a sand roughness we suggested using the logarithmic profile for the velocity distribution and evaluating the proper wall coordinate z_0 through extrapolation of this function. This makes sense since the logarithmic profile is valid down to the bed. The reason for this behavior is now understood as the result of the separation vortices. On a flat smooth bed, the vorticity has only one component in y-direction. The redistribution of the vorticity needs a cascading process to achieve the isotropic behavior in the outer flow (Frost and Bitte, 1977). During this process coupled vortices stretch and compress; however, it also needs the different vortices to interact at an angle, which requires vorticity of varying direction to be present. In the smooth case, the spanwise vorticity rises locally and thus stretches due to the mean velocity gradient. The vorticity then redirects its orientation and amplifies in magnitude (Fig. 7.5a). This process is rather slow compared to the case where vorticity is already present in all directions and can interact freely. The ω_z component, which has to vanish identically at the wall, is very weak and therefore the velocity profile in the vicinity of the wall is linear in viscous units. In the case of a rough wall, the profile is logarithmic because the separation at a grain entails vorticity of all three components (Fig. 7.5b).

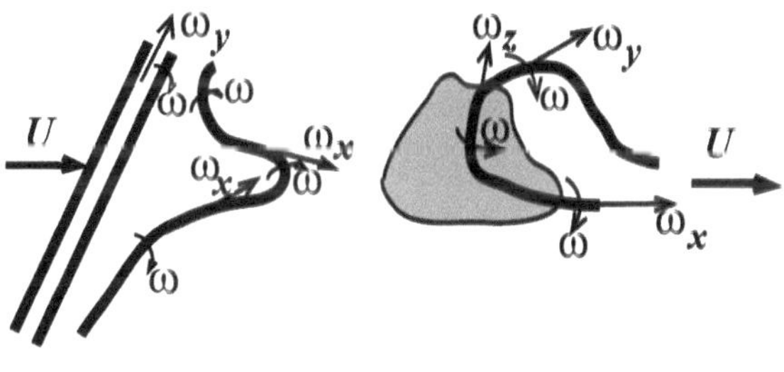

Fig. 7.5 The main vorticity lines at (a) a smooth, flat surface and (b) in the separation of a grain

Although it is not theoretically proven that a logarithmic velocity profile stands for the local isotropy of the three vorticity components, this approach generally holds for all rough beds. However, for the natural variability of the real case we are again confronted with the problem of the origin. Averaging is problematic, since the local distribution of the grains feeds back into the local velocity profile (Nowell and Church, 1979). To evaluate the velocity profile as it should be, one not only has to take the temporal mean but also to average over the bed area (see Nikora et al., 2002 and references in that paper). The consequence is that the extrapolation of the logarithmic velocity profile will no more yield a unique local quantity for the wall coordinate.

Monin and Yaglom (1971) tried solving the problem with a correction to the smooth boundary result.

$$\frac{\mathrm{d}U}{\mathrm{d}z}=\frac{u_\tau}{\kappa z}\Rightarrow\frac{\mathrm{d}U}{\mathrm{d}z}=\frac{u_\tau}{\kappa z}g\left(\frac{\Delta}{z}\right)\approx\frac{u_\tau}{\kappa(z-d)}$$
$$\wedge\quad g\left(\frac{\Delta}{z'}\right)\quad\wedge\quad z'=z-d \tag{7.11}$$

Here the function was expanded in powers of $g(\Delta/z')$ and the linear term was set to 0. From the integration of Eq. 7.11, it follows

$$U=\frac{u_\tau}{\kappa}\ln\frac{z-d}{z_0} \tag{7.12}$$

With this equation an origin elevation is fixed and shifted from the virtual origin z_0 by d upon the part of the bed consisting of the local topmost firm sediment layer, see equivalent Eq. 7.7.

This concept is very attractive for its simplicity, however, it was criticized for having little physical reasoning and therefore a series of other concepts were introduced. It is very popular to formulate d via the mean drag (Jackson, 1981); another approach used the mass conservation (De Bruin and Moore, 1985). In demand is however a law which is capable of distinguishing, e.g., a flat bed interrupted by single large elements from a tight bed of rather uniform grains. The theory should therefore make use of the separation mechanisms. One finds a first step in this direction in Nikora et al. (2002). The authors investigated how far the outer flow structures were penetrating into the rough bed. In the case of large isolated stones, these large vortical elements can penetrate down into the troughs. The tighter the cover of the bed, the less these structures penetrate. Instead, the wall normal momentum is redirected, and the structures are just transported over the bed while the fluid in between the grains is not affected too much. Nikora et al. (2002) combine this behavior with their so-called double averaging (space and time). Applying the Prandtl mixing-length theory, they use this averaged velocity $\langle U\rangle$ together with the mixing-length l and a shift of the origin d depending on the vortex-energy in z-direction

$$\frac{\tau}{\rho_\mathrm{f}}=u_\tau l\frac{\mathrm{d}\langle U\rangle}{\mathrm{d}z},\quad z\in z>k_\mathrm{s}$$
$$\wedge\quad u-\langle U\rangle\propto l\frac{\mathrm{d}\langle U\rangle}{\mathrm{d}z}\quad\wedge\quad\left(w-\langle w\rangle\right)\propto u_\tau \tag{7.13}$$
$$\wedge Z=z-d\rightarrow Z\sim O(1)\quad\therefore\quad l=\kappa Z=\kappa(z-d)$$

and with

$$\frac{\tau}{\rho_\mathrm{f}}\approx u_\tau^2\quad\therefore\quad\left(\frac{\mathrm{d}\langle U\rangle}{\mathrm{d}z}\right)^{-1}=\frac{\kappa}{u_\tau}z-\frac{\kappa}{u_\tau}d \tag{7.14}$$

Now d can be evaluated by a series of measurements. Nikora et al. (2002) have checked Eq. 7.14 in this respect for quite different roughness and found acceptably good results. This is however expected when one uses empirical results for the evaluated d. The flow separation concept was also not used in this representation. We cannot offer a better theory, but we can explain some mechanisms that can help in the development of such a new theory.

In this kind of open channel flow, vorticity can only be generated at the wall due to the no-slip condition where initially ω_y is the only component on the flat surface. Instability processes in this shear layer deform the vortex lines and any vortex loop rotated and stretched thereby producing ω_x. As we explained, the separations on roughness elements introduce vorticity of all components. Therefore, the biggest difference from the situation over flat beds lies in the relative magnitude of ω_z. Nevertheless, the vorticity in flow direction is produced just as for flat conditions, which can be observed in visualization experiments of rather weak vorticity. Weak vorticity often shows nicer results because the dye diffusion is very slow in these cases.

Due to this 3D vorticity distribution close to rough bed, the predominant component of the Reynolds stress tensor is not only in the meridian plane (2D approximation) but it also needs to be supplemented by the component related to the vertical vorticity

$$\begin{aligned} & z \rightarrow z = 0 \\ & Re_{xz} = \overline{u'_x u'_z} \quad \wedge \quad Re_{ij} = 0 \vee ij \neq xz \quad \text{smooth} \\ & \Rightarrow Re_{xz} \approx Re_{xy} \quad \wedge \quad Re_{yz} \neq 0 \quad \text{rough} \end{aligned} \tag{7.15}$$

Here we have discussed only the consequences for the mean values. Separations, however, are local phenomena and we encounter the large scouring problems at single obstacles. What is needed is a statistical description weighing these interrelations, but for this we need much better knowledge of separating flows, which we will discuss in the next chapter. We will also look at the so-called form drag, which of course can also be treated as a special roughness as it is incorporated into the sediment transport theories.

8 Flow Separation, Topology, and Vortical Dynamics

To better understand the new concepts that we will supplement to the theories of sediment transport, we have to know more about the flow separations and their topology. These are in most cases characterized by some dominant vortical structure. The vorticity becomes concentrated in tubes called the vortical skeleton of the flow. It is obvious that the dynamical behavior of these vortex tubes is most relevant for the transport since they define the areas of very low pressure needed to raise the grains up into the flow. Therefore, we have to give an introduction to all of these three elements.

8.1 Flow Separation

Flow separations on bluff bodies, such as cylinders or wings, have been treated extensively in the literature especially with respect to the early separation of boundary layer flows in adverse pressure gradients.

Flow separation generally occurs when the flow is decelerated due to the streamlines diverging right around the rear part of the obstacle. Conservation of energy requires an increasing-pressure or positive-pressure gradient. This can, for example, be the case when the geometry of the channel is altered, especially if the slope changes or when the flow encounters an obstacle. In those cases, the flow looses some of its kinetic energy, which is stored in the low-pressure cores.

We start with a sketch of a separation of 2D flow undergoing the previous deceleration

$$\frac{\mathrm{d}p}{\mathrm{d}x} > 0 \quad \therefore \quad \frac{\mathrm{d}u}{\mathrm{d}x} < 0 \tag{8.1}$$

Proceeding from the continuity and the boundary layer equation with the usual boundary condition

$$\begin{gathered}\frac{\partial u_x}{\partial x} + \frac{\partial u_z}{\partial z} = 0 \quad \wedge \quad u_x \frac{\partial u_x}{\partial x} + u_z \frac{\partial u_x}{\partial z} = u_\delta \frac{\mathrm{d}u_\delta}{\mathrm{d}x} + \nu \frac{\partial^2 u_x}{\partial z^2} \\ \wedge \quad u_x(x,0) = 0; \quad u_z(x,0) = 0; \quad u_\delta(x,\delta(x)) = u_\delta(x)\end{gathered} \tag{8.2}$$

and by integration of the continuity equation over the boundary layer thickness, we get

$$\begin{gathered}\int_0^{\delta(x)} \frac{\partial u_x}{\partial x}\mathrm{d}z = -\int_0^{\delta(x)} \frac{\partial u_z}{\partial z}\mathrm{d}z = -u_z(x,\delta(x)) + u_z(x,0) \\ \therefore \quad \int_0^{\delta(x)} \frac{\partial u_x}{\partial x}\mathrm{d}z = -u_z(x,\delta(x))\end{gathered} \tag{8.3}$$

Now we obtain a relation between the curvature of the velocity profile and the pressure gradient at the wall, assuming that this pressure gradient is about the same throughout the boundary layer

$$\nu \frac{\partial^2 u_x}{\partial z^2} = \frac{1}{\rho}\frac{\mathrm{d}p}{\mathrm{d}x} = -u_\delta \frac{\mathrm{d}u_\delta}{\mathrm{d}x} \tag{8.4}$$

The velocity profiles shown in Fig. 8.1 reflect special points along Eq. 8.4, and it is shown that the flow does not separate in c1, which is where the pressure gradient becomes zero, and therefore the criterion Eq. 8.1 is violated. The separation is delayed due to the time it needs to affect the outer flow. This is a second part of the criterion; the velocity gradient becomes zero shortly thereafter. The velocity gradient is identical to ω_y, and we reformulate the separation criterion Eq. 8.1 by requiring that the vorticity disappears at the wall, respectively $\tau_\mathrm{w} = 0$, and the flow is weakly unstable $\mathrm{d}u_\delta/\mathrm{d}x = 0$,

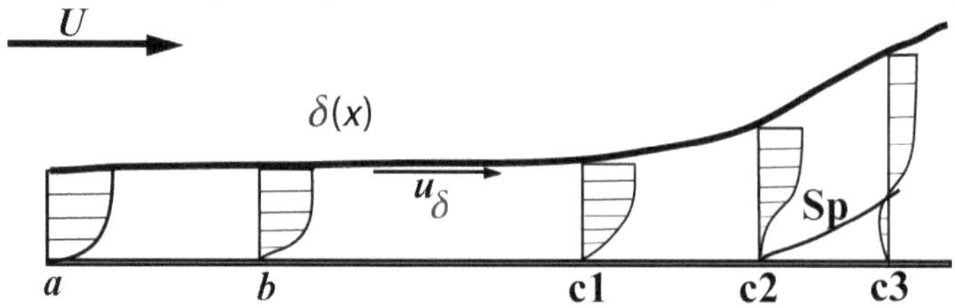

Fig. 8.1 The velocity profiles of a separating boundary layer flow and with its separation layer, the separatrix-plane Sp. The flow is characterized by its five profiles given in Table 8.1

Table 8.1 The first and second lines show the pressure and velocity gradients, respectively, at locations corresponding to Fig. 8.1. The third line gives the curvature of the velocity-profile at this location. The fourth line gives a measure for the vorticity at the wall

a	b	c1	c2	c3
$\frac{\mathrm{d}p}{\mathrm{d}x} < 0$	$\frac{\mathrm{d}p}{\mathrm{d}x} = 0$	$\frac{\mathrm{d}p}{\mathrm{d}x} > 0$		
$\frac{\mathrm{d}u_\delta}{\mathrm{d}x} > 0$	$\frac{\mathrm{d}u_\delta}{\mathrm{d}x} = 0$		$\frac{\mathrm{d}u_\delta}{\mathrm{d}x} < 0$	
$\frac{\partial^2 u_x}{\partial z^2}\Big\vert_\mathrm{w} < 0$	$\frac{\partial^2 u_x}{\partial z^2}\Big\vert_\mathrm{w} = 0$			$\frac{\partial^2 u_x}{\partial z^2}\Big\vert_\mathrm{w} > 0$
Konvex	Turning point		Konkav	
$\frac{\partial u_x}{\partial z}\Big\vert_\mathrm{w} > 0$	$\frac{\partial u_x}{\partial z}\Big\vert_\mathrm{w} > 0$	$\frac{\partial u_x}{\partial z}\Big\vert_\mathrm{w} > 0$	$\frac{\partial u_x}{\partial z}\Big\vert_\mathrm{w} = 0$	$\frac{\partial u_x}{\partial z}\Big\vert_\mathrm{w} < 0$

In the meridian plane of Fig. 8.1, the separation appears at a point (c2). In the real 3D flow, the separation happens along a line along the bed. A separatrix at which a plane starts separates the whole flow field into two volumes. This plane is called separatrix-plane, it is a separation layer on which $\overline{u}_\perp$ is zero everywhere. In general, flow separations subdivide the available volume into subspaces, which allow using geometrical descriptions within their boundaries; we talk about the topological behavior of the flow field. These geometrical laws are needed when using separation elements within a theory of the sediment transport.

8.2 Basics in Topology

The separatrix-plane in Fig. 8.1 does look as it would end up in the bulk of the flow, which actually contradicts its definition. To separate the volume into two subspaces, Sp has to end up on the bed or to roll up the entire volume. The first case generates a closed separation bubble, Fig. 8.2a. The second case describes a roll up as shown in Fig. 8.2b generating an open separation bubble, which will be discussed.

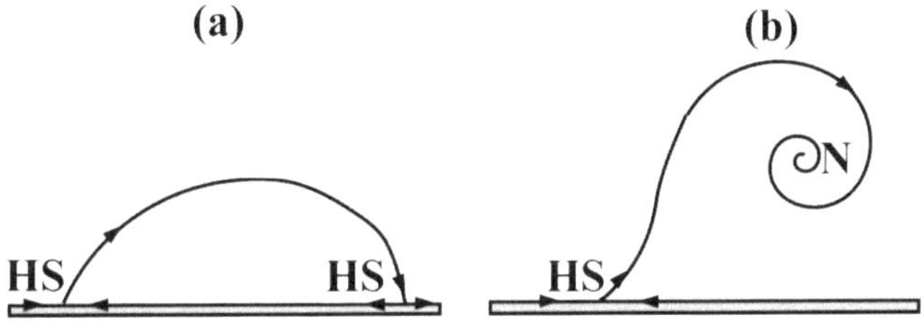

Fig. 8.2 A separatrix-plane forming a closed separation bubble (a) and rolling up the whole space in a spiraling node (b). The topological specifications are given in the text

In Fig. 8.3, we give a sketch of the flow topology in natural coordinates. The topology is given by the separatrix-planes subdividing the space. The linearized equations result in a set of singular points representing the local topology as sketched.

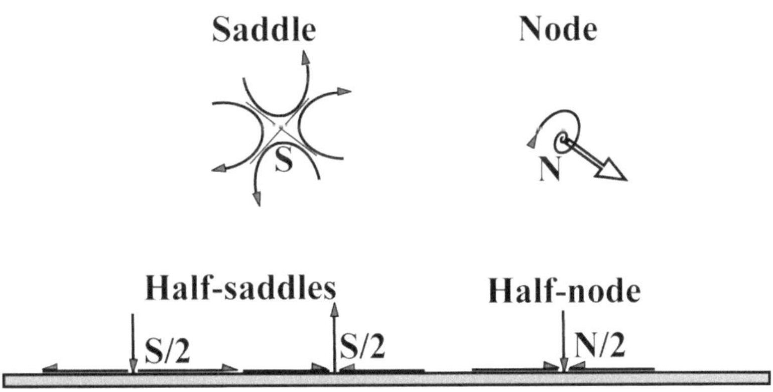

Fig. 8.3 The basic topological structures. The saddle points *S* is defined by a stagnation point in space or a half-saddle HS at the wall. The eigenvalues of *S* are real. The node N, or half-nodes HN at the wall, additionally show their streamlines spiraling near the node. Then the point has two conjugate-complex and one real eigenvalue. The sign of the real eigenvalue results from the continuity equation so that for a converging spiral, $Re(\mathrm{N}) > 0$ and for a diverging spiral $Re(\mathrm{N}) < 0$

In a 2D representation, saddle-points *S* are locations where streamlines converge to or diverge from. They are the classical stagnation points, which show up as a stagnation line in 3D. If the saddle lies on a solid surface, we call it a half-saddle HS. The sum of the three eigenvalues is a measure for the source strength, which will be zero for an incompressible fluid.

Nodes, or half-nodes HN at the wall, are points characterized by all streamlines showing circular spirals to or from the center called focus. They mark a depression. Also here, the separatrix-plane separates the space so that no streamline crosses it. Details can be found in Hunt et al. (1978), Tobak and Peake (1982), Hornung and Perry (1984), Perry and Hornung (1984), Perry and Chong (1987), Bakker and Winkel (1990), Helman and Hesselink (1990), and Dallmann and Schulte-Werning (1990).

When the flow field is solenoidal, that is

$$\mathrm{div}\underline{\omega} = 0 \quad \vee \quad (\nabla \cdot \underline{\omega}) = 0 \tag{8.5}$$

it can be characterized using saddles and nodes. In any plane through the flow field, the following condition must be fulfilled (Hunt et al. 1978):

$$\left(\sum N + \frac{1}{2}\sum HN\right) - \left(\sum S + \frac{1}{2}\sum HS\right) = 1 - n \tag{8.6}$$

$$\wedge\ HN = N/2; \quad HS = S/2$$

where n denotes the so-called connectivity of the flow field equal to the number of subdivisions e.g.,

$$n = 2 \quad \therefore \quad \left(\sum N + \frac{1}{2}\sum HN\right) - \left(\sum S + \frac{1}{2}\sum HS\right) = -1$$

The work of Hunt was a further development of the topological description of the so-called wallshear-lines by Lighthill (1963), for which the following relation holds on a 3D surface:

$$\sum N - \sum S = 2 \quad \wedge \quad A\big|_{z=0} \in 3 - D \tag{8.7}$$

For a simple separation as shown in Fig. 8.1, two separatrix-planes are possible as shown in Fig. 8.2. The closed separation bubble (Fig. 8.2a) subdivides the space into two and therefore $n = 2$, and we have

$$-\left(\frac{1}{2}\sum \boldsymbol{HS}\right) = -1 = 1 - 2 = -1$$

which satisfies Eq. 8.6.

For Fig. 8.2b, we have

$$\sum N - \frac{1}{2}\sum HS = 1 - \frac{1}{2} = \frac{1}{2} \neq 1 - 1 = 0$$

which does nor comply with Eq. 8.6. The separation topology as sketched is incomplete and has to be supplemented to what is called an open separation bubble (Fig. 8.4). And again the Eq. 8.6 is not fulfilled for $n = 2$.

$$\sum N - \frac{1}{2}\sum \boldsymbol{HS} = 1 - 1 = 0 = 1 - \boldsymbol{n}$$

The reason is that the separatrix-plane does not subdivide the space into two since it does not end on both sides in a singular point. In Eq. 8.6 n is only equal 1 or, if we think of a subdivided space, the separatrix-plane must originate at the left from a virtual saddle point. This is the usual case for flows coming from the infinite.

$$\sum N - \sum S - \frac{1}{2}\sum HS = 1 - 1 - 1 = -1 = 1 - 2 = -1$$

and now the system is correct.

We did this lengthy explanation to show how useful the simple Eq. 8.6 is in constructing the right topological map, but how careful too one has to construct such maps. The open separation bubble is one of the most important features in sediment transport, as we will see.

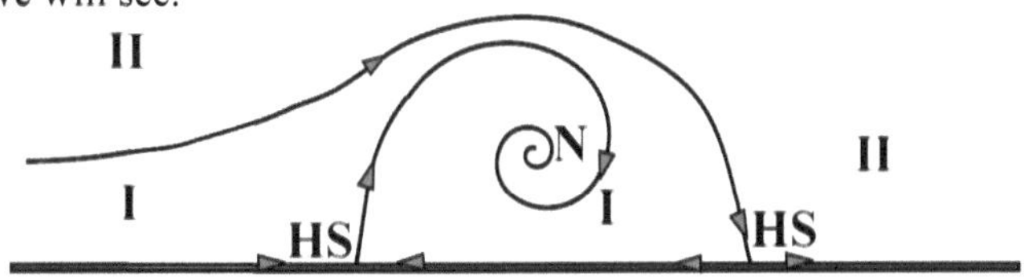

Fig. 8.4 The topological map of an open separation bubble subdividing the space into subspace I and II

Several topological structures will be discussed in Chap. 8 (Sect. 8.3). A more general representation of separations, including internal bifurcations, is published by Chong et al. (1988, 1990) who linearized the flow around critical points, relative to which $\overline{u}$ approaches zero.

Also of interest is the topological behavior of the flow interacting with an obstacle. The surface of that obstacle shall be given by natural coordinates (ξ_1, ξ_2, ξ_3) which shall locally be Cartesian, where ξ_3=0 shall define the surface and neighbored points shall be defined by ξ_1 and ξ_2. Since the surface normal velocity is zero by definition, the shear stress in direction ξ_1 and ξ_2 must be given by

$$\underline{\tau} = \mu \frac{\partial \upsilon}{\partial \xi_3}(\xi_1,\xi_2,0) = (\tau_1,\tau_2) \tag{8.8}$$

If the flow does not change too rapidly in the vicinity of the singular point (Hunt et al. (1978, 1979), the streamlines in the immediate neighborhood of the 2D surface define the local stress tensor. Given a point P on the surface of a 2D plane, the local flow can be described by a strain flow superimposed on a solid-state rotation (Batchelor 1967). A steady viscous flow in the vicinity of P can therefore be approximated by

$$\underline{\underline{\tau}} = \frac{\partial \underline{\tau}}{\partial \underline{\xi}} \underline{\xi} \quad \Rightarrow \quad \begin{bmatrix} \tau_1 \\ \tau_2 \end{bmatrix} = \begin{bmatrix} \frac{\partial \tau_1}{\partial \xi_1} & \frac{\partial \tau_1}{\partial \xi_2} \\ \frac{\partial \tau_2}{\partial \xi_1} & \frac{\partial \tau_2}{\partial \xi_2} \end{bmatrix} \begin{bmatrix} \xi_1 \\ \xi_2 \end{bmatrix} \tag{8.9}$$

With

$$J = \mathrm{Det}\left\| \frac{\partial \underline{\tau}}{\partial \underline{\xi}} \right\|$$

$$\wedge \quad \Delta = \frac{\partial \tau_1}{\partial \xi_1} + \frac{\partial \tau_2}{\partial \xi_2} \quad \text{and} \quad \Omega = \frac{\partial \tau_2}{\partial \xi_1} - \frac{\partial \tau_1}{\partial \xi_2} \tag{8.10}$$

The three defined scalars are independent of the coordinate system on the surface. Relation (Eq. 8.9) can be transformed by rotation to a new coordinate system given by η_1, η_2 via

$$\underline{\tau}' = \begin{bmatrix} T_{11} & \Omega \\ -\Omega & T_{22} \end{bmatrix} \begin{bmatrix} \eta_1 \\ \eta_2 \end{bmatrix} \quad \wedge \quad T_{11} \equiv \frac{\partial \tau_1'}{\partial \eta_1}; \quad T_{22} \equiv \frac{\partial \tau_2'}{\partial \eta_2} \tag{8.11}$$

When η_1, η_2 rotate with the main axis sytem, T_{ii} denotes the principal stress gradients.

The four components of Eq. 8.10 can be reduced to three variables T_{11}, T_{22} and Ω; and Eq. 8.10 becomes

$$J = T_{11}T_{22} + \Omega^2 \quad \text{and} \quad \Delta = T_{11} + T_{22} \tag{8.12}$$

These three values J, Ω, and Δ are independent of the frame of reference and can now be used to characterize the local properties, where T_{11} and T_{22} have to be evaluated using the system of equations (Eq. 8.12). One of these may be taken as a scale factor so that the number of parameters determining the local behavior is reduced to two J/Δ^2 and Ω/Δ, for example, if $\Delta \neq 0$ and its sign is chosen

$$J/\Delta^2 \quad \text{and} \quad \Omega/\Delta \quad \wedge \quad \Delta \neq 0; \text{sign } \Delta \tag{8.13}$$

In an incompressible, viscous fluid, the following is valid

$$u_3 = C\Delta\xi_3 \quad \wedge \quad C > 0 \tag{8.14}$$

This is a problematic definition for the separation or a reattachment. For practical cases however, it is good enough to know whether a fluid particle enters or leaves the boundary layer, which is true if the following is valid along flow-path,

$$\Delta \neq 0 \quad \text{in} \quad t \neq \infty \tag{8.15}$$

Locations where $\tau = \underline{0}$ are singular points of the vector field, their flow vicinity depends on the values of J, Ω and Δ, as shown in Fig. 8.5.

If $J < 0$, the point is a saddle point and two surface streamlines approach along the stagnation point streamlines, which mark the flow toward the critical point and away from it. The saddle point is characterized associating with an essentially 2D flow field, parallel to the obstacle surface. The separation and reattachment lines degenerate to a point on the surface. All neighboring streamlines are asymptotic to these.

If $J > 0$, the singular points on the surface are called nodes and in their vicinity, the fluid flows toward the surface for an attachment line $\Delta < 0$ or away from it for a separation line $\Delta > 0$.

A special line in Fig. 8.5 is given by

$$J = \left(\frac{\Delta}{2}\right)^2 \tag{8.16}$$

which separates the nodes from the foci. If

$$0 < J < \left(\frac{\Delta}{2}\right)^2 \quad \rightarrow \text{Sl} \in \| \text{Sl}(P) \tag{8.17}$$

all of the neighborhood surface stress lines are tangent to one line at the point, with one exception. It follows that there is one prominent line to which the other converge or diverge

The foci appear for

$$J > \left(\frac{\Delta}{2}\right)^2 \quad \rightarrow \text{Sl} \in \text{Spir.} \begin{matrix} (\Delta > 0) \text{ in} \\ (\Delta < 0) \text{ out} \end{matrix} \tag{8.18}$$

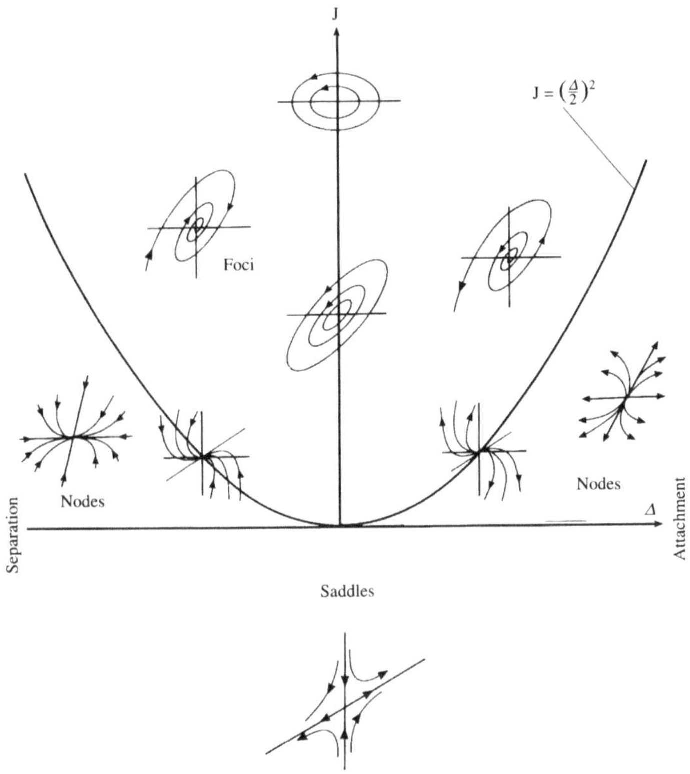

Fig. 8.5 The J–Δ plane describing the flow structure around singular points in various regions on obstacle surfaces. To define the structure univocally, J, Δ, and Ω must all be specified

The spiraling rotation of the streamlines Sl toward or away from the foci depend on the sign of Ω, or indirectly on Δ through Eq. 8.13, as shown in Eq. 8.18.

All nodes have a single streamline outside the surface, which directly aligns toward the singular point. It is directed inward for $\Delta > 0$ and outward for $\Delta < 0$, and as Lighthill (1963) pointed out the Poincaré–Bendixson theorem Eq. 8.7 must be valid. Hunt et al. (1978, 1979) applied this result to investigate the flow topology of a single obstacle, mimicking a single grain lying on the bed. From Eq. 8.7 it follows

$$\sum N - \sum S = 0 \quad \wedge \quad 2N(\infty \leftrightarrow \infty)$$

and analogous using Eq. 8.6

$$\left(\sum N + \frac{1}{2}\sum HN\right) - \left(\sum S + \frac{1}{2}\sum HS\right) = 0 \quad \wedge \quad 2N(\infty \leftrightarrow \infty)$$

As discussed earlier, in both cases 2 nodes (one source and one sink) have to be added at the infinity borders.

Chapman and Yates (1991) have formulated seven separation topologies, of which only the ones relevant to sediment transport will be discussed here.

8.3 Separation Bubbles

We now discuss the combination of the flow separation criteria with the topology in view of separations occurring at bed forms and singular elevations.

A delayed separation with respect to the point of zero pressure gradient was observed for turbulent boundary layers having moderate pressure gradients (increasing pressure) in flow direction

($\nabla p < 0 \Rightarrow \nabla p = 0 \Rightarrow$ separation delayed)

This Reynolds number-dependent boundary layer flow behavior has to be supplemented by flow separation over sharp edges or convex surfaces, where the flow separates immediately or at least with a much shorter delay. The flow encounters this situation at blocks or ridges of bed forms. We encounter sporadic separations on all scales, partly depending on the Reynolds number, so we have to distinguish between viscous dominated and turbulent separation. The viscous separations are stationary enough so that the flow field may be constructed using the topology laws given in Chap. 8 (Sect. 8.2). These are still valid for turbulent separation conditions but because of flow the dynamics, we will only be able to construct an averaged flow field. For the turbulent case, the separatrix-plane develops by instability processes of the mixing layer given the large velocity difference across the plane.

It is often a tedious undertaking to calculate the vorticity distribution within 3D separations, which is only estimated for that reason. The flow around single exposed obstacles can in good approximation be calculated using RDT (*R*apid *D*istortion *T*heory) (Hunt et al., 1978, 1991). For quasi-2D obstacles, the knowledge of the vortex skeleton contains the most information, and these vortex tubes can be rather easily constructed. We will treat the vortex mechanics in Chap. 8 (Sect. 8.4) and therefore postpone the quantitative description until used for special cases.

8.3.1 The Closed Separation Bubble

In many textbooks, this is the only topological reproduced form, and often not even the assumptions used are mentioned. Two main restrictions have always to be named: this topological picture is a 2D representation and no spanwise fluid exchange is considered. Therefore, this representation stands for a simplification of the separating flow which is a time average, which would be correct only for a stationary case. In fact such representations should be forbidden because they are misleading and in contradiction to the introduction of a topological concept since it does not give a correct picture of the flow geometry.

An idealized closed separation bubble was shown in Fig. 8.2a and reproduced again in Fig. 8.6 for two very similar forms. The situation in Fig. 8.6a represents a bubble behind a single obstacle, whereas Fig. 8.6b shows the separation over a 2D, possibly periodic, bed form (Fig. 8.6). Most closed separation bubbles occur in troughs as this was shown in Fig. 7.3.

The fluid inside the separation bubble has vorticity, which has to be fed. Along the separatrix, the normal velocity is zero and no fluid exchange is possible. At the other hand, the velocity gradient directly depends on the outer flow velocity. The velocity gradient contains vorticity, which diffuses through the separatrix, driving the circulation inside the bubble. Inside such bubbles, the flow at the bed reverses compared to the outer flow and vorticity of opposite sign is generated at the wall, damping the bubble vorticity through annihilation. When the vorticity is balanced, the separation bubble is completely stable. The circulation and the pressure difference is small. Such separations cannot move grains and do not contribute to the sediment transport, and the probability for a closed separation bubble increases with smaller outer flow velocity.

If the forcing of the separation bubble is stronger than the internal dissipation, the bubble growths in size may be swept away after reaching a certain size. When this is the case, the topology changes because a triple structure, a separation in the separation bubble, appears. The topology is analogous to the one shown in Fig. 8.7b. This happens rarely, however it is not mentioned in most of the available textbooks.

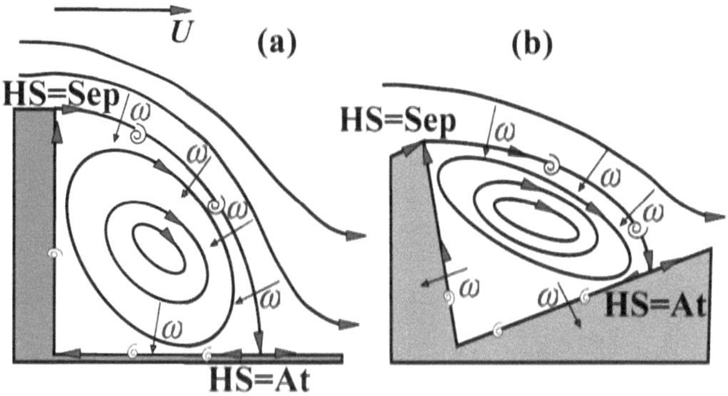

Fig. 8.6 A closed separation bubble downstream of an artificial obstacle (a) and behind a periodic bed form (b) showing an inner circulation. Both sketches show identical topology highlighting their close relation. The mechanism is explained in the text

Also when the forcing decreases, a new equilibrium will enable a closed separation bubble of smaller size. The size of the separation bubbles bears some significance since the growth as well as the shrinkage shift the reattachment point (At).

8.3.2 The Open Separation Bubble

This separation topology is most prevalent in geophysical flows. Already sketched in Fig. 8.4 and reproduced with further detail in Fig. 8.7, we are discussing the dynamics of such an open separation bubble in more detail, where we will first discuss the viscous separation and follow up with turbulent separation. Figure 8.7a is identical to Fig. 8.4, however supplemented showing also the vorticity flux *A* thin fluid layer originating from a wall boundary enters the bubble, and it not only adds fluid to it but also holds a fairly large amount of vorticity ω_y. This convective transport results in a very efficient feeding mechanism for vorticity. Hence, the vorticity within the open bubble grows much larger than for a closed bubble. For a quasi-stationary open separation bubble to be globally stable, the fluid and the vorticity have to leave the bubble at the same rate or at least in the mean for the oscillating case The inner node is of focus type, and the vorticity concentrates at the center so that the release of fluid and vorticity is achieved through a helical flow out toward the third dimension. The resulting vortex has a certain helicity density h, and its helicity H itself results from the volume integration of h.

$$h=(\underline{\omega}\cdot\underline{u}) \quad \therefore \quad H=\int(\underline{\omega}\cdot\underline{u})\,\mathrm{d}x \tag{8.19}$$

The discussion shows that open separation bubbles are naturally 3D (!) and reveal more complexity as the central cut suggests; a solution will be shown in Fig. 8.10.

Both mechanisms releasing vortical fluid can be observed, however the shedding happens on obstacles whereas the latter on bed forms. Oscillatory separations on bed forms occur only in special cases such as in very narrow channels where the sidewalls influence the flow considerably. The bubble topology during shedding is shown in Fig. 8.7b.

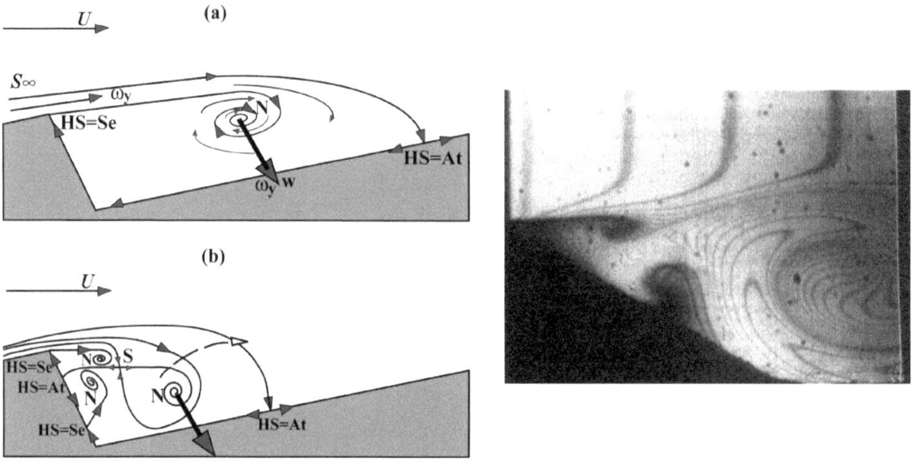

Fig. 8.7 Topology of an open separation bubble (a) in the usual form and in (b) during the shedding of the whole bubble and its photocromic visualization

The topology of the shedding mechanism has to obey the main rule in Eq. 8.6. Using the fact that three separatrices have their origin at the upstream infinite, one finds

$$\sum N-\sum S-\frac{1}{2}\sum HS=1-s$$

$$3-\left(1+3_{\infty}\right)-\frac{1\cdot 4}{2}=1-4=-3$$

representing the situation correctly. Using photochromic flow visualization, Müller and Wiggert (1989) and Müller and Gyr (1996) proved the existence of this topology during a startup-flow over a dune.

For the turbulent case, the separation is a mixing layer after the flow separates from the ridge. Mixing layers are known to be instable due roll up of the vortex sheet into discrete parallel vortices of the same sign. They grow by pairing and attach/detach in a nonstationary oscillating way during their interaction with the bed. In a 2D representation as shown in Fig. 8.8, this shows up as a simple open separation bubble, however having a complicated internal structure. Knowing about the formation of concentrated vortices with high helicity density, we can imagine how complex the full 3D topology may become. Therefore, results derived from such representation have always to be taken cum grano salis.

The velocity difference, which is of O(U), defines this separation layer. In good approximation, this layer can be thought to be very thin, and thus we can apply Batchelor's (1967) theory of an infinitely thin mixing layer. Some special details including the spanwise instability will be discussed later.

Fig. 8.8b shows the basic structure of the separating mixing layer, which is the topological equivalent to a free shear layer. Through induction, a primarily wave-like disturbance grows to a row of vortices wrapping up from the thin shear layer. Sheets comprising vortex tubes of the same sign are instable (Lamb 1879; Saffman 1992). Neighboring vortex tubes of the same sign form pairs, turning each other around while they merge through vorticity diffusion to form a single larger vortex tube. In rare cases in mixing layers triple pairing can be observed (Ho and Huang 1982). However, in the available timescale of the discussed separations they cannot merge completely usually resulting in a conglomerate of vortices within the separation bubble.

In the 2D case, if one of these vortices contacts the bed they generate a shear layer of opposite sign which sheds off, and the inflow from the outer region is interrupted. The attachment streamline oscillates over the bed with the replacement of one vortex by another, forming an open separation bubble as sketched in Fig. 8.8a. The many foci show the existence of an intensive helical flow system revealing many "holes" within the bubble. The flow is extremely 3D. In addition, some of the vortices can be released from the system as boils; we will discuss this in the context of the dunes.

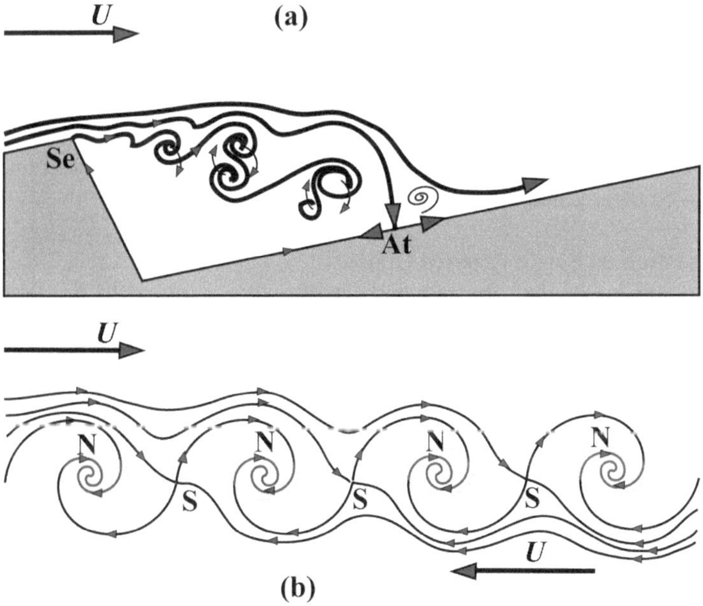

Fig. 8.8 Turbulent separation of a mixing layer type. The development and oscillation of the separatrix in (a) is shown as a result of the instability and the pairing of vortices. In (b) the essential topology during mixing of a shear layer flow is represented

8.3.3 Separations on 3D Obstacles

In Chap. 7, we introduced the roughness elements as an origin for flow separations and discussed the separation behind single grains. In Chap. 8 (Sects. 8.3.1 and 8.3.2), we discussed the topology of more or less 2D obstacles and will now expand around the topology of single 3D obstacles on the bed. We discuss two examples qualitatively, following up with quantitative data in Chap. 12.

Many details are still subject of current research. This is particularly valid for bluff obstacles because of the many difficulties involved when observing the various properties of complex time-dependent 3D flow fields. For an inviscid fluid with a uniform upstream velocity U, the flow is irrotational (rot $\underline{u} = 0$), and the velocity field $\underline{u}$ may be expressed as $\underline{u}$ = grad ϕ, where $\phi(x, y, z)$ is termed a velocity potential. For an incompressible fluid, the conservation of mass requires that div $\underline{u}$ or div(grad ϕ) = $\partial^2\phi/\partial x^2+\partial^2\phi/\partial y^2 = \Delta\phi = 0$, which—together with the boundary conditions—specifies a "potential flow" system. In this case, it imposes the normal component of u to vanish on the boundary. Generally, the potential flow approach provides the smoothest and simplest flow pattern possible over a complex obstacle shape. For an incompressible fluid, this flow pattern adjusts instantaneously to changes in U if the latter varies in time.

In real flows of viscous fluids at the defined conditions and large Reynolds numbers, the observed solution differs from the potential flow case. The reason lies in the no-slip condition, which additionally requires the surface-parallel component of the velocity to vanish on the surface. The developing boundary layer not only displaces the outer potential flow but also influences the location of the separation lines. Large variations of the separating flow regions may occur depending on the obstacle shape. When the face of the obstacle is sufficiently steep also upstream separation occurs. The downstream separation is practically always accompanied by an extensive wake region. This implies that there are substantial fluid volumes where the flow pattern is affected by boundary processes and singular points. Generally, the flow outside these regions is closely approximated with the potential laws. However, this is of little practical use since the shape of these regions and the interchange of external fluid with them must be known before ϕ can be determined.

8.3.3.1 Separation on Single Exposed Grains

Some grains are larger than the rest and therefore are exposed to the flow as if they would sit on a much smoother bed. Such a grain is represented here by a cubic element of the size (h, a, b) sitting on a flat bed (Fig. 8.9).

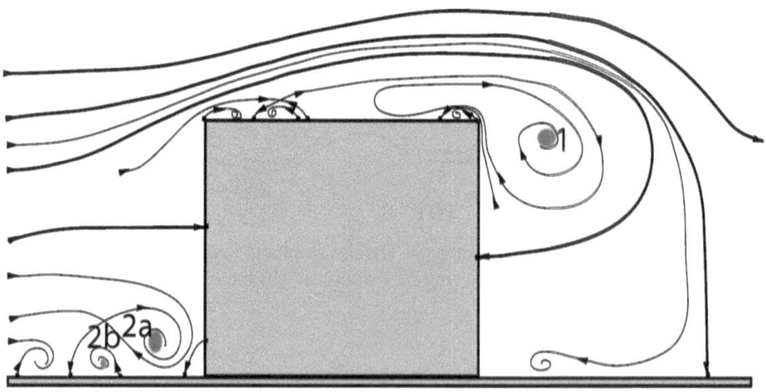

Fig. 8.9 A sketch of the instantaneous flow topology around a cubic obstacle sitting on a flat bed as seen in a central cut. We call special attention to (1) the vortex forming on the lee-side of the cube and (2a and 2b) the vortices forming the upstream separation. The singular points on the cube surface are marked with a black dot (•)

This example shows how much work is still needed to integrate such topological aspects into a statistical theory of sediment transport. One remark shall further elucidate this situation. A new theory needs to integrate the effects step-by-step beginning from the single grain topology to interacting obstacles. Much research was driven by chemical engineers investigating flow interactions of organized obstacles, such as cylinders, mainly with respect to heat transfer problems.

The sharp edges of the cube shown in Fig. 8.9 are very important for the separation topology. Here, the lee side separation causes flow recirculation generating up-flow and down-flow across the rear surface. The lee vortex (1) is large in diameter, but its importance is rather small. Its vorticity is weak because the rolled up fluid layer had little vorticity to begin with. In high contrast are the side vortices (2a and 2b) being rather small but containing a rather high amount of vorticity originating from the large velocity gradient close to the wall. Being so close to the wall therefore causes a local low-pressure field with high rotation resulting in scouring around the front and both sides of the cube. Vortex 2(a) and Vortex 2(b) have opposite signs annihilating some of their vorticity through diffusion. Vortex (2a) has more energy and absorbs (2b) further downstream, so typically a so-called horseshoe vortex survives (Schofield and Logan 1986).

Again the three-dimensionality of this separation is more complex as it seems when looking only at the central cut since the sharp side edges have their own topological features. In the following section, we present a smooth single separate obstacle, where the 3D representation of the flow topology becomes simpler.

8.3.3.2 The Topological Structure of a Flow Encountering a Sandy Elevation.
Often one encounters smooth sandy elevations protruding from the bed surface. In the central plane cut, the separation topology shows the classical open bubble as introduced in Fig. 8.6. This obstacle is called a "polynomic"-hill, whose main three-dimensionality results from the fluid being displaced to the sides. Especially at lower Reynolds numbers, this fluid will organize to a rather stable structure, however for very high Reynolds numbers the topology is closely resembled by potential flow superimposing the vorticity as concentrated vortex tubes being of similar topology as the very viscous result. This approach is used for investigating the scouring on bridge pillars and similar constructions.
In Fig. 8.10, we reproduce a 3D view of the topology of a polynomic hill.

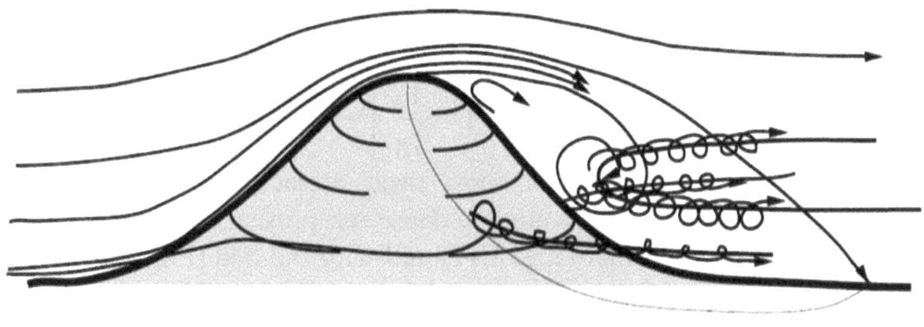

Fig. 8.10 The 3D visualization of the separation flow topology over a polynomic hill. On the hill the separatrix is marked

Further to the discussion of Fig. 8.10, the 3D nature is reflected and classified by the so-called wall shear lines, in the simplest case of the given example this shows up as the so-called "owl-face" as we will find it in Fig. 9.17(b). Although the slope of such hills can be as high as 45°, their smooth 3D character allows the potential flow to escape around the sides. The horseshoe vortices start rolling up only after the flow has passed the widest part of the obstacle and lie within the separation bubble.

8.4 Vortex Tubes and Vortex Interactions

As we saw, through the focusing mechanism vorticity is fed into and transported away along the vortex tubes. These tubes form what we call the vortex skeleton of the structures. If the dominant part of the vorticity is located within these tubes, we assume that a simplified flow field could be constructed using this approach. With the introduction of coherent structures came the expectation that the flow field could be divided into subspaces defined by these coherent structures having their unique interaction with the sediment. Following through with an integration over all the subspaces, one could yield the total transport. To a certain extent, coherent structures are part of the turbulent flow; however their dynamic behavior and interactions are too complex to achieve a general breakthrough of this concept. The situation changes when one looks at quasi-stationary vortex skeletons being more stable and even of much higher intensity. Nevertheless, it has to be cautioned that such vortex tubes do not belong to the classical turbulent structures since they are not the result of a statistical process but of a localized feedback system given by the flow and the obstacles.

Using the vortex skeleton concept requires knowledge about the interaction laws of such tubes with each other and a wall. We refer the reader interested in vortical dynamics to the books of Saffman (1992) or Green (1995) or the classic literature by Lamb (1945) or Batchelor (1967). A very original book was written by Albering (1981).

With the vortex skeleton assumption of concentrated vortex tubes, the Biot-Savart's law allows a complete description of the flow field. When the location and the strength of the vortex tubes are known, an integration of Eq. 5.24 results in a velocity estimate at point P when Q is any point on the vortex tube with the strength Γ and $\underline{r}$ is the vector distance between the two points. The field description results from the simplified form through evaluation of Eq. 8.20 at discrete grid points,

$$\mathrm{d}\underline{u}\big(\underline{x}(P),t\big)=\frac{\Gamma}{4\pi}\frac{\left[\mathrm{d}\underline{x}(Q)\times\underline{r}\right]}{r^3} \qquad (8.20)$$

$$\therefore\quad \underline{u}\big(\underline{x}(P),t\big)=\frac{\Gamma}{4\pi}\int\frac{\left[\mathrm{d}\underline{x}(Q)\times\underline{r}\right]}{r^3}$$

It is of advantage to understand the most basic interactions since the visual imagination often helps developing new models. Here we present and discuss the interaction of vortex tubes with each other (Fig. 8.11) and with the bed (Figs. 8.12 and 8.13).

On several occasions, we encountered longitudinal vortices being more or less parallel to each other and fairly aligned with the flow direction so that ω_x is the main vorticity component. They originated from separation, e.g., the mentioned side vortices

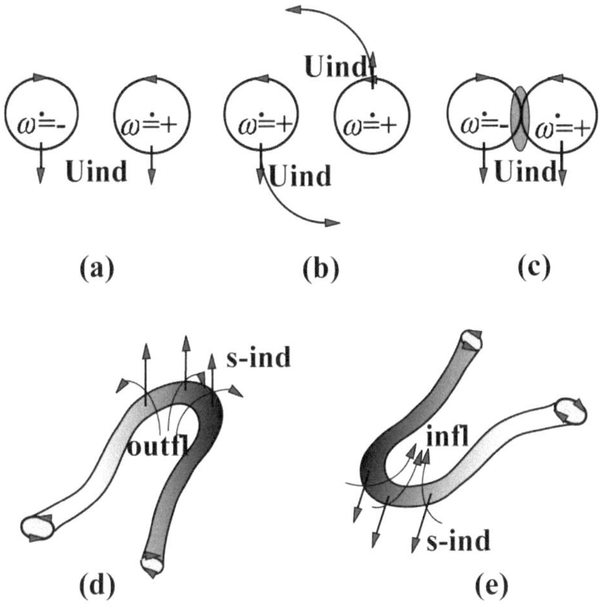

Fig. 8.11 A compilation of the most basic vortex–vortex interactions. (a–c) straight parallel vortex tubes. (d and e) Strong self-induction at vortex bends. See the text for explanations of the dynamical behavior

at obstacles additionally being enhanced by the mean shear. Two possible scenarios may evolve, depending on the respective orientation of the vorticity vectors. When they are antiparallel, as sketched in Fig. 8.11a, the induced velocity from either vortex onto the other moves both tubes parallel in one common direction, perpendicular to the flow and the connecting line of their centers. At an obstacle, as shown in Fig. 8.10 for example, the inner and also the outer tubes induce velocities, which drive both vortices toward the wall. The opposite case, when the vorticity is arranged parallel, the induced velocity rotates the vortices around each other (Fig. 8.11b). Such an interaction often ends up by a pairing process since vorticity diffuses over the area originally occupied by the two turning vortices. When vortex tubes of opposite sign approach close enough for the vorticity to diffuse from one to the other, the total vorticity is reduced through cancellation and the vortices loose strength.

Depending on the externally imposed deformation filed, vortex tubes deform into vortex loops of large curvature so that the self-induction becomes dominant. For a channel flow, these loops occur at vortex tubes in spanwise direction ($\omega_y \neq 0$). This important mechanism is responsible for the initial lift up of the loop, see Fig. 8.11d or for its downward motion Fig. 8.11e, respectively. In a channel flow, the forward looking vortex loop can rise up to the water surface. On breakup, it shows up as a pair of vortex cups or as a boil. Fluid is pumped upwards together with the loop, transporting low-velocity fluid toward faster-moving regions. The backward facing loop will be driven toward the bed, where this vortex loop will split into two tornado-like vortex branches possibly attaching themselves to the wall. The volume between the loops will be accelerated, bringing fluid of high velocity toward the bed. The combination of the low-pressure core and the momentum flux is one origin of scouring.

If one could freshly seed the bed, the locations of these vortex interactions on the wall would display a negative image, removing sediment along the patterns visualized by the wall shear lines (Fig 8.12a). We observe bifurcations, positive for separations and negative for attachments where the separation lines again are the intersection of the separatrix-planes with the bed.

In Fig. 8.12b, the so-called mirror concept is explained. It is an artificial construction originally used for nonviscous fluids. The disappearing velocity at the wall determines the boundary condition; to achieve this condition, one introduces a velocity field that compensates the induced velocity by the vortex interacting with the wall. This is the case if one replaces the wall by a vortex, which is compensating the action of the first one. This is a vortex whose properties are one of the original vortex reflected at the wall. The wall acts like a mirror. For a vortex line with a vorticity $\underline{\omega}$, the nonconvective part of the velocity with which the vortex in a distance $2a$ of the two centers is passing over the bed is given by its circulation κ:

$$\kappa = \frac{1}{2}\omega a^2 \rightarrow u = \frac{\kappa}{2a} = \frac{\omega a}{4} \tag{8.21}$$

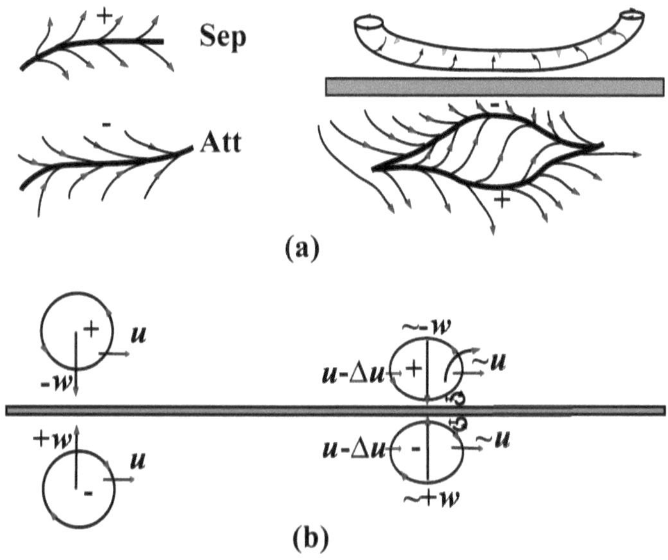

Fig. 8.12 The behavior of vortices interacting with the bed. (a) showing the wall shear lines and (b) an explanation of the mirror concept

For this non viscous case, also the pressure can be calculated

$$p(x) = \Pi - \frac{\kappa^2 \rho}{a^2}\cos 2\theta \cos^2\theta \tag{8.22}$$

Also, for a vortex line parallel to the wall the resultant force of the bypassing vortex on the bed is given by

$$\frac{\kappa^2 \rho_f}{a^2}\int_{-\pi/2}^{\pi/2} \cos^2\theta \cos(2\theta)\, a \sec^2\theta \,\mathrm{d}\theta = 0 \tag{8.23}$$

For the viscous case, the no-slip condition must be fulfilled. The induced flow at the wall produces a local velocity field of opposite vorticity (Fig. 8.12b). Given the right Reynolds number, this secondary vorticity can actually absorb most of the original energy and reflect the original vortex. At the closest approach, annihilation acts and the vortex cross section distorts.

These mechanisms shall be used to explain the interaction of vortex rings with the bed because vortex rings are a good model for vortex loops which often occur in the context of vortex skeletons and separations (Fig. 8.13).

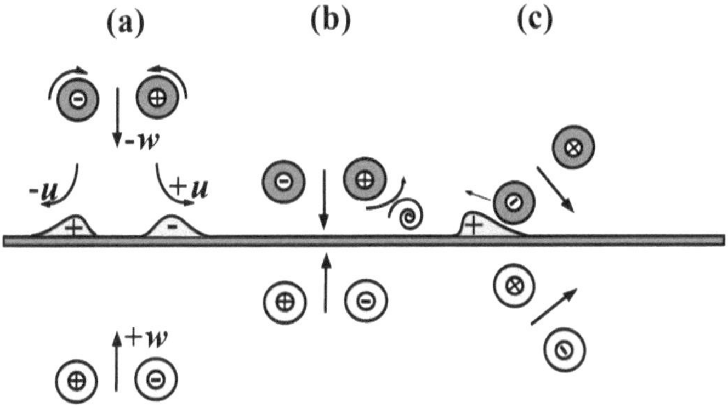

Fig. 8.13 Vortex rings interacting with a wall using the mirror concept for the explanation given in the text

The classical textbook case shows the interaction of a circular vortex ring approaching the wall perpendicularly, this is shown in Fig. 8.13a. The mirror concept explains why the vortex ring in approaching the wall is opening the main radius and its velocity component $-v$ is decreasing since it moves toward a stagnation point. The spreading of the vortex produces a radial velocity, which shows up in the central cut as a $+u$ and $-u$ component. This induced flow is a boundary layer flow with a vorticity of opposite sign compared to the one of the interacting part of the vortex ring. The strength of the vortex ring becomes reduced by annihilation and the spreading, and the secondary vorticity in the new boundary layer flow can become comparable in strength. Its induced velocity on the vortex ring can lead to bouncing of the ring as explained in Fig. 8.13b. See for example for such vortex rings Munro and Dalziel (2003), or for simulations see Doligalski et al. (1994) or Walker et al. (1987).

The perpendicular inflow is a very special case, the usual configuration is a vortex ring inclined by an angle to the wall (Fig. 8.13c). The branch closer to the wall reacts first and is stronger influenced by the mirror vortex. The so-induced boundary layer starts to develop earlier than its counterpart. The process becomes asymmetric and the vortex ring becomes elliptical, it gets deformed in the third dimension with respect to its axis, and fluid is transported in φ direction, which means an inner helical flow begins. This is usually ending in a breakup of the vortex ring. The same can be said for the interaction with the water surface producing two vortex branches showing up at the interface as naps. Visualizations show that these branches are "attached" to the interface, as explained in Fig. 3.7, in form of a tornado-like flow with a high-underpressure core, most important for the sediment transport.

9 Fine-Sand Dynamics

In Chap. 7, we discussed the importance of roughness properties and local separations for the sediment transport. The roughness of the bed, however, is prescribed by its grain composition. A size classification of the material involved makes sense, since this procedure also has the advantage in aiding to find the right "cooking recipe," which is necessary for the principally different transport mechanisms.

A bed made of fine sand is one of the most investigated cases from laboratory channels. This way the Froude similarity for rivers with rather large gravels can be modeled. The results showed that turbulence plays a minor role; however, separations are very important. Whenever this is the case, the existing results show considerable scatter.

We now treat the transport behavior of fine sand meaning a bed composed of grains, which, when flattened, would act as a hydraulic smooth surface. The grain diameter when scaled with wall variables should always be within the viscous sublayer, Eq. 2.42a based on Eq. 2.34

$$k_s^+ \leq 5 .$$

Hydraulically smooth means that the roughness does not influence the friction factor and it is not important which grains are involved. Therefore, one can use sand of uniform grain size as long as the above requirement is satisfied. However, one has to be careful since an increasing discharge also affects the physical size of the grains fulfilling the above criterion, which causes a decrease in scale, and every grain size exceeds the height of the viscous sublayer at a certain load. This is important for the description of the different transport types and bed forms with an increasing load.

We want to take the opportunity to remark on a popular misconception that the flow zone close to the bed at hydraulically smooth condition is termed "laminar" instead of "viscous." It shall be insisted that "laminar" cannot be right. Using a microscope, one clearly sees intermittent motions changing direction and velocity, which are modulated by the outer flow.

Beginning with a flattened bed, one expects the developing instability structure to be the same as for the flat wall, where we showed the development of coherent structures. Hence, for the sediment transport description under these circumstances, the coherent structures mechanism is dominant. We note, that results found for the fine-sand transport cannot be extrapolated to a situation with rougher conditions since the scales of the coherent structures and their creation mechanism require smooth wall conditions.

9.1 Stable Beds and Incipient Motion

We start from a sand-bed with an artificially flattened surface in a 2D open channel flow. With the present understanding, this bed should be stable up to a critical wall shear stress given by the Shields criterion postulated in Chap. 3 (Sect. 3.1). Using the grain diameter as length scale, we determine a critical Froude number (of the grain)

$$\mathrm{Fr}_{*\mathrm{c}} = \frac{\tau_{\mathrm{wc}}}{g(\rho_{\mathrm{s}} - \rho_{\mathrm{f}})k} = \frac{u_{\tau\mathrm{c}}^2}{g\rho' k} = f(k^+) \tag{9.1}$$

For a fine sand of $k^+ = 5$, incipient motion of the bed load occurs when the critical grain based Froude number exceeds $Fr_{*c} \approx 0.022\ / 0.04$ (Shields diagram Fig. 3.1). The upper value is valid for a stable bed after initial self-organization, which accounts for the initial rearrangement of some unstable grains originating from the preparation of the flat bed. The stability increases with the decreasing grain size so that, e.g., for $k^+ \approx 1$ critical Froude number has already quadrupled to $\mathrm{Fr}_{*\mathrm{c}} \approx 0.08$. This higher stability also allows the development of a steeper velocity profile with a larger shear gradient at the wall. This underscores the extreme stability of fine material where, e.g., lake-chalk cannot be transported in laboratory experiments needing an outer disturbance to start the transport. In an experiment one could show that a single water drop falling on the water surface can by its pressure field, which is not of turbulent origin, destabilize the complete bed.

Also, at incipient motion, the transport does not start moving single grains but as a common motion moving most grains within a given wall area to form a cloud. The cloud motion is induced by a coherent structure. The viscous tornados with their limited low-pressure potential are not yet capable of raising the grains, nor are the longitudinal vortices energetic enough to cause the transport. Grass (1971) postulated the incipient motion to origin from the so-called sweep events while they interact with the bed. Schmid (1985) confirmed this statement experimentally, measuring the velocity field using an LDA system, while simultaneously filming the sediment motion using a synchronized camera. The incipient motion was totally correlated with the sweep events (Fig. 9.1).

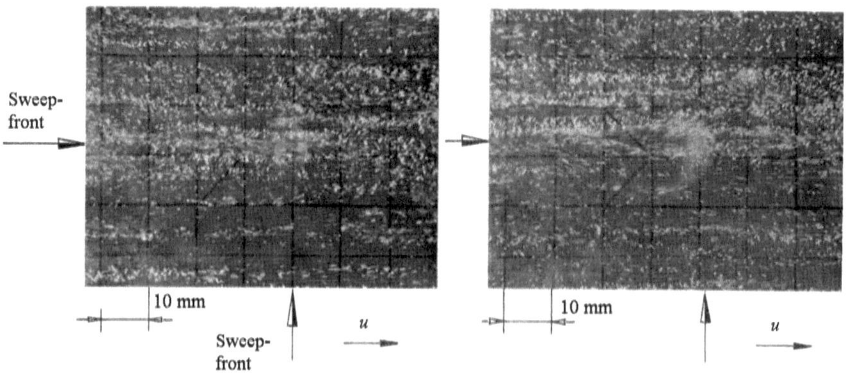

Fig. 9.1 The influence of a sweep event on the incipient motion on a flat surface. (Data: u_τ= 18.9 mm/s, ν= 1 mm²/s, Re = 39000) (Schmid, 1985)

Visualizations, like Fig. 9.1, show that a jet like flow toward the wall is highly retarded due to the interaction (friction) with the bed. The developing front vortex due to this retardation is the one with an extreme vorticity and therefore under pressure.

The mechanistic model constructed from data is shown in Fig. 9.2. Here, the sweep was thought as part of a Λ-vortex system, a mechanism first shown in Fig. 6.4.

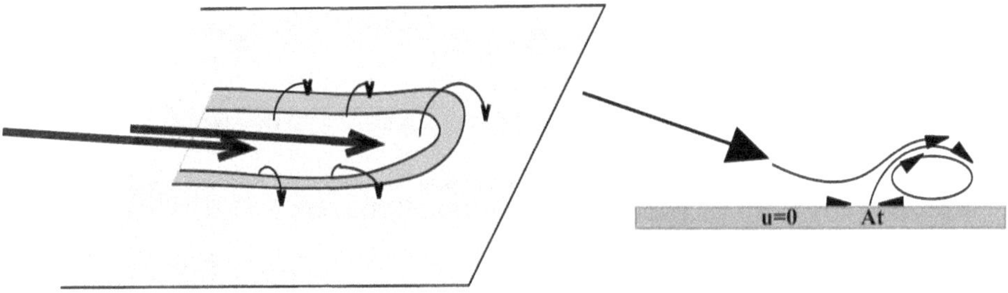

Fig. 9.2 The origin of the strong front-vortex due to the retardation of a sweep event and the formation of longitudinal vortices by the convection of this front-vortex. The attachment point At is dynamically shifted

Direct numerical simulations of turbulent boundary layer flows (Robinson, 1992) showed a similar picture, where the location of lowest pressure was located between the ejection and sweep events which were identified by their velocity signatures in the boundary layer flow. This is compatible with the sweep model insofar, as the front vortex is at the downstream end of a sweep event, and its influence can be calculated by using Eq. 6.87. Therewith one can estimate the pressure distribution of such a vortex in a first approximation (Fig. 9.3).

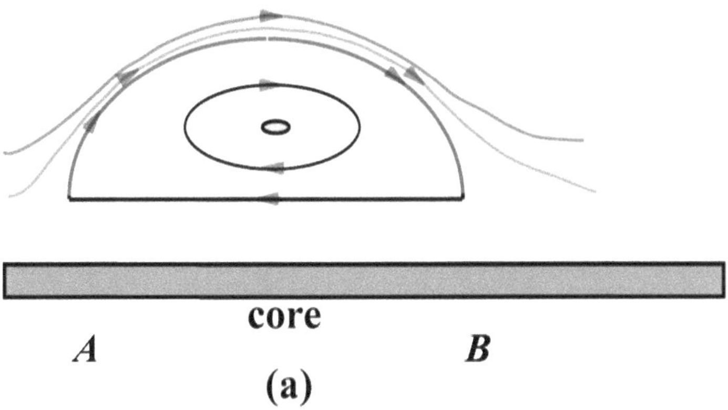

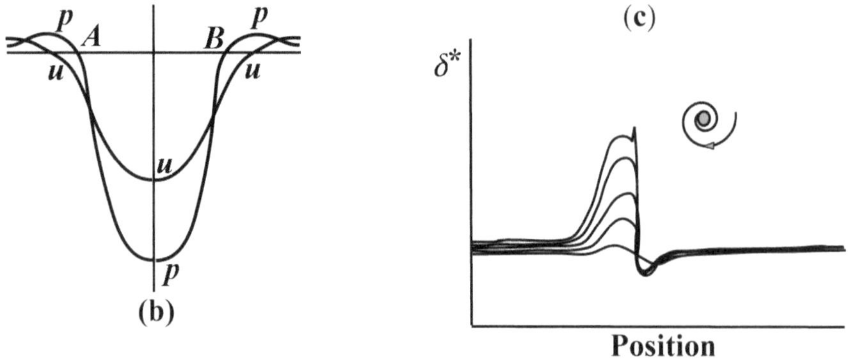

Fig. 9.3 The reaction of a bypassing vortex (a) on the wall-near boundary layer flow and the deformation of the displacement thickness δ^*, (b) The pressure field, and (c) the displacement

Figure 9.3a shows a central cut through a simplified vortex bypassing over the bed. The dimension of its core is given by the distance A–B. Figure 9.3b shows the velocity and the pressure at the wall in a coordinate system moving with the vortex. In Fig. 9.3c, the reaction of the displacement thickness δ^* is plotted showing how the vortex mixes up the boundary layer.

The most provoking flow pattern for an incipient motion is a sweep producing a critical low-pressure field, first lifting and transporting and later depositing the grains further downstream. The mechanism is sketched in Fig. 9.4 and a quantitative estimate can be found in Chap. 10 (Sect. 10.3).

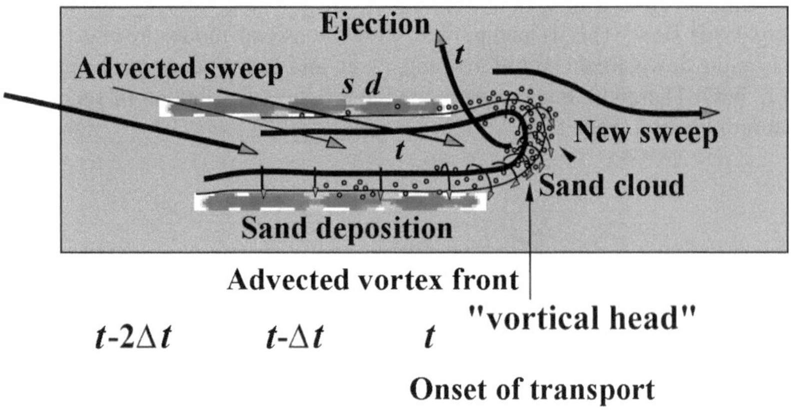

Fig. 9.4 The incipient motion as induced by a sweep event at the bed. The retardation produces a front-vortex capable of raising grains through its low-pressure A cloud of suspended particles is produced. The vortex is convected with the flow forming two longitudinal vortices of small circulation from which the suspended particles settle in accord with the sign of the vorticity in these side-vortices. The ejection is generally not able of raising particles, because they only exhibit a small pressure difference in the core

For most cases, the only real active element is the front-vortex of the decelerated sweep, whereas the other vortex elements are important as a guiding system controlling the deposition pattern. A longitudinal stripe pattern can be recognized. It is important to see that this event is not due to a raising and depositing by the longitudinal vortices as often mentioned in the literature. This concept associates the initiation of sediment transport for fine-sand beds with the turbulence of the flow, where this aspect was neglected in the classical theories. The transported volume for a single transport event will scale with the sweep structure and is therefore limited in size.

It is interesting to observe the incipient motion to stop after an initial rearrangement, showing a stable bed configuration being not the striped but what Schmid called an orange-peel surface, see (Fig. 9.5). The highest elevations are separated by roughly 10–20 mm or a distance of 100–200 viscous units. The elevations have a width of 2.5 mm or 20–50 viscous units, their height however is only of the size of a grain diameter of 0.02 mm. This is the most stable bed configuration corresponding to the highest measured critical Shield parameter for the incipient motion.

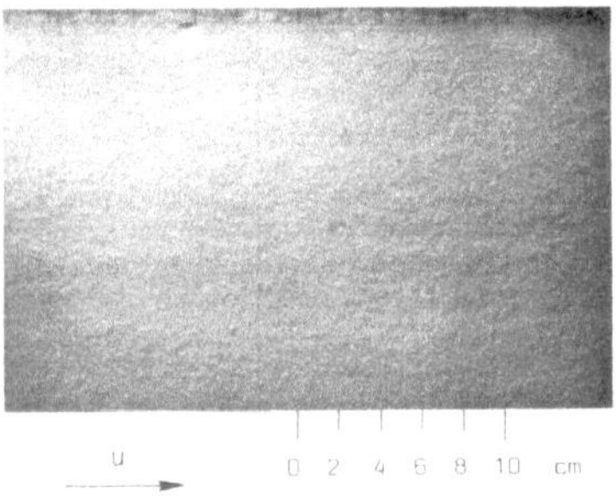

Fig. 9.5 A typical orange-peel bed configuration. ($u_\tau = 11.1$ mm/s, $\nu = 1$ mm^2/s, $Re_k = 2.21$, $Fr_k = 0.058$)

The most important part of this process is its self-organizing behavior. The flow sediment transport interaction produces a stable configuration and the model needs to reflect this possibility. By the transport mechanism shown in Fig. 9.4, the bed starts to deviate from the flat configuration, and the deposited material produces elevations; the ongoing process however does not redistribute those elevations again. These, in turn, interact as disturbances with the instability process, and the modified coherent structure behaves under-critical, resulting in a stabilized bed. This altered instability process is not yet understood, however, we have some hints pointing to how this mechanism might work. Rempfer et al. (2003) investigated the case where the wall can react as a compliant wall to pressure fluctuations of the outer flow and showed the extremely subtle reactions of the stability. Therefore we will be modest and argue based on averaged measurements.

Schmid (1985) measured the intermittency of the sweep events and showed that the intermittency factor i_4 increases close to the wall as the bed is forming an orange-peel configuration

$$i_4 \equiv T_4 / T_{total} \quad \wedge \quad T_4 \rightarrow t \ni \left((u', w') \in Q_4 \right) \tag{9.2}$$

With the intermittency, also the number of sweep events increases, although the momentum transport to the wall remains constant (Fig. 9.6).

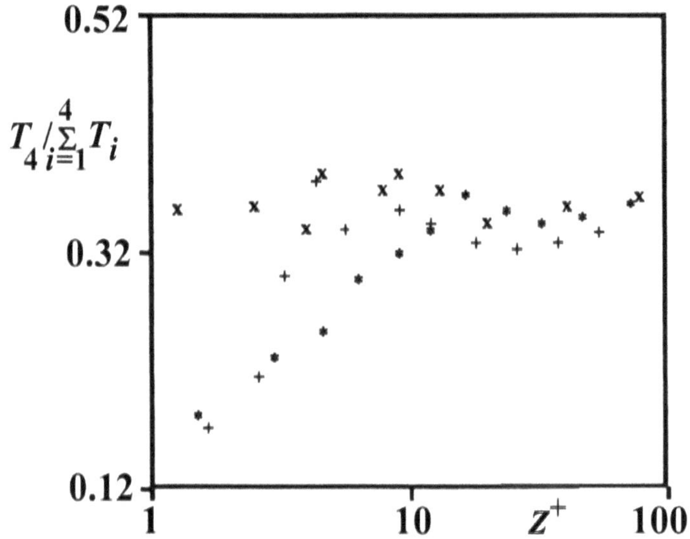

Fig. 9.6 The intermittency factor of the sweep events as a function of their wall distance. For a smooth wall at Re = 29200 (*), an orange-peel configuration at Re = 2 0400 (+), and a natural riblet configuration at Re = 27100 (x).

We can therefore argue that more but weaker sweep events are present, which supports the maximum friction factor theorem as mentioned in Chap. 4 (Sect. 4.2.2).

An indication for the behavior of such bed forms was already presented by Schlichting (1936), although Schlichting's investigations were performed using much rougher elements. He measured the energy dissipation of a bed of spheres of diameter d, which have been deposited at flat smooth surfaces at a distance s of the centers, by varying s. The highest energy dissipation he found for s was s = 1.4d, which is compatible with the distance of the highest elevation in an orange-peel configuration.

9.2 Sediment Stripes as a Bed Form

With increasing u_τ, a modified type of incipient motion develops from the same mechanism, also showing a striped pattern at first. Schmid (1995) measured the scales of the stripes just after they developed and found

$$\lambda_{xs}^{+} \approx 760 \quad \& \quad \lambda_{ys}^{+} \approx 90$$

These scales are very similar to the sweep structure dimensions known from turbulent boundary layer flow close to the wall, which also supports the model for incipient motion. The developing stripe pattern is reproduced in Fig. 9.7.

Reflecting to Fig. 9.6, also the transport was observed to be still intermittent with a constant intermittency factor throughout the boundary layer, the sweeps are essentially aligned in flow direction. In the beginning, this streaky structure is weakly stable, the crests of the depositions 'undulate' in spanwise direction being continuously reshaped. However, after some time they stabilize, canalizing the sweeps along the developing stripes. Now, the self-organization is subject to a feedback mechanism, where the instability is similar but not exactly the same as for the flat plate.

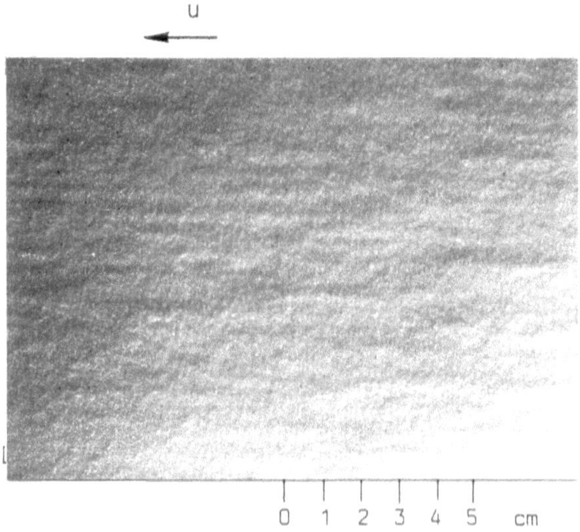

Fig. 9.7 The longitudinal sand-stripes of a bed formed by a flow of $u_\tau = 13$ mm/s, $\nu = 1$ mm²/s, $Re_k = 2.56$, $Fr_k = 0.052$.

The behavior of surfaces with a striped structure of the scales of coherent structures has been investigated exhaustively since they have drag-reducing properties, which we call riblets. However, the most efficient drag-reducing riblets have a much shorter spanwise interval (Hage et al., 2000; Bruse et al., 1999; and Bechert et al., 1997; and literature therein). It is noteworthy that the idea of a drag-reducing wall roughness was stimulated by the discovery of the coherent structures. Liu et al. (1957) speculated that a riblet structure of the same scale as the coherent structures should be drag reducing, and in fact Gyr (1999) showed that scales as we encounter them in the stripy bed form were drag reducing (Fig. 9.8).

The striped structure has many similarities with artificially produced riblet surfaces, and shall therefore be called natural riblets. Looking at the riblets theory of Luchini et al. (1991), one understands the canalization mechanism for the sweeps as follows. The crest elevation restrains the spanwise velocity fluctuations and therefore the flow instability occurs always in the valleys and has less effect in the lateral direction. Using the convection velocity of the sweeps and their lifetime, one can estimate an average length L_{st} of the stripes

$$L_{st} = u_c T_{sw} \tag{9.25}$$

As Fig. 9.8 reveals that it is a weak effect, nevertheless existent within a friction velocity range of 11.5 mm/s $\leq u_\tau \leq$ 12.5 mm/s. Although the effect is small, the result is important as it confirms the principal action mechanisms.

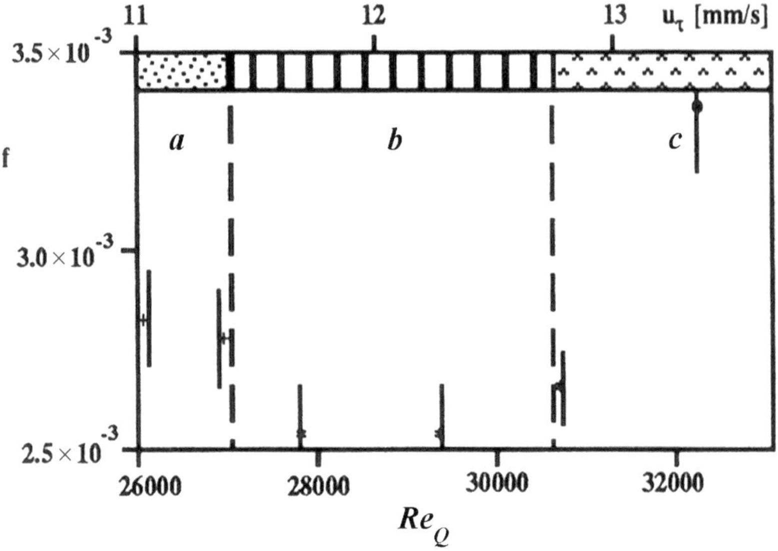

Fig. 9.8 The friction factor *f*, $f = 8u_\tau^2/u_Q^2$, as a function of *Re* for the three ranges: (a) an orange peel configuration, (b) natural riblets, and (c) an arrowlike pattern

9.3 The Arrowhead-like Bed Forms

The striped structure ends abruptly with further velocity since a single deposit in one of the valleys destabilizes the weakly stable coherent structures. After a very short time an arrowhead-like bed form develops (Fig. 9.9). Also the transport has changed now and became more continuous, however, still displaying intermittency. In this state, other flow structures besides the discussed coherent structures become more important.

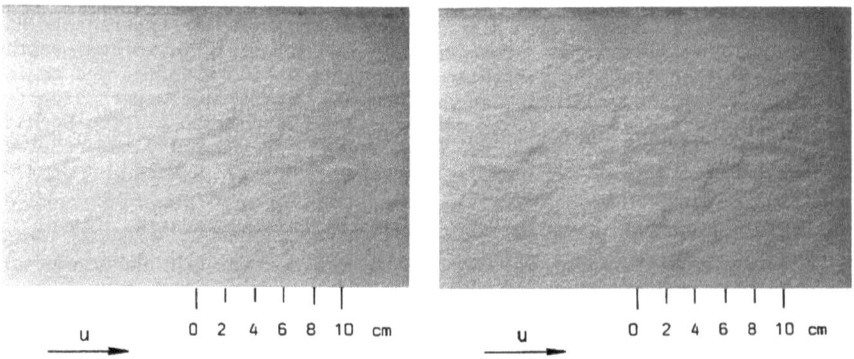

Fig. 9.9 Arrow-head like bed forms that developed at $u_\tau = 13$ mm/s, $\nu = 1$ mm^2/s, $Re_k = 2.58$, $Fr_k = 0.052$. First stadium at left and in the form they transform into pre-ripples at the right

This pattern was already found by Gyr and Müller (1975) and was interpreted as being formed by shear waves. The forms have the typical full aperture angle of 25°.

This reflects exactly the value found theoretically by the instability calculations of Benney and Lin (1960) and Lin and Benney (1961). The same result could also be derived from the instability theory of Landau (1944).

To rephrase, the bed forms follow the coherent structures as they develop with increasing load, and there is no doubt that they are subject to feedback from the accumulated incipient sediment transport on the instability processes of the viscous sublayer. However, the question remains whether this will be still the case with further increase of the flow.

9.4 The Ripple Formation

With further increase of the flow rate, ripples start developing. Their formation however need not follow the previously described pattern, but they can also develop suddenly when the flow rate increases abruptly. Ripples forming under those conditions are called spontaneous ripples.

The main property of the ripples is their periodic, 2D spanwise structure as they would be small dunes.

Since measurements show that the burst cycle still exists, one can ask whether it is the reason for the development of ripples. Additionally, one often assumes a second instability process for the flow and the sediment transport, which we will discuss at greater length within the context of dune formation. For the ripple formation, rather empirical descriptions were given in the literature as, for example, by Simons and Senturk (1977).

In this regime, the sediment transport is rather continuous and a layer of high-sediment concentration develops close to the wall. Liu (1957) and Hardtke (1979) proposed the idea of a developing Kelvin-Helmholz instability at its outer edge, reflected in the erosion as well as in the deposition process. However, such instability should have been measured, which has not happened until now. Such a model would also need to adapt a new rheology. Liu and also Stehr (1975) supplemented this concept by postulating that the ripple formation is due to the laminar-turbulent transition of the viscous sublayer. However this layer was never laminar, even before ripples form, and also the transition frequency cannot be found in the developing bed form.

The instability method used to define the laminar-turbulent transition was now also applied to investigate the ripple formation. One calculated the growth rate of an infinitesimally small 2D wave disturbance, however supplemented by a transport equation for the slow changes of the bed configuration by the transport process. It is common practice to use a most simple transport equation for this purpose, such as the Exner relation

$$\frac{\partial Q_{\mathrm{s}}}{\partial x} = (1-n)\frac{\partial h}{\partial t} \tag{9.4}$$

This relation represents a pure mass conservation, Exner (1925), where h is the bed elevation and n the porosity of the sediment. The results of this instability calculations are still used (Kennedy, 1969; Richards, 1980; and Engelund and Fredsoe (1982). Such a description is a posteriori since the number of parameters involved always allows by a fitting process to identify the found periodicity. Only Kennedy tried to tie the instability to a water surface interaction. However, since the ripples develop independent of the water depth, a different mechanism must exist.

In Sect. 9.3, we found that the ripples develop from an arrowhead pattern and therefore the initial assumption of a 2D disturbance must be wrong. We encounter the very difficult problem of how a 2D pattern develops from a 3D perturbation. Raudkiwi (1963) postulated a selective redeposition of the sand, where the transport was governed by shockwave-like effects initiated by the lee area of the upstream grains. He found that this rheological approach, which was very similar to the Kelvin–Helmholz instability, was not sufficient. He therefore argued that the separation process controls the equilibrium state given by a width–length relation. However, he did not compile his ideas in the formulation of a mathematical theory. Raudkivi (1982) still proposed in his book that turbulence plays a minor role in the ripple development. Grass (1970, 1971) came to a similar conclusion using the topological results found by Allen (1971), who investigated the development of ripple-like structures on a plate of gypsum exposed to a flow. This experiment cannot be applied to a feedback system of an erodible bed described here; however, it gives a good perception of the topology involved.

9.4.1. The Separation Concept

Williams and Kempt (1971) submitted the idea that ripples form from a striped bed structure, when the crest height is about four viscous units leading to separation. They analyzed this separation to initiate a reproduction mechanism based on the depression topology of Allen (1971). The authors also assume continuous sediment transport with an overlapping factor of $n = 0.2$, Eq. 3.43, as defined by Grass. This was the first attempt of combining a separation concept and statistical elements of turbulence.

Jackson (1995) gives a ripples classification and confirms Schmid's result (1985) that ripples can appear only for

$$k_s^+ \leq 13 \quad \rightarrow \quad k_s^+ \leq 12.5\,(\text{Schmid}, 1985) \tag{9.5}$$

We now return to the original question of the development of the 2D structure. Hardtke (1979) argues a lateral coupling of the separation vortices to be responsible for the 2D organization. This is only half the truth, and we will analyze this mechanism further.

The sharp criterion in Eq. 9.5 itself is a first hint that a special mechanism is involved in the ripple formation. Let us recapitulate: we started from a hydraulically smooth wall defined by Eq. 2.42a. The hydraulic surface roughness is not affected by the redistribution of sand grains when $k_s^+ < 5$. The developing bed configuration as such is a new roughness, especially when separation occurs at the crest. Then the momentum exchange of the flow with the surface results from pressure differences over the obstacles. This concept was applied to postulate a triple decomposition of the wall shear stress (Eq. 3.137). The transition between a viscous dominated friction and a sand-roughness occurs at $k_s^+ \approx 5$, so what is the meaning of the criterion $k_s^+ < 13$, Eq. 9.5?

For a bed of this roughness, the coherent structures still exist to a certain extent but are somewhat altered from those described in Chap.6 (Sect. 6.2.5). While criterion Eq. 9.5 is reached without major changes in the coherent structures, we assume that these are not contributory. Williams and Kempt (1971) thought in these terms when they postulated that a higher roughness depresses the stripes formation, which is a necessary step in the ripples development. In contrast to this, Gyr (2003) postulated that the criterion should discern between viscous and turbulent dominated separations. Doing so allows ripples to form as long as the separation shear layer is viscous dominated in contrast to the turbulent separation typical for dune formation. The topological distinction of the two different types of separation is shown in Figs. 8.7 and 8.8.

9.4.2 The Relation Between Ripples and Coherent Structures

It is known from experiments that ripples develop when the transport close to the wall is high enough, criterion Eq. 9.5 is fulfilled and separations on preripple forms occur. In addition to these three basic suppositions the experiments show two mechanism of ripple formation. The first one as indicated in Sects. 9.1 and 9.2 where the ripples develop slowly from the arrowlike pattern as the velocity of the flow is increased. The second mechanism is spontaneous, when the flow was increased very rapidly. Also here, one observes the formation of short-lived striped bed form. Then later in time, diamond shaped pattern form like the arrow pattern, however only on a larger scale. The investigations of Williams and Kemp (1971) and Schmid (1985) dealt with the first mechanism, whereas the investigation of Hardtke (1979) focused on the spontaneous case. Raudkivi (1982) used Hardtke's model, however, giving a wrong topology for the vortex tubes field.

Over slight elevations ($h^+ \approx 2k_s^+ \approx 6$–$8$), the sweeps deviate laterally and the flow can separate on their lee side leading to scouring and thus enhancing the separation process. The separations form essentially the same way over the diamond shaped pattern. The individual ripples lose their identity when they join and build up essentially 2D laterally extending ripples. Once this ripple form has built up, the transport decreases, although the wall shear stress is larger now. Therefore, with a larger amount of sediment in suspension, the decrease in transport must be due to a decrease in the bed load. It reveals itself in the slow propagation of the ripple sediment body itself. In other words, the form and the propagation velocity of the ripples, v_{ri}, describes a major part of the sediment transport mechanism. This velocity is an order of magnitude smaller than the flow velocity.

$$v_{\mathrm{ri}} = \frac{q_{\mathrm{s}}}{hg\left(\rho_{\mathrm{s}} - \rho_{\mathrm{f}}\right)\left(1-n\right)} \tag{9.6}$$

The transport is still highly correlated with the sweep events and the separation is more an interruption than a transport mechanism, therefore the sweep scale should show up in the scale of the ripples. In fact there is a high correlation between the two length sales (Table 9.1).

Table 9.1 Comparison between the length scales of the burst events and the preforms of the ripples

Length scale	Burst	Pre-ripples
λ_x^+	810	835
λ_y^+	96.5	94.5

The length scale in x direction remains also after the 2D forms have developed, and we assume the coherent structures to be still the main transport mechanism. This view is also supported by the fact that the ripples only exist until bed forms completely interrupt the sediment transport. However, as soon as grains are transported over a larger distance due to an outer noncoherent structure, the ripples disappear. This introduces the local intermittency as another variable of the model and reminds us that several theories are based uniquely on this coherent structure argument. Instead of deriving the periodicity by a rather problematic stability theory, one now can predict the periodicity via the coherent structures. The connection to the scales is given by

$$13.1 \leq v_{\text{rix}}^{+} = \frac{\overline{\lambda_x^{+}}}{T_{\text{P}}^{+}} \leq 14.4 \tag{9.7}$$

The scales and transport mechanism are reproduced in Fig. 9.10. The quasiperiodicity of the sweeps reflects on the bed forms because they find an anchor on the ripples crests, and thus impose their length scale. So the "sweep-length" becomes the wavelength of the ripples, Gyr and Müller (1996).

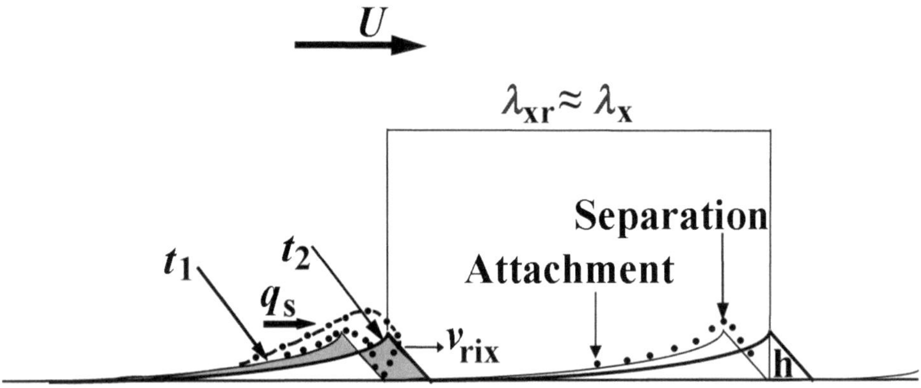

Fig. 9.10 2D representation of the ripple-propagation over a time interval. Two grain trajectories on the left side show grain transport and deposition in the separation zone due to the recirculation flow field existing therein. On the right, the flow attaches to the bed and develops a new boundary layer separating at the crest. The flow attachment point is the location of highest shear and where grains start to move first. Shown is the coincidence of the two length scales. The shape moves with the velocity given by Eq. 9.7

9.4.3 The Ripple Formation Mechanism

The ripple formation is initiated by feedback mechanisms, which also support the self-organization to its final stable appearance (Gyr, 2003). These mechanisms are reproduced in a series of figures coupling to the separation processes, helping in the understanding of self-organization.

The first bed form develops from the arrowhead configuration as soon as the flow detaches at the lee side of the elevation, producing a lateral vortex (Fig. 9.11). This vortex does not only scour through its low-pressure field but also forces an interruption in the transport. Since the arrowhead pattern developed due to the coherent structures,

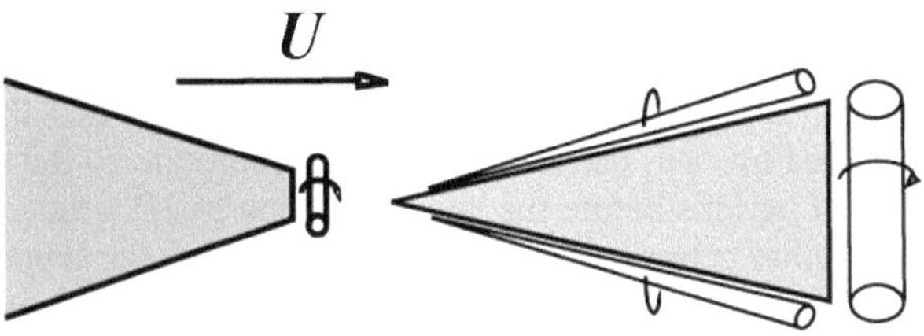

Fig. 9.11 The arrowhead pattern develops a separation at its arrow pointed end, where the lee vortex develops scouring and growing laterally producing a pre-ripple form mirror imaging of the original bed form (Fig. 9.9)

These forms propagate due to the sediment transport as it is steered by the topology of the flow field; this is shown in Fig. 9.12.

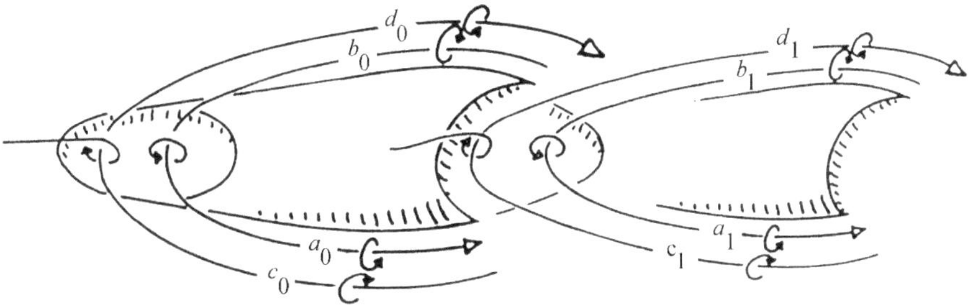

Fig. 9.12 The pre-form of a ripple propagates due to the induced flow field characterized in topology by Allen (1971) for a depression, shown here by the scour of the lee side vortex tube

This flow is very similar to the flow around a flush mounted half sphere shown in Fig. 9.13, which is a good representative for a single grain.

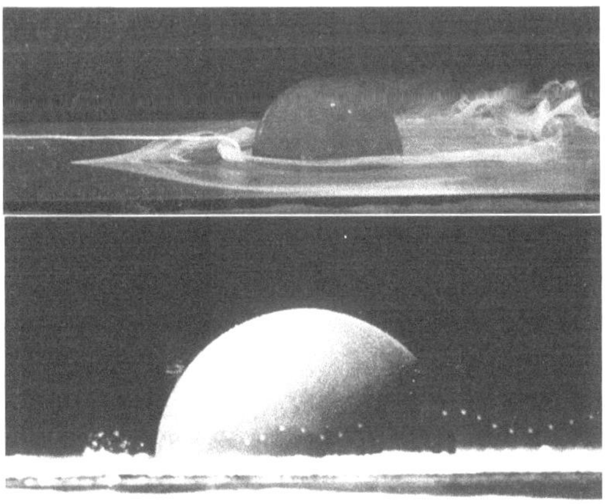

Fig. 9.13 The flow around an embedded half sphere. The first image shows the horseshoe-like vortex system by injection of color to the boundary layer very close to the wall shortly upstream. The second picture shows particle moving in such a field

its length scale already correlates with the transport intermittency. Such developing bed forms would not only reproduce but would also grow in width and ultimately become 2D. This is Hardtke's (1979) model, who assumed additionally that the crest vortices would coalesce. Nature however, takes another way in reproducing the sediment bodies over quite a number of periods before the lateral merging to 2D structures occurs. This behavior reflects the rather complex flow and its separations at these pre-forms.

Schmid (1985) reproduced naturally developing pre-forms by fine tooling and casting so that these forms could be investigated in further detail; the representative form in Fig. 9.14 has the relevant dimensions given in millimeter.

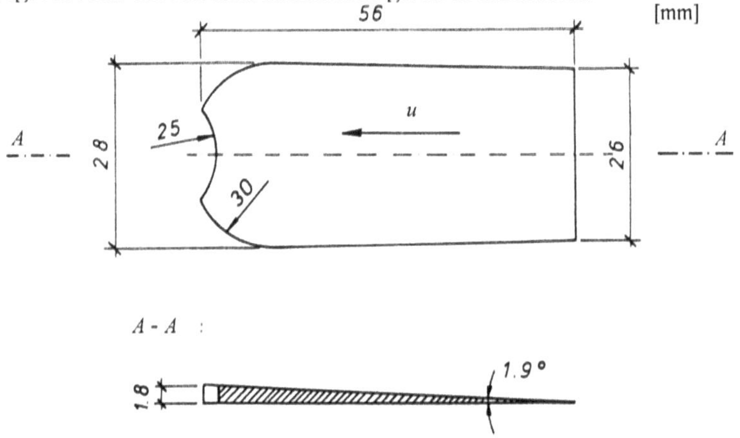

Fig. 9.14 The cast of a naturally developed pre-ripple form as used for flow measurements

We first reproduce the evaluated flow topology around a pre-ripple in Fig. 9.15 to gain an overview of the separation mechanism.

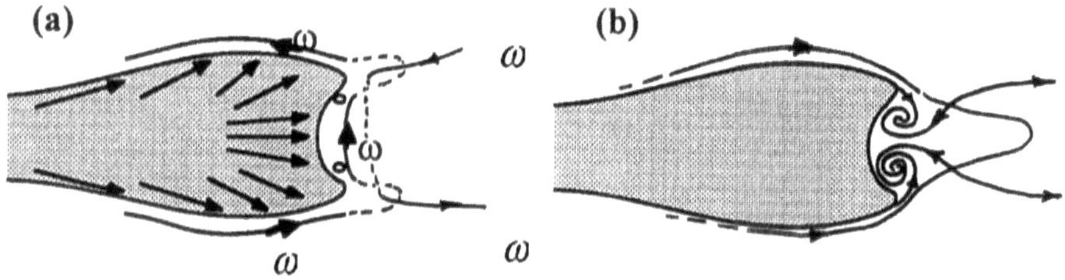

Fig. 9.15 The topology of the flow around a pre-ripple as shown in Fig. 9.14. (a) Shows the main vortex skeleton of the flow, and (b) the pattern of the wall-streamlines showing the typical "owl face"

After passing over the pre-ripple, the flow separates at the lee edge producing the crest vortex embedded in and stabilized by the open separation bubble. Because of the obstruction of the sediment body, the fluid also displaces laterally increasing the strength of the side vortices with increasing elevation angle. This part of the vortex skeleton is highly dependent on the mean shear. The side vortices become strongly helical and tend to be transported in flow direction. However, originally they were connected to the lee vortex. This connection must break producing a rather complex vortex system. In this situation, the flow also separates at the rear vertical edges of the body producing two vertical vortices, best seen in the wall-stream lines as in Fig. 9.16 b. The lee vortex and the side vortices are of opposite signs in the convective zones just downstream the pre-ripple.

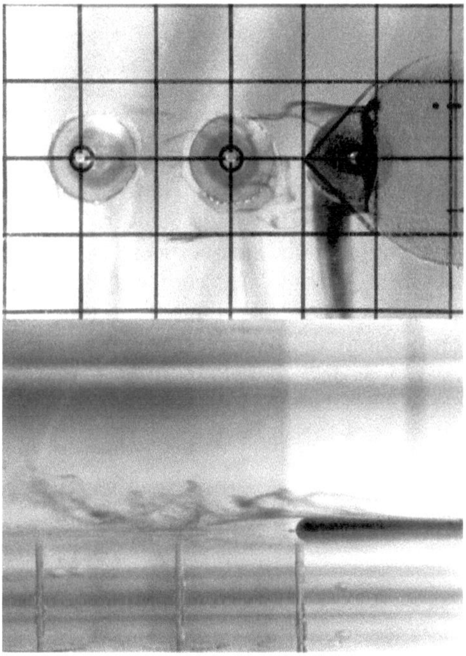

Fig. 9.16 Visualization of the flow over a pre-ripple as shown in Fig. 9.14. Flow is from left to right respectively just reversed. The left image shows the vortex skeleton in lee of the form. The right image shows the open separation bubble with turbulent outside flow. The grid mesh has 10 mm spacing

Under these conditions, the stronger vortex usually absorbs the weaker one, while the remaining vortex has the sign of the stronger element usually being the side vortices. For this reason, these pre-forms propagate over long distances having a narrow width compared to their length, showing a half moon shape elevation in the upstream half and a similar sized cavity in the downstream half of the ripple. The visualization confirming this viewpoint is given in Fig. 9.16.

With this process in mind we can now explain the development of 2D ripples. This is the case when neighboring forms start to influence each other. The process is shown in Fig. 9.17.

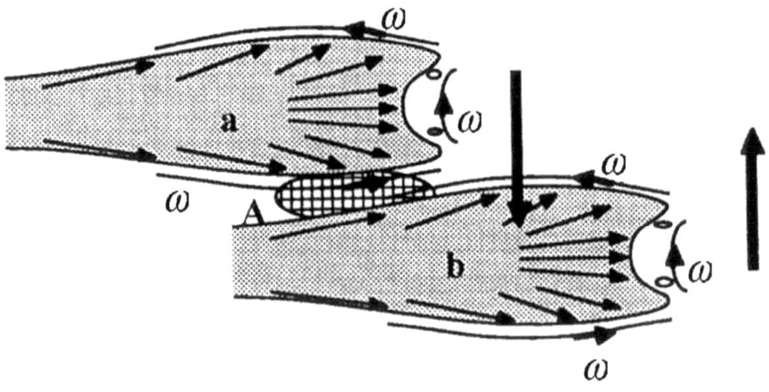

Fig. 9.17 A sketch of neighboring pre-ripples beginning to interact resulting in a 2D ripple. The side vortices annihilate in between the forms strengthening the remaining separation vortex

As soon as two pre-ripple forms approach laterally, their interacting side vortex system weakens by the diffusive annihilation. The remaining vorticity is weakened, since the interacting vortices have opposite signs. This happens in the crosshatched area labeled A in Fig. 9.17. A damping of the side vortices disturbs the equilibrium between the circulation of the lee vortex and the side vortices. The lee vortex starts spreading laterally, since now its transport is less limited in this direction. This leads to coalescence of neighboring forms to 2D ripples, which can be nicely observed in the experiment (Fig. 9.18).

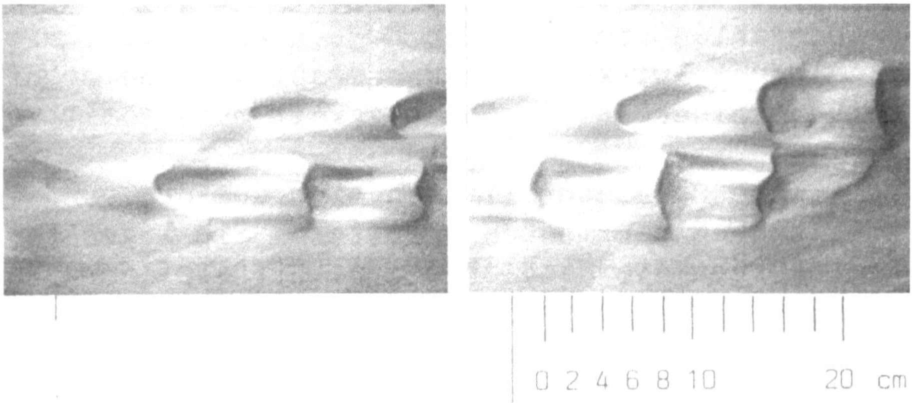

Fig. 9.18 The 2D merging of neighbored pre-ripple forms to ripples by the interaction of the vorticity between them. A time sequence

However these 2D bed forms are not stable, they again disintegrate in 3D forms because the topology of the open separation bubble starts dominating the local flow field. Most of the rolling up lee-vortex vorticity is transported out of the separation bubble via the spanwise helical flow. With increasing ripple width, this helical flow becomes stronger since vorticity and mass are fed to the bubble over larger area. In a channel flow, the lateral growth ends at the sidewalls where the flow field reorients in flow direction. This intense vortex tube erodes sediment along the sidewalls, an effect observed as scolding of the side borders from the redirected separating vortex. At a certain aquired ripple width, the lateral transport cannot rid the added vorticity and the vortex tube becomes instable. We can treat the instability of such vortex tubes in different ways, commonly one investigates the wavelike deformation of the vortex core. The result is a wavy pattern of the crest. An example is given in Fig. 9.19.

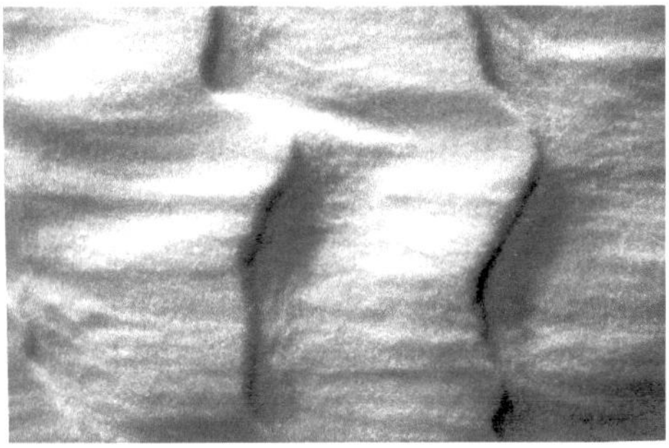

Fig. 9.19 A spontaneously produced ripple and its transition to a 3D form. The longitudinal stripes of sweep transports are visible on the ripples surface. A wavier ridge forms from the straighter 2D ripple. The image shows an area of 145×100 mm after 125 s of established flow from a flat bed at u_τ = 16.8 mm/s, $\nu = 10^2$ mm^2/s, $Re_k = 3.53$ and $Fr_k = 0.087$

For a simulation during the ripple development using Biot-Savart's law, an estimate for the circulations of the vortex skeleton need to be known.

Here in Fig. 9.20, we show the results measured in the symmetry plane for the pre-ripples as shown in Fig. 9.14. As the dimensions indicate, it is a very flat sand body, twice as long as wide, and exhibiting in a half-moon ridge on the lee side. The open separation bubble feeds through a thin part $\hat{\delta}_0^+$ of the boundary layer flow containing most of the vorticity. In our case, $\hat{\delta}_0^+ = 8$. The amount of fluid Q^+, to be transported away laterally by the helical lee side vortex is best estimated from the inflow. Evaluating at the cross-section α after the step, we found $Q^+ \approx 2600$.

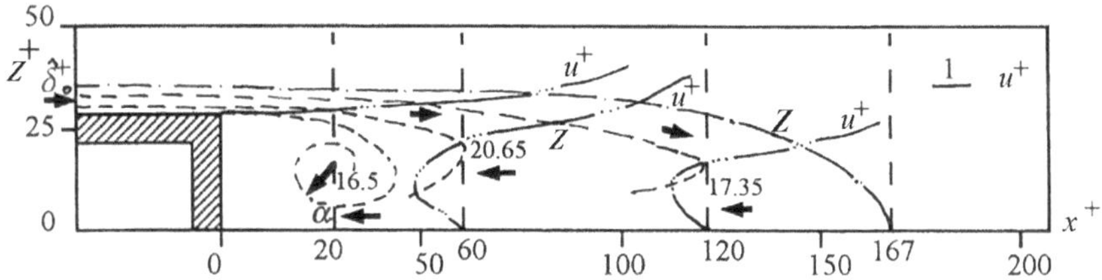

Fig. 9.20 The velocity profiles and maximum velocity vectors in the central plane behind the pre-ripple form of Fig. 9.14. The separatrix is sketched as dash-point line

Dividing this estimate for the discharge by the entire bubble volume of $V^+_B \approx 15\times10^4$ yields the average residence time of the fluid in the separation bubble. We estimated $t^+ \approx 60$, a rather long time. The measured values for the circulations around the three main vortex tubes are for the lee vortex, the side vortices, and the vertical vortex

$\Gamma^{+} \approx 550,\ 2 \times 550,\ 110$ respectively. In our case, it is evident that the lee vortex is in equilibrium with the side vortices, which has been postulated for a stable pre-form as the one investigated was the cast of a naturally formed stable body.

9.4.4 Alternative Concepts of a Ripple Formation

Although we are convinced that the mechanism described in Sect. 9.4.3 is correct, we will supplement the discussion with an alternative concept first introduced in leading paper of Bagnold (1941). It is probably more important for the ripple and dune formation in air instead of water. His concept was based on the argument that a suspended grain accelerates during its flight time while removing momentum and energy from the flow. When such a grain later hits the bed, it transfers enough momentum to suspend another grain. If the momentum of the suspended grain is high enough, a chain reaction is possible. Anderson (1987, 1990) underlined these ideas with a theory based on a separation of the different transport processes, where a part of the suspended grains had received much of energy by the bombardment whereas the second category of grains has just enough energy to roll on the surface. The first category of grains holds the process alive. There is little interaction between the two populations and therefore the ripple formation is due to the local differences in the transport regimes due to the second mechanism.

The periodicity correlates with the transport length of the bed-load transport and not with the jump distances as assumed by Bagnold. The linear stability theory predicts a wavelength of about six times the mean jump distance of the low-energy particles. Andreotti et al. (2002) observed that those grains transported by the so-called "reptating" motion of the particles move about a distance of $100d_s$ in one jump and further continue to roll on the sand surface after landing for a few times this distance in accordance with the linear theory but for the wrong reason.

Yizhaq et al. (2003) expanded this investigation using non-linear stability theories and, proceeding from Exner's Eq. 9.4, found good agreements with measured Aeolian ripples.

Bagnold's treatment was also used for the explanation of dune development and made the stability theory very popular. As we have seen for the ripple formation in water, the observations contradict such a model, and one has to ask whether the forms develop differently in water and air. Both fluids being Newtonian, the forms should be similar for a given Reynolds number, if they are due to turbulent flow structures. However, this comparison is not simple since the momentum transfer mechanisms differ strongly in both cases. For air flow, the momentum transferred through the grain impact on the bed becomes dominant. Therefore, in air the saltation mechanism was used to model the transport (Anderson, 1987, 1990), where turbulence is used as a nonstructured phenomenon.

Lacking a well-defined physical concept, an empirical description was tried for the sediment transport using mean values found as average over the wavelength. This averaging smoothed out the local velocity variations resulting in an unrealistic profile. The approximation given in the literature is

$$\frac{\overline{u}}{u_\tau} = 2.5 \ln 11 \frac{h}{k_s} \qquad (9.8)$$

where h is the water depth above the highest elevation, and the grain roughness k_s is equivalent to the height of the transported bodies. Besides the local velocity variation, also the suspension load and the transport regime of the sediment bodies are of major importance, which shows how problematic such assessments are. One can adjust the two parameters k_s and κ (Zanke, 1982), however, these prediction are unsatisfactory. Earlier descriptions did not distinguish between the grain sizes in a physical sense, and therefore the empirical formulations missed the essential differences.

Without a clear physical concept, one cannot distinguish between ripple and dunes and therefore no classification is possible. This gap was bridged by empirical formulations or/and definitions. Using the stream power Eq. 4.7 or 4.10, Simons et al. (1964) plotted

$$P \propto \tau_w U \tag{9.9}$$

the stream power per unit area as a function of the grain size d_s. For a quartz sand of 2.8 mm grain size, they found an abrupt break in the slope at the value $P = 1.0$ Wm^2, which he defined as the criterion distinguishing ripples from dunes.

Instead of d_s, Hill et al. (1967) used, d_s^+ and used a criterion function

$$g\frac{\overline{d_s^3}}{\nu^2} = f_r \tag{9.10}$$

Using a grain size distribution they found the transition between ripples and dunes again by a sudden change in slope at $\overline{d_s^+} \approx 9.5$ for $f_r \approx 200$ by plotting f_r versus $u_\tau \overline{d_s} / \nu$, Znamenskaya (1962, 1969) plotted the Froude number over the ratio of the stream and settling velocities and mapped the different areas empirically to the diverse bed forms. Use of the settling velocity introduced a new physical idea, which is relevant for extremely fine sand where most is transported as suspension.

To summarize, the different authors used the following models to deduce the transitional grain diameter separating ripples from dunes.

Bogardi (1965, 1974) used

$$gd_s / u_\tau^2 \Leftrightarrow d_s$$

$$u_\tau \left[1.65 g \rho_f \left(\rho_s - \rho_f\right)\right]^{1/2} \Leftrightarrow d_s \tag{9.11}$$

Liu (1957) used

$$u_\tau / u_v \Leftrightarrow d_s^+ \tag{9.12}$$

Chabert and Chauvin (1963) used

$$\mathrm{Fr}^* \Leftrightarrow Re^* = d_s^+ \tag{9.13}$$

van Rjin (1984) used

$$\mathrm{T} = \tau_w / \tau_{wc} - 1 \Leftrightarrow d_s^+ \tag{9.14}$$

Chabert and Chauvin tried using Shields' representation and found the limit for the ripple forms in agreement with Eq. 9.5. van Rjin required two criteria together, where $d_s^+ < 10$ and $T < 3$ resulted in moderate agreement. A series of additional empirical criteria was published, however, with diverging results. Nevertheless, these empirical theories are valuable, since often they are based on grain distributions not fulfilling our criterion Eq. 2.34, but being within the limits of Eq. 9.5.

Kennedy (1969), who first formulated bed forms theoretically, acknowledged that his theory couldn't distinguish between ripple and dunes. He agreed that they are quite different and thus must be of different origin. The concept of coherent structures can explain the difference, which is a major advancement for the new theory based on turbulence and topology.

9.4.5 The Flow Resistance of Ripples

With the ripples established as the new bed form also the drag has changed compared to the flat bed or pre-ripple shapes. In the classical procedure, the drag was derived from the wavelength averaged velocity profile. This approach was consistent with the separation of the wall shear stress components as shown in Eq. 3.137. Several authors found this to be problematic in case of fine sand and added a fictitious value τ_{su} for the energy, which is used by the suspended material, and Eq. 3.137 can be written as

$$\tau_w = \tau_v + \tau_r + \tau_f + \tau_{su} \approx \tau_w = \tau_r + \tau_f \tag{9.15}$$

The contribution from viscosity and from the suspensions can be neglected. Together with the ridge, the adjoining separatrix locally subdivides the bed into two parts. An active part of the bed begins on the upstream face at the attachment separatrix and ends at the ridge. A second, more passive part follows from the ridge to the attachment separatrix of the next ripple and acts mainly as a form roughness by its pressure distribution. Within the separation bubble, we also have a recirculation drag, however, this contribution is smaller by an order of magnitude. Neglecting this part, the energy dissipation relation averaged over one wavelength is

$$\frac{\overline{u}^2}{u_\tau^2} \approx \frac{\overline{u}^2}{u_{\tau,r}^2} + \frac{\overline{u}^2}{u_{\tau,f}^2} + \frac{\overline{u}^2}{u_{\tau,su}^2} \tag{9.16}$$

or in form of an energy slope I,

$$\tau_w = \rho_f g H I \quad \therefore \quad I \approx I_r + I_f + I_{su} \tag{9.17}$$

This representation also lacks physical reasoning. The geometry is given by a relative slope σ and a relative ripple height h_{rel}.

$$\sigma_r = \frac{h}{\lambda_r} \quad \text{and} \quad h_{rel} = \frac{h}{H} \tag{9.18}$$

The relative slope σ_r is well approximated by τ_w/τ_{wc}, which varies considerably since the ripple periodicity depends also on the grain distribution and the boundary layer flow.

$$d_s^+ < 4 \quad \Rightarrow \quad \frac{\lambda}{d_s} \approx ¢ / d_s^+ \tag{9.19}$$

Yalin (1979) set the constant equal to 2250, but other experiments showed that its value varies between 1000 and 2500 (van Rjin, 1984).

Engelund (1996b) evaluated the ratio of the total wall shear stress to the portion due to the sand roughness. He used the Shields parameter and calculated the relative energy slope I_f due to the separation zone via an energy dissipation using the Carnot equation applied to a Bernoulli equation. Starting with the Bernoulli equation

$$\frac{1}{g}\int_1^2 \frac{\partial u}{\partial t}\mathrm{d}s + \frac{u_2^2 - u_1^2}{2g} + (z_2 - z_1) + \frac{p_2 - p_1}{\rho_f g} = 0$$

$$\wedge \quad H = \frac{u^2}{2g} + z + \frac{p}{\rho} = ¢ \tag{9.20}$$

$$\wedge \quad \mathrm{d}H = udu + gdz + \frac{\mathrm{d}p}{\rho_f} = 0$$

where z_1 and z_2 are the highest and the lowest elevation of the bed forms, respectively. The Carnot equation for the expansion loss is

$$\Delta H_f = \frac{C_E (u_1 - u_2)^2}{2g} \tag{9.21}$$

where C_E is the expansion loss coefficient. Using the discharge density q_f Eq. 9.21 can be rewritten as

$$\Delta H_f = \frac{C_E}{2g}\left[\frac{2q_f}{2H - h} - \frac{2q_f}{2H + h}\right]^2 \cong C_E \frac{u^2}{2g}\left(\frac{h}{H}\right)^2 \tag{9.22}$$

$$\vee \quad I_f = \frac{\Delta H_f}{\lambda} \cong \frac{C_E}{2}\frac{h^2}{\lambda H}\mathrm{Fr}^2$$

In this representation, the energy slope depends on the relative slope and height (Eq. 9.18) of the sediment body, the square of the Froude number and C_E needed to be evaluated. By measuring pressure in the separation bubble, C_E could be estimated. But if the separation bubble can be evaluated as shown in Figs. 9.16 and 9.22, it is advised to integrate Bernoulli's equation directly along a streamline on the separatrix plane.

9.4.6 The Importance of Ripple Formations

Before we close this chapter on ripples, we will shortly summarize. We do this, although these forms are not really important, since they are the only bedforms that can be completely understood via turbulence and the viscous dominated separation. They are prototypes for explanations going beyond such assumptions and are good conceptual configurations to which one can tie new ideas using similar topological concepts. In nature, ripples are rare and if they exist, they are commonly found as substructures on the much larger dunes.

In an established ripple formation, the lowest pressure occurs just beneath the lee vortex in the separated region, where wall shear stress is directed upward the rear face of the ripple. This can cause a sediment transport in this direction and it is certain that the lee vortex causes the fairly steep grade of the lee faces. Usually, τ_w is small on the lee side of the ridge and therefore the rear slope of the sediment body adjusts itself to the angle of repose Φ, Chap. 1 (Sect. 1.2.2.2). This behavior already shows that averaging over a ripples' length is not very meaningful and that the separation of the wall shear stress contributions Eq. 9.15 is problematic.

The pressure field contribution that keeps the flow in equilibrium cannot be calculated by using Bernoulli's equation, since it would require integration along both streamlines acting as the inner and outer separatix (Fig. 9.20). However, the pressure drop can be estimated well using only the outer separatrix but the velocity on this separatrix must be known.

One very important result of this investigation was the explanation for the vorticity field self-organization as shown by the vortex skeleton of the separating flow.

9.5 Dunes of Fine Sand

Ripples are the result of viscous dominated separations of the open bubble form. Once established, further increase of the flow will deform ripple (Fig. 9.19) and the new intermediate 3D configurations again organize to larger 2D forms called dunes. They appear when the open separation bubble becomes turbulent one as shown in Fig. 8.8. The geometric scaling of the form and periodicity are the result of turbulent separations and not anymore due to the instability process of the boundary layer. Bent and Best (1996) described these differences, characterized the flow parameters, and provided a good literature summary, however, without being able to give a consistent distinction between the two states, which is possible only with topological concepts.

One of the important mechanisms is the interaction of the separation with the outer flow. For ripples, we found the local pressure produced by the quasi-stationary separation bubble to be significant. The pressure field decays by the reciprocal of r, the distance from the dynamical source of the pressure, as mentioned in Chap. 5 (Sect. 5.2.1). Bennet and Best postulated, therefore, that a bed of ripples could influence a zone of $z/H \leq 0.3$ only, which is a good approximation for rivers with ripples, however not for sea floors for example. At dune ridges, the flow separates as a turbulent mixing layer, which penetrates much further into the outer flow (Müller and Gyr, 1983, 1986). The separation is unstable in the sense of a Kelvin–Helmholtz instability as mentioned by Müller and Gyr and confirmed by Nezu and Nakagawa (1993), Bennet and Best (1995), and Baas and Best (2002), these instabilities are leading to vortices, which grow and amalgamate while convected away from the separation edge. Therefore, wavy vortex tubes in spanwise direction, which behave as described in Chap. 8 (Sect. 8.4), characterize the separation, where it is also explained how boils and the scouring is produced (Fig. 8.11 d and e). Decisive for this separation mechanism is that at least in rivers the flow over the whole depth becomes influenced. Especially, also the water surface starts to interact, in other words, surface waves and separation mechanisms become coupled, and it is hard to say what is the origin of the later on forming of dunes. The waves can be the initial disturbances, if the pressure field of the waves can influence the entire flow, however, if the dunes origin from the smaller bed forms, it is rather so that the waves become forced by the separations. But one thing is clear; as soon as they interact, a feedback mechanism will hold the process in a quasi- stationary state to which more or less stationary dunes belong, which move by sediment transport.

9.5.1 The Feedback Mechanisms

The feedback mechanisms between the sediment transport and the flow is essentially the same as for the ripples. The sediment transport produces a sediment body sitting on the bed and producing a separation of the flow, which is for the sediment transport a singularity and the transport becomes discontinuous, the result is a periodic bed form that propagates. This is the essence of the feedback system which in detail is however, much more complicate, and we will now try to explain it.

Let us start on the ridge of the dune where the flow separates. The criterion for a separation was given in Chap. 8 (Sect. 8.1) and as can be seen from Table 8.1 or Fig. 8.1 be given by

$$\tau_{w} = 0 \tag{9.23}$$

In other words, the wall near flow changes its sign; the downward flow becomes negative that means the flow is directed versus the crest. However, since $\left|-\tau_{w}\right| < \tau_{wc}$, practically no sediment transport is involved with this flow. On the other hand, the sediment transported toward the ridge by the stream upward flow starts to sediment into the separation can be taken in the mean bubble where it settles, however, not by forming a natural angle of repose because the fall out is more like a rain. The lee-side slope of the dunes, however, is one of the natural angle of repose, since it forms intermittently by an avalanching process of bed material (Ha and Chough, 2003).

In addition, the vortices participating in the instability process of the mixing layer can influence the settling process selectively. The sediment becomes collected in clusters and is falling out as a cloud or as a veil (Meiburg, 2003; Meiburg et al., 2000; Wallner and Meiburg, 2002; Martin and Meiburg, 1994). So with the proper definition for the mixing layer, the sediment transport can be simulated.

These processes within the separating vortices are similar to those already mentioned for the interaction of a turbulent suspension flow for which the flow structures and the grains are much smaller and therefore more dynamic. We recommend the following list of references for further study; Squires and Eaton (1990), Elghobashi and Truesdell (1993), Truesdell and Elghobashi (1994), Maxey et al. (1997), and Sundaram and Collins (1999). The influence of the wall near turbulence on suspensions was treated especially well by Pan and Banerjee (1996), or the influence of heavy particles on a fully developed channel flow by Kulik et al. (1994) and Kiger et al. (2003). Druzhinin (1995) made a series of investigations to further understand the interaction mechanisms by observing the suspension behavior at stagnation points, areas of constant vorticity or, e.g., in Stuart vortices. One of his results was that for small Stokes numbers, the two-way coupling reduced the vorticity in the center of the vortex, whereas the strain in a hyperbolic stagnation flow increased. Since all real processes are nonstationary, the investigations were extended in this direction (Saffman, 1961; Chen and Chung, 1995). Dimas and Kiger (1998) extended the linear stability investigation and found that an increasing Stokes' number for a given mass concentration damped the instability, however, for special relations of Stokes' number to mass concentration secondary instabilities also occurred in the low-frequency range.

The large number of references given shows the complexity and difficulty involved describing only the transport to and the sedimentation within the separation bubble. For those who would like to simulate these processes, this literature will be helpful. In general, one can say that the sediment settles with a smaller slope than the natural angle of repose.

Now let's discuss the attachment area as seen in Fig. 8.8. The mixing layer ends up with the outer separatrix attached to the bed and having the topology of an open separation bubble with many internal nodes and saddles. The vortex skeleton is just as complex since some of the lateral vortices are already paired and have deformed cores and continue to form vortex loops, which are so important for the further development of the structures. As we know the forward directed loops rise whereas the backward directed ones descend, the open bubble becomes highly 3D deformed, especially at location where a loop approaches the bed (Fig. 8.12) or touches the bed or the surface. At such location the vortex core opens and ends up in tornado-like attached vortices as described in Chap. 8 (Sect. 8.4). In Fig. 8.13, it is also shown how the vorticity at least can be estimated. In other words, we have the tools to simulate the mechanism and therewith the sediment transport, although the mechanics is so complicated that it will be impossible to describe it in an analytical form.

The main conclusion is the important result of a highly fluctuating attachment location, which is essential also for the sediment transport because the attachment location separates the processes. We have no other choice than time averaging, which is problematic because an attachment point far downward is coupled with a low resuspension. An averaging procedure needs to include all these correlations. For practical purposes, however, a much simpler averaging method is used. It is geometrically possible since the mixing layer thickness increases linearly with distance from its origin, see Landau and Lifschitz (1991). The result is a conical linearly thickening layer containing the large vortex instabilities as sketched in Fig. 9.21.

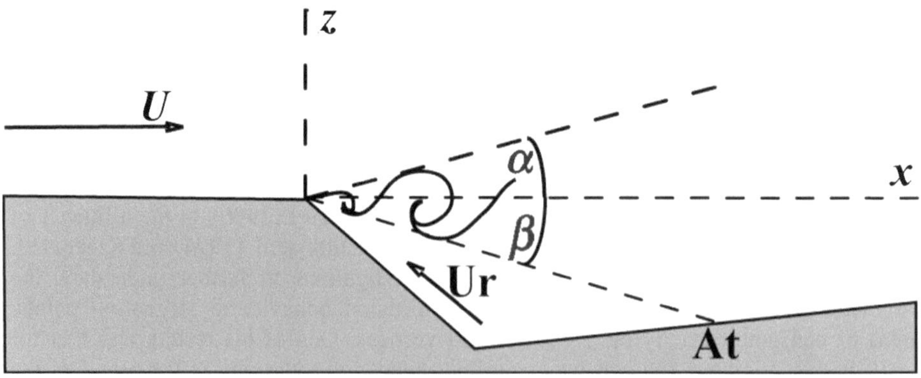

Fig. 9.21 Schematic of a mixing layer in its temporal mean as for example on the lee side of a dune

This representation is valid only for $U_x / \nu > 4\times10^5$ since it assumes self-similarity. This *Re* number is reached for large dunes but not for those originating from ripples, and such averaging does not apply. A further problem is the determination of the upper and lower dispersion angle α and β, respectively. They cannot be determined theoretically but need experimental evaluation where the main parameters are the ridge-angle and the ratio of U/U_r. Landau and Lifschitz state, e.g., that for $U/U_r \approx 30$ and a scarp of 90°, resulted in $\alpha = 5°$ and $\beta = 10°$ and a pressure difference between the outer and inner two turbulent areas of $p_1 - p_2 = 0.03\rho_f U^2$.

The 2D mixing layer behind a splitter plate without sediment is discussed in, e.g., Tennekes and Lumley (1972), who measured the dispersion angle to be ± 3.4° from the symmetry plane. For dunes, however, we recommend using the values given by Landau and Lifschitz. A good estimate for the attachment location for dunes is

$$x_{At} \approx h / tg10° = 5.67h \tag{9.24}$$

This was calculated assuming both a horizontal mixing layer separation and horizontal bed at the attachment area. The leeward dune slope was thought to be 90° since not the angle but only U_r must have the correct value. The evaluated length measured from the crest to the attachment point agrees well with the measurements of McLean et al. (1996)

$$\Lambda_A \approx 5 \div 6h \tag{9.25}$$

The attachment area shows very active flow and sediment resuspends due to several mechanisms, of which the tornado-like low-pressure cores and the horizontal redirection of the outer flow are strongest. Close to the reattachment area, the decelerated flow develops a front vortex as discussed for the ripple formation and sketched in Fig. 9.2. The resulting pressure field shows highly fluctuating depressions, which are responsible for the observed sediment transport (Cherry et al., 1984; Hudy et al., 2002). Most of the suspended grains carry on in the turbulent boundary layer flow developing from there on to the crest of the next dune. Here, the coherent structure as described in (Sects. 9.1 and 9.2) exist, where they can cause the formation of ripples as substructure.

The separation and attachment subdivide the transport mechanisms and therefore dunes cannot be defined by a single length scale. In addition to their periodicity Λ, one also needs Λ_{At}, the length of the separation bubble. The length scale ratio Λ/Λ_{At} is a measure for the bed exposure to the erosion process. In addition to the two length scales, we need the maximum length scale for the fall out from the mixing layer, Λ_{sm}. When Λ_{sm} is larger than Λ_{At}, only a part of the transported grains fall out to the rather quiet region of the separation bubble whereas those transported further than the separation bubble will not settle any more. In this case, we have very intensive suspended load transport finally resulting in the disappearance of the dunes. This is actually observed, so we understand the principal mechanisms. It is desirable to quantify the description, probably by classification via the different length scales. To do so it is valuable to have as much information as possible, which we receive best from the literature.

9.5.2 The Development of Dunes as Described in the Literature

Let's start with the question, how the wavelength Λ can be described as a function of the water depth H? Yalin (1964) empirically found this wavelength dependency of the dune to be proportional to the water depth

$$\Lambda = 2\pi H \tag{9.26}$$

(This is of course not applicable for wind dunes, since H is an undefined value in this case. The transport mechanism is different and for airflow we do not have an upper ceiling as in a water channel flow.)

The second characteristic value for the dune is its height h, for which the literature results stray considerably; some results are summarized in Table 9.2.

Table 9.2 Several equations for the relation of h/H for dunes

Author	Result	Assumptions
Knorez (1959)	$\left(\frac{h}{H}\right)_{max} = \frac{3.5}{\ln\frac{H}{d_s}+6}$	$u >> u_s$
Yalin (1964)	$\left(\frac{h}{H}\right) = \frac{1}{6}\left(1-\frac{\tau_{wc}}{\tau_w}\right)$	
	$\left(\frac{h}{H}\right)_{max} < \frac{1}{6}$	$\tau_w >> \tau_{wc}$
Nordin and Algert (1965)	$\left(\frac{h}{H}\right)_{max} < \frac{1}{3}$	
Znamenskaja (1965)	$\left(\frac{h}{H}\right) = f\left(\frac{u}{u_c},\frac{H}{d_s}\right)$	Plot
Allen (1968)	$h = 0.0086H^{1.13}$	
Führböter (1967)*	$h < \frac{2}{\Phi\Psi+1}H$ $\left(\frac{h}{H}\right)_{max} = 0.4$	For triagonal sediment-body Φ-form coefficient, Ψ-transport exponent for U $\Phi \approx 1$ for long banks $\Psi \rightarrow 4$and $u >> u_c$

Gill (1971)**	$\frac{h}{H}=2\Psi\alpha\frac{\tau_w}{\tau_w-\tau_{wc}}$ $\frac{h}{H}=2\Psi\alpha\frac{\tau_w}{\tau_w-\tau_{wc}}\frac{1}{1-Fr^2}$ $\left(\frac{h}{H}\right)_{max}=0.25$	Kinematical ansatz Dynamical ansatz $\alpha=0.5$, $\Psi=4$; triangonal form

*The height h and the dune form define the form coefficient Φ, and we find the velocity profile using potential theory. Through Bernoulli's equation, we obtain the upper boundary condition since the water surface is lowered over the ridges. In this form, the dunes act as an additional roughness. Führböter (1980) expanded his theory to dunes for not predetermined H. In this case, the self-organization is caused only through interaction of flow and sediment transport. He found

$$\frac{h}{H}=\frac{2}{2\Phi\Psi+1} \quad \text{instead of} \quad \frac{h}{H}=\frac{1}{\Phi\Psi} \tag{9.27}$$

**Gill (1971) postulates an equation in which he considers the derivatives of both the roughness coefficient and the velocity profile, however they are unknowns. The most striking result is, however, that for a unity Froude number $Fr=1$, the dunes disappear

One can use this list also to make rough estimates without being concerned about the sediment transport.

Raja and Sony (1976) published a purely empirical equation, Steer (1975) introduced one based on a boundary layer, and Zane (1976) combined these results to formulate a range,

$$0.15 < h/H < 0.3 \tag{9.28}$$

Another important dune measure is the steepness (h/Λ). For this characteristic value van Ruin (1984) gives the function

$$\frac{h}{\Lambda}=0.015\left(\frac{H}{d_s}\right)^{0.3}\left(1-e^{-0.5T}\right)\left(25-T\right) \tag{9.29}$$

$$\wedge \quad T=\frac{\tau'_w-\tau_{wc}}{\tau_{wc}} \quad \left(\tau'_w \rightarrow (\text{eq. } 9.15)\right)$$

These empirical results are all very useful for certain simulations, but the physical concepts only enter with respect to the choice of parameters. However, Eq. 9.26 is a clear hint that the dune length is correlated to the surface waves. Kennedy (1969) used this assumption to gain understanding of the transport mechanism with best results. For this reason, it shall be discussed here in detail.

He assumes that the flow can be given by a velocity potential ϕ and by a double sinusoidal boundary, the surface wave as limitation of the free surface and the sandy bed form having the same frequency as the boundary condition, as shown in Fig. 9.22.

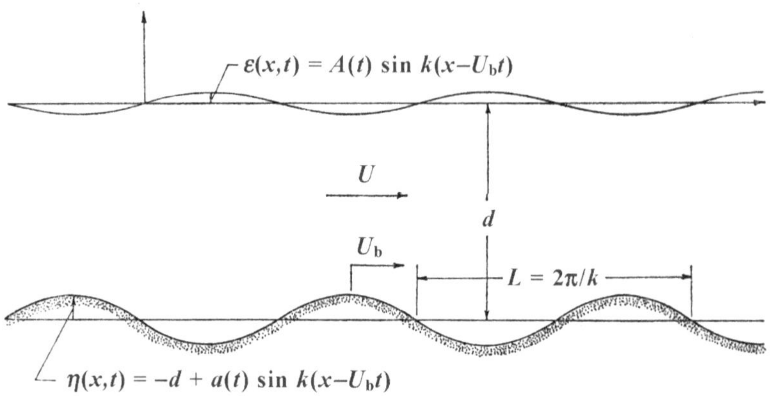

Fig. 9.22 The flow and boundary condition of the open channel flow with dunes as formulated by Kennedy (1969) from Kennedy (1969)

For this flow the Laplacian equation is fulfilled and given by

$$\begin{aligned} &\frac{\partial \xi}{\partial t}+U\frac{\partial \xi}{\partial x}-\frac{\partial \phi}{\partial z}=0 && z=0 \\ &g\xi+U\frac{\partial \phi}{\partial x}+\frac{\partial \phi}{\partial t}=0 && z=0 \\ &\frac{\partial \eta}{\partial t}+U\frac{\partial \eta}{\partial x}-\frac{\partial \phi}{\partial z}=0 && z=-d \quad \wedge \quad d=\overline{H} \end{aligned} \tag{9.30}$$

The velocity potential in accordance with Eq. 9.30 is given by

$$\begin{aligned} &\phi(x,z,t)=UA\left[\frac{1}{\mathrm{Fr}^2 kd}\cosh kz+\sinh kx\right]\cos k\left(x-U_b t\right) \\ &\mathrm{Fr}=U\sqrt{gd}, \quad k=2\pi/\Lambda, \quad L=\Lambda\,(\text{Fig.}\,9.22) \end{aligned} \tag{9.31}$$

where A is the amplitude as shown in Fig. 9.22.

Assuming that the transport is slow enough ($\frac{\partial A}{\partial t} << UkA$), we can attach a coordinate system to the moving dunes and thus have a stationary flow. Then, the relation between the two wave amplitudes is given by

$$a(t)=A(t)\left[1-\frac{1}{\mathrm{Fr}^2 kd}\tanh kd\right]\cosh kd \tag{9.32}$$

and the phase angle between surface wave and the dune is

$$\begin{aligned} &U^2>\left(g/k\right)\tanh kd \rightarrow \theta=0 \\ &U^2<\left(g/k\right)\tanh kd \rightarrow \theta=\pi \end{aligned} \tag{9.33}$$

Contrary to a traveling wave, the dune body moves by the sediment transport on its surface. Kennedy decided to use a transport relation based on the continuity equation

$$\frac{\partial T}{\partial x}+\frac{\partial \eta}{\partial t}=0 \quad \wedge \quad T=q_{s}b \tag{9.34}$$

where b is the width of the channel. This description reflects Kennedy's potential flow assumption fairly well, and it should work quite well outside the separation bubble. It becomes very problematic if there is a separation mechanism involved. What is needed is a transport equation based on the flow local conditions, and this requires a known sediment transport description. However, every description based on an independent wall shear stress is incompatible with a potential flow theory, and therefore Kennedy used the empirical theory of Allam et al. (1966)

$$T(x,t)=m\left[U-u_{c}+\frac{\partial \phi(x-\delta,-d,t)}{\partial x}\right]^{n} \tag{9.35}$$

where m is a dimensional coefficient and n is a dimensionless exponent. The transport rate is related to the predicted velocity at a mean level of the bed z = –d. The most interesting value in Eq. 9.35 is δ, which Kennedy called the ignorance factor. Formally, it is the phase shift between the two waves and corresponds to the lag of the dune body with respect to the surface wave. Without this term, the potential theory predicts only stable beds. These assumptions are in agreement with an observed phase shift between the local bed elevation and the local shear stress, which is $\neq 0$ or $\neq \pi$ (Benjamin, 1959; Engelund and Hansen, 1966; Iwasa and Kennedy. 1968). Since the local sediment transport depends on the local flow conditions, the phase shift of the shear stress must contribute to $d=\overline{\boldsymbol{H}}$. Also the concentration of the transported sediment lags behind the maximum shear stress, and the flow needs time to adjust again. The opposite is true if we have a deceleration. We ascertain a transport relaxation time, which can also be expressed in a relaxation length.

If we expand Eq. 9.35 as binomial series, we can describe the phase relation of the flow and the transport, which can now be used for a stability investigation. The results of this investigation are reproduced in abbreviated fashion in Table 9.3, which are sorted sequentially so that the upper limit of one class matches the lower limit of the next configuration.

Table 9.3 Bed-form criteria

Bed forms	Equations	Phase relations	Direction of motion
Ripples + dunes upper limit.	9.36	$jkd = 3\pi/2$	Downward
Transitive forms	$\mathrm{Fr}^2=(j/kd)\tanh kd$	$jkd = 3\pi/2$	Upward
Flat bed	$\mathrm{Fr}^2=(j/kd)\tanh kd$	$jkd = \pi$	
Antidunes	9.36	$jkd = \pi/2$	Downward
Antidunes		$jkd < \pi/2$	Upward

The bed forms as they are evaluated by a stability investigation based on the configuration as given by Fig. 9.22. The relations are presented through the critical Froude numbers as functions of the phase relation $j = \delta/d$. With increasing importance of the transport relaxation length, the implicit Eq. 9.36 becomes valid for dominant *kd*

The relations are presented through the critical Froude numbers as functions of the phase relation $j = \delta/d$. With increasing importance of the transport relaxation length, the implicit Eq. 9.36 becomes valid for dominant *kd*

$$\mathrm{Fr}^2 = \frac{1 + kd \tanh kd + jkd \cot jkd}{(kd)^2 + (2 + jkd \cot jkd)\, kd \tanh kd} \tag{9.36}$$

In spite of the simplified flow field, the evaluated results are in good agreement with the experiments and Kennedy's theory can be used for simulation purposes.

The circumstance that fairly high-relaxation values of about j = 5–6 were needed for the matching to the diverse bed forms was surprising to Kennedy. It was this result that stimulated him to postulate that for bedforms with high-relaxation length also suspension load transport must be considered, whereas those with a short-relaxation length should be expressible through bed load transport only. This prediction is probably correct but must be discussed in further detail, because, in our opinion, Kennedy probably drew the wrong conclusion. He recognized that his theory couldn't discern between ripples and dunes, and he stated that ripples must be the result of bed-load transport whereas dunes are created mainly by suspended load transport.

A remarkable point of this theory is its capability to describe dunes in rivers as well as wind-generated dunes in a desert when the flow domain extends to the unlimited half space. Setting

$$\mathrm{Fr} \to 0 \because g \to \infty, \quad \frac{\partial a}{\partial t}(0) \to \max\,? \tag{9.37}$$

we find

$$\begin{gathered} 2kd = (2 + jkd \cot jkd) \sinh 2kd \\ \therefore 2 + k\delta \cot k\delta = 0 \quad \therefore \quad \Lambda = 1.24\delta \end{gathered} \tag{9.38}$$

Kennedy was aware that his theory was incomplete and needed a physical interpretation and evaluation of d (=H). We will follow with an attempt to solve the problem by introducing separations instead of only phase shifts.

9.5.3 An Attempt to Find a Statistical Theory for Dunes

We purposely used the word attempt here since too many parameters are involved, e.g., when one considers separation, and the formulation of an analytical solution for dune development becomes impossible. Several assumptions and simplifications must be made even for this averaged representation. Therefore, it will never be free of criticism, however statistical information is needed to run simulations.

Kennedy and several other authors start their stability investigation from an initially sinusoidal bed deformation. Especially, a phase lock between the deformation wavelength and the surface wave was presumed, although we know that it is not the case if the source of the deformation does not origin from the surface wave. We therefore have to show how a disturbance grows to become sinusoidal. Julien (1995) showed how a large enough disturbance, like a degenerated ripple, could grow to become a sinusoidal bed form. Here we reproduce his ideas because they contain the main elements of a feedback mechanism.

For a 2D channel flow it is assumed that

$$\begin{aligned} &\tau_{xx}, \tau_{yx}, \tau_{yz}, \tau_{zz} \approx 0; \\ &\overline{u} = \left(\overline{u}_x(z), \approx 0, \approx 0\right); \\ &\mathrm{rot}\underline{\overline{u}} \approx \frac{\partial \overline{u}_x}{\partial z} \end{aligned} \tag{9.39}$$

And the Bernoulli equation

$$E = \frac{p}{\rho g} + \frac{u^2}{2g} \tag{9.40}$$

The result is a simplified form for the equations of motion which are now given by

$$\begin{aligned} &g\frac{\partial}{\partial x}\left[\frac{p}{\rho g} + \frac{\overline{u}_x^2}{2g}\right] = g_x + \frac{1}{\rho}\frac{\partial \tau_{zx}}{\partial z} && : x \\ &\frac{\partial}{z}\left[\frac{p}{\rho} + \frac{\overline{u}_x^2}{2}\right] = g_z + \overline{u}_x\frac{\partial \overline{u}_x}{\partial z} + \frac{1}{\rho}\frac{\partial \tau_{xz}}{\partial x} && : z \end{aligned} \tag{9.41}$$

If $\partial\tau_{xz}/\partial x = 0$, the pressure distribution remains hydrostatic and Eq. 9.41 can be integrated over the mean water depth $d = H$, which results in

$$p = \rho g(d - z)\cos\theta \tag{9.42}$$

In a uniform flow u_x and p are constant in x-direction; so it follows that the left side of Eq. 9.41(:x) is zero and by integration we find

$$\begin{aligned} &\tau_{zx} = \rho g(d - z)\sin\theta \quad \propto z \\ &\because z = d \rightarrow \tau_{zx} = 0 \\ &\wedge \quad z = 0 \rightarrow \tau_{zx} = \tau_{\mathrm{w}} = \rho g d\sin\theta = \gamma d S_0 \end{aligned} \tag{9.43}$$

In a nonuniform flow, Eq. 9.41 has to be replaced by

$$\underbrace{-\frac{1}{\rho}\frac{\partial \tau_{zx}}{\partial z} = g\sin\theta}_{\text{uniform}} \underbrace{- g\frac{\partial}{\partial x}\left(\frac{p}{\rho g} + \frac{u^2}{2g}\right)}_{\text{non-uniform}} \tag{9.44}$$

flows

Let's assume a local disturbance of height Δz in an under-critical flow. On the upward (up) and downwards (down) side of the perturbation. Compared are two location on both side, where the flow depth is H_n. From a specific energy diagram, it is shown that the small disturbance Δz causes a decrease in specific energy when approaching the perturbation, thus E and τ_{zx} are characterized by

$$\text{(up)} \qquad g\frac{\partial E}{\partial x}<0 \;\wedge\; \frac{\partial \tau_{zx}}{\partial z}\Uparrow, \quad \left(\frac{\partial \tau_{zx}}{\partial z}\right)_{\text{nunif}} > \left(\frac{\partial \tau_{zx}}{\partial z}\right)_{\text{unif}} \tag{9.45}$$

$$\text{(down)} \qquad g\frac{\partial E}{\partial x}>0 \;\wedge\; \frac{\partial \tau_{zx}}{\partial z}\Downarrow, \quad \left(\frac{\partial \tau_{zx}}{\partial z}\right)_{\text{nunif}} < \left(\frac{\partial \tau_{zx}}{\partial z}\right)_{\text{unif}}$$

Near the bed, the velocity term in Eq. 9.44 can be neglected and Eq. 9.44 simplifies

$$-\frac{1}{\rho}\frac{\partial \tau_{zx}}{\partial z} = g\sin\theta - g\frac{\partial H}{\partial x} \quad \because \quad z \approx 0 \tag{9.46}$$

The consequence is that upward of the disturbance, the wall shear stress increases and erosion occurs whereas downstream the flow eventually separates and sedimentation results. This is the basic mechanism by which a dunelike sediment body forms.

We see from Eq. 9.46 that for $\partial H/\partial x > \sin\theta$, the shear stress will reverse sign. The integration over the entire water depth will result in a negative wall shear stress τ_w when $\partial H/\partial x$ is large. This means that downstream of the disturbance, there may be a region with $\tau_w < 0$ where separation occurs.

The main restriction is that the disturbance has to be present over the entire width of the channel. This is exactly the situation over a ripple bed, and therefore Julien's description agrees with the observations.

The problematic part of Julien's theory is the initial coupling of the small ripple disturbance with the free surface deformation since $\partial H/\partial x$ should be negligible in this case. In fact, ripples form independent of the water surface deformation, and we know that the transition to dunes is caused by a change in the separation topology and not by a coupling mechanism with the free surface. What is needed is a theory, which further entails the details of separation.

A sediment body displaces the flow and is additionally characterized by the formation of the vortex skeleton. With increasing flow velocity, additional instability elements like Kelvin–Helmholtz vortices appear as soon as the separation changes from a viscous to a turbulent dominated form. The new system must produce bed forms of longer wavelength. This analysis contradicts the investigation of Nasner (1974), who found that dunes can increase in only height but not in length, a result that became the used standard in the relevant literature. Nasner's result is probably correct as soon as the dunes are coupled with the surface waves because here the dune length is forced upon by the standing wave, as in Kennedy's theory. However a nonstationary process with changes in the vortex skeleton must occur to achieve this state.

An additional problem lies in the sinusoidal wave assumption as shown in Fig. 9.22 being in contradiction with the observation that dunes exhibit a triangular shape with a long luff side and a rather steep lee side. This discrepancy does not appear significantly in the stability investigations however, because the separation bubble follows the sand body and they together modulate the main flow with a sinusoidal-like disturbance. This is the case when the outer separatrix attaches at the bed (Fig. 8.8a). Such a model alone is not sufficient since it does not contain the intermittency of the sediment transport due to the bed forms. It therefore seems obvious to compare δ with Λ_{At}, however δ now has a new meaning. It is not anymore a relaxation length but a distance after which the sediment transport is constant, the so-called saturation length. Formally the new δ determines the phase shift of Eq. 9.35. From Bernoulli's law, the flow has its lowest pressure over the ridge, where the flow separates, and the free surface should be lowest just at that location. Since the luff side inclines, the flow overshoots resulting in phase shift with respect to the dunes, which is small however compared with the interruption of the erosion. In addition, since the pressure difference between the outer flow and the separation bubble is small [see Sect. 9.5.1, the feedback system with the free surface occurs rather late.

However, the locations at which the mixing layer instabilities merge are rather stable, like in a mixing layer. Acoustic forcing through a feedback of the sound wave could be the probable reason for this behavior (Kiya et al., 1997), who introduced the so-called control factor m having an optimum value of $m = 1$ for the fundamental mode and, or $m = 2$ for the first harmonic. A standing acoustic wave is created on the separation bubble. The consequences of this feedback are that the gravity wave couples with the "dune-wave" as given by Airy's law, and consistent with Kennedy's theory, the following speed of propagation of the wave can be calculated

$$u_g = \sqrt{\frac{g}{2\pi\Lambda} th \frac{2\pi H}{\Lambda}} \tag{9.47}$$

A theory based on the flow separation also sustains the argument that one has to distinguish between sediments having a large or a small fraction suspended in the flow. In a pure bed-load regime, the sediment erodes between the attachment and the ridge and deposits in the separation bubble. If the deposition is incomplete and material remains in suspension for more than a wavelength, the accumulation of transported material ends up in a new status, a flat bed with a very high transport. When we have mainly suspension transport the process becomes extremely complicated, since the sedimentation is distributed over the whole channel and affected by local disturbances, such as the separation bubbles and the instability vortices, as described by Meiburg.

Around the attachment area, most of the transported sediment erodes, where a smaller part recirculates versus the luff ridge but most sediment accelerates versus the next ridge within the newly developing boundary layer flow. One can only imagine how complex such a process may become, however there is hope that one understands several of the partial mechanisms, enabling the design of a simulation program. The important mean scales are shown in Fig. 9.23.

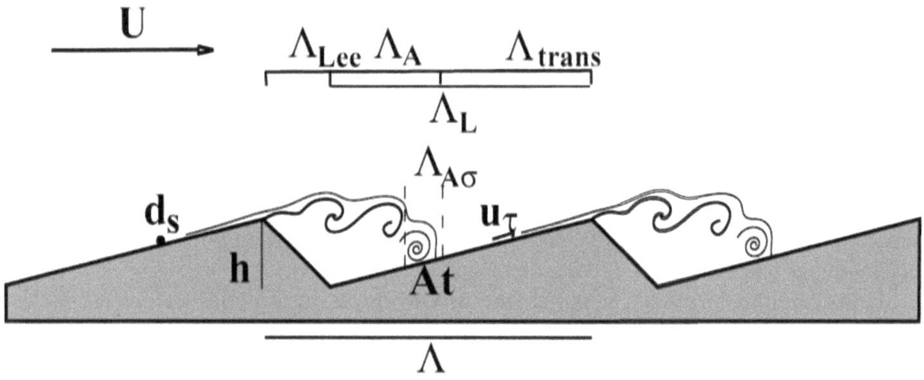

Fig. 9.23 The main parameters for dune development: dune height h and length Λ, and length of the luff Λ_L and lee sides Λ_{Lee}. Λ_A corresponds to the length of the separation bubble, which is the distance between the ridge and the attachment At. From this we derive $\Lambda_{trans} = \Lambda - \Lambda_A$, the lenth over which the bed erodes with its characteristic wall-shear velocity $u_{\tau w}$. The attachment region At is characterized by the attachment line moving within a given range $\Lambda_{A\sigma}$. Here most of the sediment suspends due to strong tornado-like vortices. The grain distribution is given by its size d_s and density ρ_s

It is important to recognize that the scales are mean values and can only be used in a statistical sense. The separation mechanism has been described in Fig. 8.8 exhibiting a separating and reattaching mixing layer with a Kelvin–Helmholtz instability. That means a quantity like Λ_{At} should be introduced as a statistical distribution. This interpretation allows weighing the contributions from different sediment transport regimes. Such a theory, however, does not explain the growth mechanism of dunes.

A transport for that regime can be calculated using the continuity equation in combination with Eq. 9.26.

9.5.4 A Vortex Shedding Theory

A separation bubble generally sheds vortices at a frequency f_{det}, which we will take as our starting point for the shedding theory. The shedding frequency is thought of as the natural instability of the Kelvin–Helmholtz wave. We know that this occurs only in very rare cases since these instability vortices roll up to an open separation bubble of very complex topology. Although the energy is not transported away by the vortices, we assume the separation to be of shear layer type. Like the separation around an obstacle being characterized by a Strouhal number, the separation bubble can be seen as a self-sustained resonator with by the vibration frequency f_{det}, Levi (1983)

$$f_{det} = \frac{1}{2\pi}\frac{U}{\Lambda_A} \tag{9.48}$$

derived from the so-called "universal Strouhal law" as it provides $1/2\pi$ as a theoretical value for the unit Strouhal number. Occasionally, one experimentally finds indicators for a second-harmonic component, which is fully consistent with Kiya's control factor $m = 2$. To include all possible harmonics, Eq. 9.47 can be generalized as

$$f_{\text{det}} = \frac{m}{2\pi}\frac{U}{\Lambda_{\text{A}}} \tag{9.49}$$

Verbanck (2004) used this concept to relate the separation with the averaged energy dissipation as given by the energy slope S_{E}. The conservation of head in a given river reach is a positive function of the intensity of the topography-forced gravitational process

$$S_{\text{E}} \propto f_{\text{att}}^{+\alpha_1} \tag{9.50}$$

where α_1 is an arbitrary (positive) real number exponent, and f_{att} is the characteristic frequency of the nonturbulent gravitational process induced by the bed-form topography. This frequency is reflected also in the water-surface profile from interaction with the gravity waves as described by Airy's law Eq. 9.47.

The vortex generation will be maintained if and only if the necessary energy input is provided to the system and a high-vortex shedding-frequency is equivalent to a large energy loss from flow separation. Accordingly, we can write

$$S_{\text{E}} \propto f_{\text{det}}^{+\alpha_2} \tag{9.51}$$

Since both Eqs. 9.50 and 9.51 have to be satisfied, they can be combined to a single one

$$S_{\text{E}} \propto \frac{f_{\text{det}}^{\alpha_2}}{f_{\text{att}}^{\alpha_1}} \tag{9.52}$$

and dimensional consistency imposes that α_1 and α_2 must be equal given by α. Equation 9.50 should logically be written as

$$f_{\text{att}} = \frac{\sqrt{\frac{g}{2\pi/\Lambda} th \frac{2\pi\bar{H}}{\Lambda}}}{\Lambda} \Rightarrow f'_{\text{att}} = \frac{\sqrt{\frac{g}{2\pi/\Lambda} th \frac{2\pi\bar{H}}{\Lambda}}}{\Lambda_A} \tag{9.53}$$

Equation 9.52 with Eqs. 9.49 and 9.53 ends up in the ratio $\Lambda/\Lambda_{\text{A}}$ as in the previous chapter, and this shows that this ratio is the essential value for describing the sediment transport by dunes, however we cannot predict it. Verbanck (2005) propagates a minus 10/3 law based on empirical data

$$S_E = \left[\frac{f_{\text{att}}}{f_{\text{det}}}\right]^{-10/3} \quad or \quad \frac{m}{2\pi} = S_E^{0.3}\text{Fr}^{-1};\, m \geq 1 \tag{9.54}$$

with m a control factor, which reflects the complex feedback-control loop process active in the separation cell immediately downstream of the bed-form crest. However, it does not contain a description of the transport.

9.5.5 Asymptotic Behavior of Dunes

In this chapter, we described the formation of dunes by a variety of processes, however, most of the time we are interested in the final equilibrium bed form and its possible existence should prove an asymptotic behavior of the interaction process. Kennedy (1969) tried to predict the asymptotic behavior from his theoretical framework and the empirical laws for dune forms, compiled in Table 9.2. He recognized that this cannot be done in the frame of his theory, but it is evident that the dune development ends when

$$jkd = k\delta = 2\pi \tag{9.55}$$

If Eq. 9.55 holds, neither additional sedimentation nor erosion can occur, which does not imply that the dune is not migrating. Already Engelund and Hansen (1966) concluded this by introducing

$$\eta(x,t) = \eta(x - U_b t) \tag{9.56}$$

into the transport Eq. 9.34. One integration then demonstrates that for constant-amplitude bed waves, the local transport rate varies linearly with the local bed elevation, hence the maximum and minimum of the two functions coincide.

Equation 9.56 indicates the significance of the continuity equation. It is assumed that the mean transport can be described by the migration of the dune alone,

$$\overline{T} = \frac{U_b}{\Lambda}\int_0^{\Lambda}\eta(x)dx \tag{9.57}$$

$$\Rightarrow \overline{T} = \overline{T}_b = \frac{U_b}{\Lambda}a_0\int_0^{\Lambda}(1 + \sin kx)dx = U_b a_0$$

where a_0 is the equilibrium amplitude assuming bed-load transport only. The key value for this interpretation is U_b, which can be evaluated through a binomial expansion of Eq. 9.35 (Alam et al., 1966). By eliminating U_b, one finds the relation

$$\frac{2a_0}{\Lambda} = (\beta / n\pi)\frac{U - u_c}{U}\frac{\tanh kd - F^2 kd}{1 - F^2 kd \tanh kd} \quad \wedge \quad \beta = \frac{\overline{T}_b}{\overline{T}} \tag{9.58}$$

where n is the power law exponent of Eq. 9.35, as well as the slope in the logarithmic representation of

$$\overline{T} \to U - u_c \tag{9.59}$$

This representation given by Alam et al. (1966) also needs empirical correlations to fit the data. The results agree well with the measurements, especially for $\beta \to 1$, that is, for rougher material moving by bed-load transport. Kennedy evaluated the following quantities for a closed rectangular channel flow using Eq. 9.38

$$\begin{aligned} U_b &= n\overline{T}k\frac{U}{U - u_c}\coth kd \\ \frac{2a_0}{\Lambda}(\beta / n\pi) &= \frac{U - u_c}{U}\tanh kd \end{aligned} \tag{9.60}$$

and for a half space flow

$$\begin{aligned} U_b &= n\overline{T}k\frac{U}{U - u_c} \\ \frac{2a_0}{\Lambda} &= (\beta / n\pi)\frac{U - u_c}{U} \end{aligned} \tag{9.61}$$

It is worthwhile to repeat that all these explanations are based on bed-load transport but the significant temperature sensitivity indicates that suspension loads or at least sedimentation processes versus rolling motions are important as well (Al-Shaikh, 1961; Burke, 1966; and Franco, 1968). This result underscores the need for a separation based transport theory of dunes.

With changing temperature and viscosity, particles of different size can settle in the separation bubble, and the dune propagation will be affected. The even smaller suspended particles of the outer flow do not participate but being carried by the mean flow as described and shown in Chap. 3 (Sect. 3.3).

The main difference of the separation based theory, when compared to the phase shift approach, is the clear distinction of the completely different transport mechanisms acting in the two regions Λ_A and Λ_{trans}.

Using the separation concept, we will now introduce physical criteria and classifications. We first look at an asymptotic state using the continuity Eq. 9.56, and it follows that the transport in a co-transported coordinate system is zero,

$$q_s = q_{strans} - q_{sA} \quad \Rightarrow \quad q_s \to 0 \tag{9.62}$$

which means the whole transport can be described by the dune propagation as given by Eq. 9.57.

If Eq. 9.62 ceases to hold, e.g., $q_{strans} > q_{sA}$, the asymptotic state no longer exists, the dune erodes, and the bed-form approaches a flat bed, with a very high-sediment transport in the near wall layer. Here, the fluid should be assumed non-Newtonian for the description of the fluid stresses.

The opposite case, e.g., $q_{strans} < q_{sA}$ should result in dune growth, however, this ends up in a contradiction since the accumulating material must first be eroded, in other words q_{sA} has the meaning of a deposition capacity $\hat{q}_{sA}$, which differs from Eq. 9.62. The criterion for dune growth now becomes

$$q_{strans} < \hat{q}_{sA} \tag{9.63}$$

As a consequence, the dune can only approach its equilibrium state, Eq. 9.62, through adaptation of the length scales. What happens physically?

For the stationary dune, all of the eroded material has to be deposited completely in the separation bubble and the dune propagates without a change in form. The sedimentation occurs throughout the separation bubble, peaking close to the crest. In case of a grain distribution, one observes a separation by size and since the dune propagates over the finer sand deposition, the sediment body becomes stratified. The maximum scouring happens around the attachment area, where practically no deposition occurs.

When the dune shape is not in equilibrium, e.g., $q_{strans} > q_{sA}$, erosion decreases the height and the slope of the luff side, and the dune will shrink as well when the deposition capacity is exceeded. For $q_{strans} < q_{sA}$, the slope height and scale of the separation bubble will increase until it matches the erosion by which the dune shape and size approach an equilibrium state.

This description requires no interaction with the surface wave, however, if such an interaction starts, the dunes produce a standing wave pattern, which stabilizes the separation process. Since now the length scale Λ can only vary within a small range, the dune adapts its height together with Λ_A until the equilibrium is reached. This development is in accordance with the observations but the quantification of these processes needs still additional research, where simulations could achieve some of the possible progresses. For this reason, we quoted so much empirical data.

The explanation for the quasi-2D appearance of dunes is still missing. For now, one could argue just as for the ripple formation, where the open separation bubble releases fluid and vorticity toward the *z*-direction by transport through lee-vortex, and its interaction with neighboring bed-forms. However, the mechanism is different in its details. The Kelvin–Helmholz instability forms spanwise vortices (Fig. 8.8), which pair up generating a wavy core. We know from vorticity dynamics (Fig. 8.11) that forward facing loops raise fluid and produce boils whereas the backward bended loops are scouring downwards. The spanwise instabilities of successive vortex cores are shifted half a wavelength equal to a phase shift of π (Bernal, 1981; Jimenez, 1983; and Jimenez et al., 1985), so the scouring happens exactly where the preceding vortex just deposited its material, thus two-dimensionality is explained.

We now know the processes, and it has become evident that an analytical description including all the details is outside our capacity. Modeling these physical insights of known elements, simulations can be constructed based on the knowledge we have from the real system.

9.6 Antidunes

We encountered already the antidunes in Kennedy's theory see Table 9.3. This form of sediment bed and sediment transport occurs, when the dunes grow (Eq. 9.63) so that the flow over the dune crest becomes critical and the disturbance cannot move downward. In that case, we have a strong sediment transport but the bed forms travel upstream. The separation mechanism is too complicated to be used and the wave theory with a phase shift is appropriate. Usually the antidunes grow further until the surface wave breaks. We then observe a sudden release of the antidunes resulting in the formation of a flat bed with high transport and the growth of new instabilities to antidunes. The instability process is still unknown in this case, but it has been speculated that it is induced by a shock wave in the non-Newtonian transport layer.

Verbanck (2005) goes a step further in considering the antidune standing wave as the final alluvium bed form. If the stream power is very high, the antidunes are the natural forms (Gilbert, 1914; Kennedy, 1963), and they possess some very astonishing features because the water surface deformation stays in phase with the sediment bed profile. Compared to other bed form types, the streamline curvatures are especially significant. Also, the antidunes have marked amplitudes but cannot really be described as significant protrusion reducing the cross section. Verbanck (2004) noticed that for the upper alluvial regime, the transformation from the upper-stage plane bed to the antidunes standing-wave bed condition actually coincides with a decrease in the friction factor. Some other explanations exist based on assumptions that the suspended bed-load material inhibits the turbulence (Galland, 1996; Garg et al., 2000; Hsu et al., 2003). One of the consequences of this turbulence damping is the locally reduced shear stress allowing the bedform to grow in height and maintain itself (Nelson et al., 1993).

10 Mixtures of Medium Grain Sizes

Riverbeds composed of fine sands as treated in Chap. 9 are rather rare. The sediment usually has a wide grain size distribution and consists of various materials. Classical sediment transport concepts must then be extended to include new aspects such as flow separations as well as feedback concepts which explain the self organization of the sediment in bed forms.

In classical descriptions, the transport is estimated for individual size fractions of the sediment and then integrated over the entire distribution. Such a representation does not consider bed forms. In addition, when the various fractions of sediment are transported in different modes, a sorting process occurs. Fine sand is transported preferentially, whereas less of the coarser fractions are moved. This selective transport mechanism can only be neglected if the deficit in fine sand can be compensated by a feeding from upstream. However, this is not the case along the river, as long as the bed is the only source of the sediment. This illustrates that even sediment transport without a change of the flow conditions can become unsteady.

Sorting must occur when only part of the sediment is transportable. When all sediment fulfills the transport criteria, the sorting process becomes very complex, and we can expect a large spectrum of sorting processes. In other words, one would have to classify the various sorting processes. In the following section, we treat a limited number of cases that are relevant for the construction of simulations.

10.1 Armoring

In Chap. 4 (Sect. 4.4), we mentioned that sediment transport is especially easy to characterize for an armored bed, as it is a rare case in which a steady state is achieved by a self-organizing process. It is also a very useful configuration that can be artificially generated to produce a stable bed.

10.1.1 State of the Art

We speak of an armored bed when the grains covering the bed are not transported anymore and protect the layer beneath from erosion, Gessler (1971). In other words, the grain distribution beneath the armoring also contains fine material. The evolution of the armored bed is the result of a self-organizing process. The sorting occurs due to the transport of the fine material which is moved, whereas the coarser one remains in place. The covering layer then becomes rougher and rougher until the sediment transport stops completely. This process occurs even when the discharge does not change, although the velocity profile does because of the increasing roughness of the bed. The logarithmic profile approaches to the wall. In other words, we are dealing with an asymptotic process, which in detail may be very complex and interactive but which results in a stable state.

The sorting process must be described by introducing additional elements because the surface layer contains a portion of fine sand which in accordance with the transport criteria should be moved.

A first attempt to explain this contradiction was made by Coleman (1967), who introduced a so-called exposition height for isolated grains. The idea is that larger grains are not easily moved when embedded, but when they are more exposed to the flow, a higher force acts on them, and they are even easier to move than some of the finer material. Since an exposure of individual grains is not uncommon, we discuss this explanation further.

The exposition height P was defined as the height above the mean bed surface. Since the single outstanding grains are contributing to the mean height too, Coleman suggested using a conditional averaging procedure. The deficiency is that the mean becomes somehow arbitrary. However, when the same criteria are used for all cases, the result is consistent. The corresponding results were found by Fenton and Abbott (1977) and are represented in Fig. 10.1 by using the dimensionless wall shear stress

$$\Theta_c = \Theta\left(\tau_{wc}\right) \quad \ni \quad \Theta = \frac{\tau_w}{g\left(\rho_s - \rho_f\right)d_s} \tag{10.1}$$

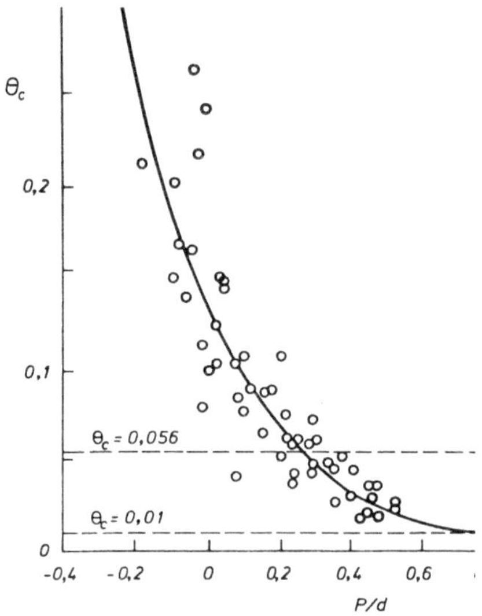

Fig. 10.1 The dimensionless critical wall shear stress Θ_c as a function of P/d

The results show that the strongly exposed larger grains are easily eroded as a consequence and the surface layer becomes smoother, which is just the opposite of what one would expect if fine sand is transported preferentially. Since this kind of smoothening occurs too, we recognize that one has to be very careful in the interpretation of armoring.

Let's go back to the original idea by considering the ratio of individual and the mean grain size d_i/d_s. In accordance to this classification the critical wall shear stresses were introduced as τ_{wc} for the mean grain size and τ_{wi} for the individual ones.

For $\tau_{wc} > \tau_w > \tau_{wi,}$ only strongly exposed grains can be transported. The bed is stable and is in the state described by the exposition-height model. Some large grains may roll on the bed.

For $\tau_{wc} < \tau_w < \tau_{wil,}$ the entire sediment surface is in motion with the exception of very large individual grains, which remain in place. The armored bed is the result of the separation and its vortex skeleton on the large grains, as sketched in Fig. 8.9. The horseshoe vortex on these elements is scouring around it with the effect that the large grains start to slide into their own depressions, whereas the fine material is still transported.

Raudkivi (1982) pointed out that for an efficient scouring process, a wide distribution of the grain sizes must exist. He empirically determined that for the armoring process to occur due to separations on the large elements, they have to be about six times as large as the smallest, or $(6d_s)^3 \approx 200$ times heavier. The observed armoring process occurs also for a narrower distribution, which means that several mechanisms must be involved.

A significant contribution was given by Einstein's description of the sediment transport (1942, 1950). His main concern was not the transport of the largest grains, but their influence on the small ones. He recognized that some fine material on the lee side of larger grains is not moved. This was the basis for his concept of describing the transport of grains of different sizes, and he introduced a hiding factor ξ. This factor is a function of the individual grain size d_s, the roughness k_s defined by the mean grain size, and the wall shear stress $u_{w\tau}$. The physical interactions are so complex that he needed a series of interrelated equations to describe it Eq. (3.27). In that equation, δ' is the limit of the buffer zone within which turbulence as well as viscous influences have to be considered. The thicker this layer is, the smaller the forces on the grains in the bed become, because for coarse material, $k_s >> \delta'$, and X tends to 1(Fig. 10.2). Therefore, one finds

$$\xi = \xi\left(d / 0.77 k_s\right) \tag{10.2}$$

and ξ tends toward 1. Finer material can be deposited behind larger ones, see (Fig. 10.2).

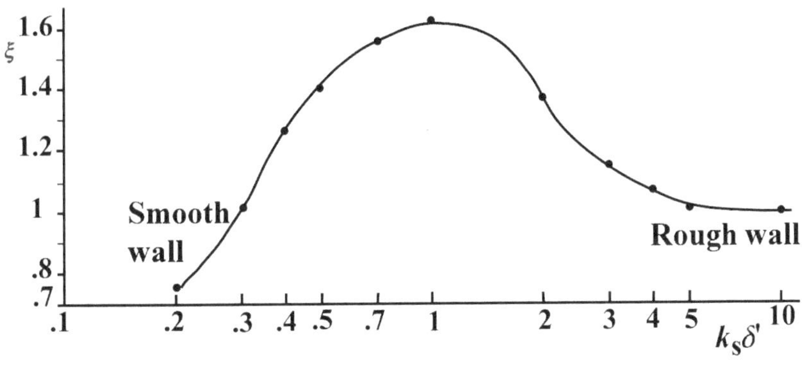

Fig. 10.2 Einstein's hiding factor ξ as a function of a relative roughness k_s/δ'

This mechanism explains the armoring process, but one has to bear in mind that Einstein's result was found for rather fine material, and the system of equations contain too many interactions to be functionally described.

Nevertheless armoring is a state without sediment transport that means that the transport over all classes of grains ceases. On the basis of this knowledge, we try now to formulate a mechanism based on flow separations and its micromechanical processes of how a layer of fine material is formed just beneath the surface layer, the so-called fine-sand anomaly.

10.1.2 A Hypothetical Armoring Mechanism

On river beds with sufficiently coarse sediments, the flow will separate on larger grains. The separation is characterized by a horse-shoe vortex, which starts on the upstream face and consists of two side branches, which combine to form a wake in the lee of the particle (Figs. 8.9, 8.10, 9.12 and 9.13). On the basis of this concept, the armoring can be described by a process as described below.

Initially, fine material is transported but not the coarser one. The surface becomes rougher since the fine grains between the larger ones are eroded. A sorting occurs without any motion of the larger grains.

With the increasing exposure of the larger grains, the flow separation on them becomes more noticeable. The horseshoe vortex starts to scour, and fine sand is deposited behind the large elements.

Some of the large grains sink deeper due to their own scouring actions, while others become instable, roll away, and stop at a preferential location.

The engraving of large grains is only a transitional state since fine sand around them will be removed again. Exceptions are those grains, which have already large grains as neighbors.

A stable bed surface is the result of a decreasing influence of the separation vortices. This is the case when large enough neighbors are close by. In that configuration neighbored horseshoe vortices are damped by vorticity annihilation.

Scouring and engraving will always be present, and this also explains the fine sand anomaly. The rearrangement of grains is the main contributor to this anomaly. The moving larger grains will preferentially settle just behind another larger one. Since fine material has been deposited there, the settling grains sit on this fine material and become more stable because of the higher friction due to the increase in the number of contact points of the grains. The dependence of the stability on the number of contact points was introduced in the classical theory of Bagnold (1941) by using a concentration c, which is equivalent to a porosity of the support when defining the wall shear stress

$$\tau_w = c(\rho_s - \rho_f) g d_s tg\alpha \tag{10.3}$$

$tg\ \alpha$ is the friction factor between the layers and approximately the natural angle of repose. For the wind transport, Bagnold gives the value $ctg\ \alpha \approx 0.4$.

In addition, the newly deposited large grains will have good protection from the oncoming flow in the lee of the preceding grain.

10.2 Turbulence Dominated Sediment Transport

Turbulence dominated sediment transport is possible for fine sand mixtures as discussed in Chap. 9. The turbulent forces acting on a grain were investigated by several experiments. Müller et al. (1971) studied this aspect on single grains exposed to a vortex passing by. These authors as well as Grass (1970, 1971) found that the underpressure areas moving with the flow are producing the dynamical lift forces required to raise a particle. Grass also showed that these areas are correlated with decelerating sweeps (Chap. 9, Sect. 9.1). The pressure distribution on the bed is nearly identical to a superposition of the one for a flow over fine sand and the pressure field originating from the separation vortices on the larger grains by using statistical methods to evaluate this pressure distribution. The papers cited below could be useful for designing a numerical model.

Arndt and Ippen (1968) investigated the pressure difference, which is reqired to produce cavitation on the bed. Since the vapor pressure is precise pressure reference, these measurements are of high value. They found

$$(\Delta p)_{\mathrm{cav}} \approx 16\tau_{\mathrm{w}} \tag{10.4}$$

Emerling (1973) found that the mean pressure fluctuations at the wall are three times as large as the mean wall shear stress, and that the peak values in a turbulent flow are about six times as large (Eq. 3.15).

These results are similar and differ only with respect to the roughness of the test beds.

If one postulates that the lift force is equivalent to the weight, as this was done by Müller et al. (1971), one finds Eq. 3.16 or

$$\frac{\tau_{\mathrm{w}}}{g(\rho_{\mathrm{s}} - \rho_{\mathrm{f}})d_{\mathrm{s}}} = \frac{2}{3c_{\mathrm{max}}} = \Theta_{\mathrm{max}} = \Theta_{\mathrm{c}} \tag{10.5}$$

With the two values $c_{\mathrm{max}} \approx 16$ and 18, we get $\Theta_{\mathrm{max}} \approx 0.037–0.042$. With Eq. (10.5), which corresponds to the Shield's curve, one can calculate the grain that can just be moved. The agreement is good, but one has to notice that for wind transported material Eq. 10.3, the incipient motion is by a factor 10 smaller then for water due to the density difference and grain–grain interaction.

10.3 Sediment Transport Dominated Separation

The transport which occurs when separations occur on bed forms was outlined in Chap. 9, but there is also a sediment transport due to the separation on single roughness elements. This kind of transport is important for coarse grains. This can be seen by comparing the size of an area of underpressure due to coherent structures and the size of a grain. For roughnesses of a size of $d_{\mathrm{s}}^{+} > 20$, the concept of coherent structures will fail as well since it is based on the instability process on a smooth bed, as discussed in Chap. 3 (Sect. 3.2.4). In other words, we are confronted with the old question of how to describe a rough surface as discussed in Chap. 7 (Sect. 7.4). The scouring along the sides of a grain and the deposition of fine material in its wake can be described for single grains exposed in the flow but also the larger particles are arranged and therefore exposed as a group to a local flow defined by the complex interaction of the separation flows on all the participating grains of this heap. This could be taken in consideration by a statistical description based on given roughness type. An analytical one could be best, but this cannot be expected since the systems are too complex. But it can be advantageous to have some empirical data of this kind to construct better models.

We discussed several interactions of flows around elevations, in Chap. 9 (Sect. 9.1), and showed that Schlichting (1936) tried to find statistical relations of the drag as a function of the spacing of the roughness elements. In the meantime, a lot of efforts were made to characterize various surface roughnesses, mainly in the literature for chemical engineers, who need this information for constructing reactors.

The usual approach for an open channel flow with a rough bed is to adapt a modified logarithmic velocity profile, e.g., Cebeci and Smith (1974),

$$\frac{U}{u_\tau} = \left[\frac{2.303}{\kappa}\ln\left(z^+\right) + C_1\right] - \frac{\Delta u}{u_\tau} + \frac{\omega}{\kappa} f\left(\frac{z}{\delta}\right) \tag{10.6}$$

In the first parenthesis, we find the Prandtl–Karman term for a flat bed, which is supplemented by the second term, which stands for the decrease of the velocity due to the roughness, whereas the third term stands for Coles (1956) law of the wake. This equation is usually simplified by using k_s and omitting Coles correction for $z/h \leq 0.15$. By using von Karman's constant as $\kappa = 0.4$, we get

$$\frac{U}{u_\tau} = 2.5\ln\left(30\frac{z}{k_s}\right) \tag{10.7}$$

Equations 10.6 and 10.7 represent mean velocity profiles, and the associated drag is given by Eq. 2.39.

The separation vortices are the source of the vorticity present in the mean flow, which is given by the derivative of Eq. 10.7 as

$$\omega(z) = \frac{\partial U}{\partial z} = \frac{2.5u_\tau}{z} \tag{10.8}$$

For a more analytical representation, assumptions are required. A first one would be to postulate a conservation of circulation Γ by which the vorticity displaced by the frontal area of the grain is released in the two side vortices, also called horseshoe vortices. As a first approximation, the displacement area can be represented as that by an exposed half sphere, with $r = d_s/2$, and we get

$$\frac{\Gamma}{2} = \int_{\mathrm{d}F} \omega(z)\mathrm{d}f \tag{10.9}$$

with the usual problem that for z = 0, Eq. 10.8 has a singularity. This can be overcome by introducing a two-layer configuration with an interface at z_V equal to a viscous length scale. With this split, ω can be formulated for $d_s > z > z_V$ by using Eq. 10.8, whereas for $z_V > z > 0$, a linear velocity profile Eq. 2.42a with a constant vorticity is assumed. The boundary is assumed to be arbitrarily at z_V ($z^+ = 5$), and therefore

$$z^+ = 5 \therefore z_\nu = \frac{5\nu}{u_\tau} \tag{10.10}$$

The integral Eq. 10.9 can be evaluated by using polar coordinates, and $r_s = d_s/2$ as

$$\frac{\Gamma}{2} = \int_{\varphi_v}^{\pi/2} \frac{c_1}{r\sin\alpha} r^2 \cos^2\alpha \, \mathrm{d}\varphi + \int_0^{\varphi_v} \frac{c_1}{z_v} r^2 \cos^2\varphi \, \mathrm{d}\varphi \tag{10.11}$$

$$\wedge \quad c_1 = \frac{2.303}{2\kappa} u_\tau ; \varphi_v = \arcsin\frac{z_v}{r}$$

This integral can be calculated numerically by using

$$\frac{\Gamma}{2} = c_1 r \left[\int_{\varphi_v}^{\pi/2} \frac{\cos^2\varphi}{\sin\varphi} \mathrm{d}\varphi + \frac{r}{z_v} \left[\frac{1}{2}\varphi_v + \frac{1}{4}\sin 2\varphi_v \right] \right] \tag{10.12}$$

At this point, we need additional assumptions about the vorticity distribution in the horseshoe vortex since the vorticity is mainly concentrated in the vortex skeleton. In visualizations, such as shown in Fig. 9.13., it can be seen that the vorticity released into the wake from the side zones can be neglected. Based on the same observations, it can be assumed that the horseshoe vortices are of Rankine type. That means, a vortex with a solid body rotation of the core and an outer potential vortex discribes the velocity field

$$u_\varphi(r) = \begin{cases} r\omega & 0 \le r \le r_0 \\ \dfrac{\Gamma}{4\pi r} & r_0 \le r \end{cases} \tag{10.13}$$

and the associated pressure field

$$p(r) = \begin{cases} p_0 + \dfrac{\rho}{2} v^2 & 0 \le r \le r_0 \\ p_\infty - \dfrac{\rho}{2} v^2 & r_0 \le r \end{cases} \tag{10.14}$$

The time dependence of a similar vortex was discussed in Chap. 6 (Sect. 6.2.7) and is given by the Oseen solution Eq. 6.86, respectively Eq. 6.87.

In this estimate, Γ is supposed to be known, but on r_0, any assumption is very uncertain since this quantity depends on the Reynolds number and other parameters. To provide a suggestion on the insight, how one could progress, an evaluation is made based on the visualization Fig. 9.13 for $Re_\mathrm{d} = 38800$ and $r_0 = 1/8 d_\mathrm{s}$.

With this assumption the constant value of the vorticity in the core of the lateral vortex becomes

$$\omega_x = \frac{\Gamma}{4\pi r_0^2} = \frac{4\Gamma}{\pi d_\mathrm{s}^2} \tag{10.15}$$

This value can be inserted into Eq. 10.13, which represents the velocity field. The pressure can now be calculated from Eq. 10.14, with a pressure p_0 on the vortex axis of

$$p_0 = p_\infty - \rho v_{\max}^2 \quad \text{and} \quad p_\infty = \lim p(r) \xrightarrow{r=\infty} 0 \tag{10.16}$$

This approach is for a single exposed grain, and allows an estimate of the transport of fine material sitting along the sides of such a grain by using the lift force as a reference. The small grain can be transported when

$$F_z \cong \Delta p A \tag{10.17}$$

with A denoting the area of exposition of the small grain with respect to the side vortex.

A similar evaluation can be made for the transport of fine sediment by coherent structures if we can describe the vortex produced by a sweep. As a first step, we again introduce a vortex model and the advection velocity of the core, here taken as U. The circulation can be estimated from the velocity u' and the scale of the sweep, since the entire vorticity in the sweep will be found in the front vortex as ω_y,

$$\omega_y \approx \frac{u'}{h_s} \quad \therefore \quad \Gamma \approx \omega_y h_s L_s = u'UT_s \tag{10.18}$$

Using Eqs. 10.13 and 10.14 with the above assumptions leads to

$$u_\varphi(x,z) = \begin{matrix} \left(\frac{\Gamma}{2\pi r_0^2} z + U, -\frac{\Gamma}{2\pi r_0^2}(x-Ut)\right) & \wedge (x-Ut)^2 + z^2 = r^2 \Rightarrow r \le r_0 \\ \left(\frac{\Gamma}{2\pi r^2} + U, -\frac{\Gamma}{2\pi r^2}(x-Ut)\right) & \wedge \quad r \ge r_0 \end{matrix} \tag{10.19}$$

and

$$p = \begin{cases} p_0 + \frac{1}{2}\rho_f \left(\frac{\Gamma}{2\pi r_0^2}\right)^2 r^2 & r \le r_0 \\ p_t - \frac{1}{2}\rho_f U^2 - \frac{1}{2}\rho_f \left(\frac{\Gamma}{2\pi r}\right)^2 r^2 & r \ge r_0 \end{cases} \tag{10.20}$$

$$\wedge \quad p_0 = p_t - p_0 + \frac{1}{2}\rho_f U^2 - p_0 - \rho_f \left(\frac{\Gamma}{2\pi r_0^2}\right)^2 \quad \wedge \quad \frac{\Gamma}{2\pi r_0} \equiv u_{max}$$

with the pressure p_t at the stagnation point.

Using Eqs. 10.19 and 10.20, the lift forces can be calculated the same way as for the lateral vortices.

10.4 Induced Secondary Flows

An open channel flow is defined by a free surface by two sidewalls and a bed. The aspect ratio of most rivers is so large that the influence of the sidewalls can be neglected. This simplification seems even more appropriate with respect to sediment transport since it mainly occurs in the vicinity of the bed. For a bed consisting of fine sand, this is indeed a good approximation, but when the sediment consists of different grain sizes, secondary flows develop and can become dominant even for the transport.

What is a secondary flow?

Prandtl (1926) noted that in any channel there is a change in slope from the side walls to the bed, and this corner is associated with a change of the drag force, which has to be compensated by a spanwise momentum transport. Eichelbrenner and Preston (1971) showed by using the Reynolds Eq. 2.8 that the pressure gradient acting on the outer flow is greater from the bed toward the corner than toward the smooth bed, and this difference in pressure produces a secondary circulation with a vorticity in the flow direction. Einstein and Li (1958b) came to the same conclusion by using the vorticity equation in the Reynolds representation

$$\frac{\partial \bar{\zeta}_i}{\partial t}+\bar{u}_j\frac{\partial \bar{\zeta}_i}{\partial x_j}=\bar{\zeta}_j\frac{\partial \bar{u}_i}{\partial x_j}+\frac{\partial^2 \overline{u_k'u_j'}}{\partial x_l \partial x_j}-\frac{\partial^2 \overline{u_l'u_j'}}{\partial x_k \partial x_j}+\nu\frac{\partial^2 \bar{\zeta}_i}{\partial x_j^2} \tag{10.21}$$

with

$$\bar{\zeta}_i=\frac{\partial \bar{u}_l}{\partial x_k}-\frac{\partial \bar{u}_k}{\partial x_l} \tag{10.22}$$

Gessner (1973) found by using the energy equation for the mean flow

$$\frac{\bar{k}}{\partial t}+\bar{u}_j\frac{\partial \bar{k}}{\partial x_j}=\frac{\partial}{\partial x_j}\left(-\frac{\bar{p}}{\rho_f}\bar{u}_j+2\nu\bar{u}_i S_{ij}-\overline{u_i'u_j'\bar{u}_i}\right)-2\nu S_{ij}S_{ij}\overline{u_i'u_j'}S_{ij} \tag{10.23}$$

$$\wedge\quad \bar{k}=\frac{1}{2}(\bar{u}_i\bar{u}_i);\quad \wedge\quad S_{ij}=\frac{1}{2}\left(\frac{\partial \bar{u}_i}{\partial x_j}+\frac{\partial \bar{u}_j}{\partial x_i}\right)$$

and without its turbulent supplement

$$\frac{\partial k}{\partial t}+\bar{u}_j\frac{\partial k}{\partial x_j}=\frac{\partial}{\partial x_j}\left(\frac{1}{\rho_f}\overline{u_j'p}+\frac{1}{2}\overline{u_j'u_i'u_i'}-2\nu\overline{u_i's_{ij}}\right)-\overline{u_i'u_j'}S_{ij}-2\nu\overline{s_{ij}s_{ij}}$$

$$\wedge\quad k=\frac{1}{2}\left(\overline{u_i'u_i'}\right);\quad \wedge\quad s_{ij}=\frac{1}{2}\left(\frac{\partial u_i'}{\partial x_j}+\frac{\partial u_j'}{\partial x_i}\right) \tag{10.24}$$

that only the term

$$\frac{\partial}{\partial y}\left(\bar{u}_x\overline{u'v'}\right) \tag{10.25}$$

is contributing to the secondary flow.

Using his model the well-known schematic of the secondary flow can be constructed as shown in Fig. 10.3.

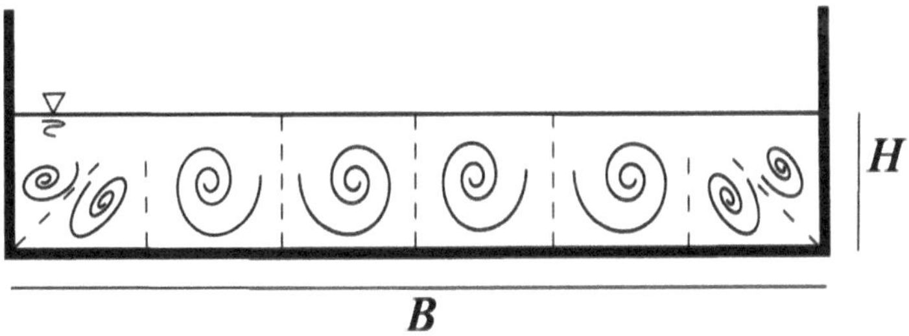

Fig. 10.3 A cross section of a rectangular open channel flow with secondary flow starting by a flow toward the corners and propagating toward mid-channel by forming quadratic cells of side length H

The flow with a component toward the corners stimulates a series of rollers that are embedded in a quadratic cell structure of a side length H. This is the case for an even aspect ratio B/H. In all other cases, the cells are elliptic and of low stability, especially when the aspect ratio is an odd number. Townsend (1976) investigated the secondary flow at the edges of a flat plate, a problem relevant for wings of aircraft. He used the Reynolds equation (Eq. 2.8) and a boundary layer approximation he published in 1961. With respect to open channels, he showed that secondary flows are the result of the non-Newtonian properties of the Reynolds Shear stress tensor (Eq. 2.9). Townsend (1976) showed that a spanwise variation of the wall shear stress could be the origin of secondary flows. This implies that the secondary flow is directed toward the wall where the wall shear stress is higher. Hinze (1967) deduced this result from the equation for the kinetic energy and confirmed it in 1973.

However, one has to be aware that vortices in flow direction are not steady; they exist in the time averaged velocity distribution only. The secondary flows consist of conical vortices attached to the bed, which grow and decay periodically with a spanwise distribution as indicated. The instanteneous velocity distribution in the conical vortices is

$$u_i(\underline{x}) = u_0(x - x_0) f_i\left[(y - y_0)/z_a, z/z_a\right] \tag{10.26}$$

with the origin of the cone at $\underline{x}$ = (x_0, y_0, 0), $u_0(x–x_0)$ is the velocity scale in a cross section at a distance $x–x_0$ from the origin, and $z_a(x–x_0)$ corresponds to the diameter of the vortex at that location. When $z_a = H$ the vortex decays and is replaced by a new one, only the value of x_0 is a variable.

For a bed consisting of uniform grains or very fine sediment, no variations of roughness is uniform, but for a grain size distribution, sorting produces roughness changes which lead to variations of the wall shear stress.

From Fig. 10.3, it follows that as soon as a local sorting occurs, the process becomes self-organized, since the transport in span direction will be enhanced, resulting in bed forms of longitudinal stripes in flow direction, i.e., with rough and smooth stripes alternating in spanwise direction. This is a self-organizing process, which shows that a treatment of the bed as a homogeneous cover is inadequate for sediment mixtures, and the transport of such sediments cannot be adequately described without introducing secondary flows. This is one of the essential limitations of the classical representation for this class of sediments.

In case the process of sorting does not start spontaneously, it will be triggered by the corner flow. When the aspect ratio is not an even number, the system needs some time to find its asymptotic form, which is determined by the corner flows. Since secondary flows are associated with mixtures containing a sufficient fraction of coarse material, it is possible to investigate the system experimentally by preparing a bed in which an alternating roughness is built in from the start. Such investigations are important as they show how strong the secondary flows are (Wang and Nickerson, 1972; Müller and Studerus, 1979, 1981; Studerus, 1982; Nakagawa et al., 1981 and Nezu and Rodi, 1985). All these investigations confirm that once longitudinal stripes of different roughness exist, the bed forms in the channel are sustained.

When a secondary flow exists, Eq. 2.8 cannot be reduced by simplifications as the one leading to Eq. 2.12 since U is not anymore a function of z only, i.e.

$$\underline{U} = \overline{u}_i = \left(U_x(0,0,z), U_y(0,y,z), U_z(0,y,z)\right) \neq 0 \tag{10.27}$$

which shows that the transport toward the wall is inhomogeneous. In the mean, the flow can still be considered a 2D and as only depending on y and z. In other words, by using the simplification

$$\frac{\partial U_i}{\partial x} = 0 \quad \wedge \quad \sqrt{u_x^{'2}} = 0 \tag{10.28}$$

and introducing the pressure as an outer force given by the slope, Eq. 2.8 can be written as

$$\frac{\partial \overline{u}_x}{\partial t} = 0 = -\overline{u}_y \frac{\partial \overline{u}_x}{\partial y} - \overline{u}_z \frac{\partial \overline{u}_x}{\partial z} - \frac{\partial \overline{u_x' u_y'}}{\partial y} - \frac{\partial \overline{u_x' u_z'}}{\partial z} + gS \tag{10.29}$$

The negative sign stands for a deceleration and the positive for acceleration. It was found that in the outer flow region all four decelerating terms are of the same order of magnitude and compensate for the acceleration due to the slope. Close to the wall one finds however that the first and the fourth term on the right are dominant and of opposite sign. The fact that the two terms compensate each other confirms that a momentum exchange between the vertical advection and the turbulent contribution is present. The momentum exchange due to the two remaining terms is only about 5–10% of gS. The inhomogeneity of the rolls over the height is another proof that the roll model is a virtual one and resulting from the averaging of conical vortices.

The rolls are driven by the spanwise variation of the momentum transport toward the bed, which is compensated by a momentum flux in spanwise direction as illustrated by the results, as shown in Figs. 10.4, 10.5, 10.6, and 10.7.

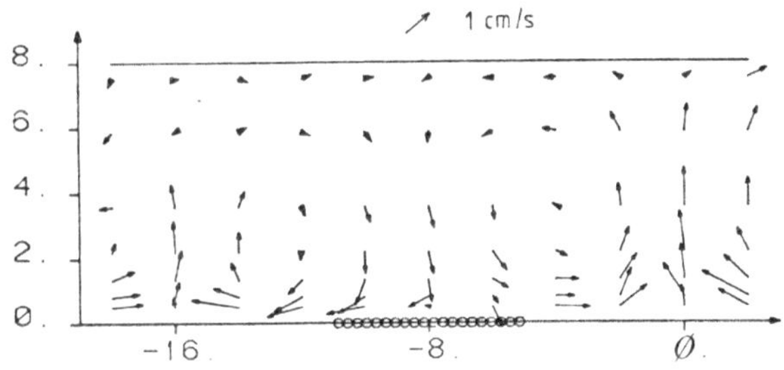

Fig. 10.4 The mean velocities $<u_y>$ and $<u_z>$ in an open channel flow with alternating roughness in the spanwise direction. From Studerus (1982)

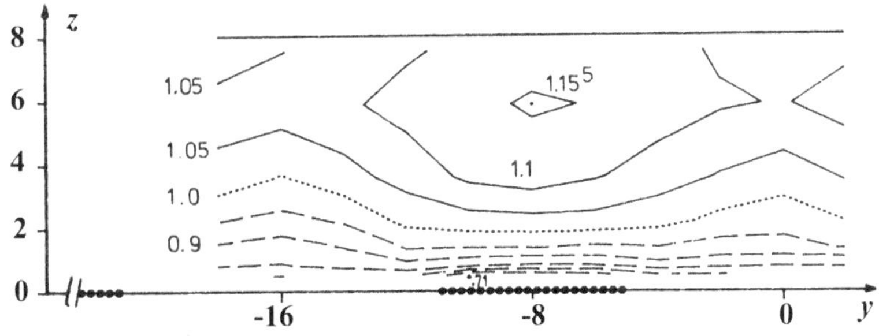

Fig. 10.5 The iso-velocity lines in flow direction normalized with U, $<u_x>/U$. From Studerus (1982)

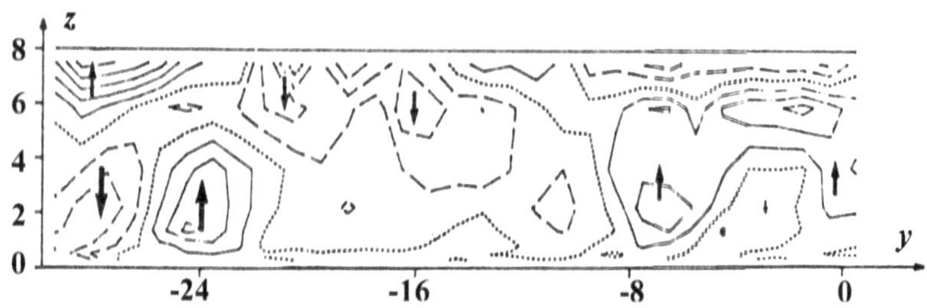

Fig. 10.6 The iso-lines of the vertical velocities $<u_z>/U_x$ in an open channel of 25 m × 0.6 m at x = 4.5 m in intervals of 0.5% of U_x. Dotted line for $U_z = 0$, solid lines for positive, broken ones for negative. From Studerus (1982)

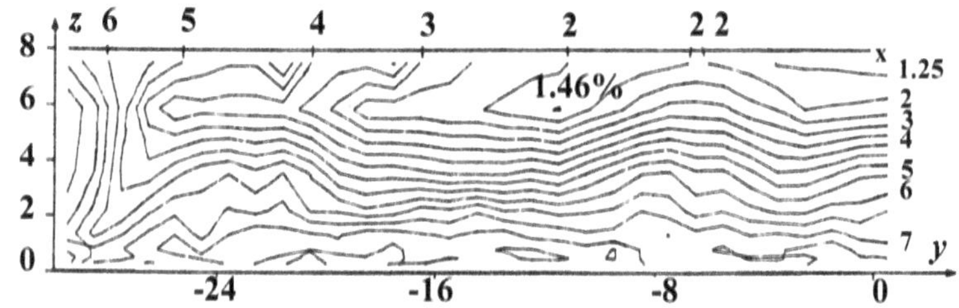

Fig. 10.7 Turbulence intensity in an open channel of 25 m × 0.6 m at x = 4.5 m in intervals of 0.5%. From Studerus (1982).

A complete description of the secondary flow should relate Eq. 10.29 to a specification of the roughness. To achieve this, the shear stresses may be split into advective and turbulent contributions by using experimental evaluated values

$$\begin{aligned} \tau_{xx} &= -\rho_f \overline{u}_x^2 - \rho_f \overline{u_x'^2} \\ \tau_{yx} &= -\rho_f \overline{u}_x \overline{u}_y - \rho_f \overline{u_x' u_y'} \quad \Rightarrow \quad \rho_f \overline{u}_x \overline{u}_y >> \rho_f \overline{u_x' u_y'} \\ \tau_{zx} &= -\rho_f \overline{u}_x \overline{u}_z - \rho_f \overline{u_x' u_z'} \quad \Rightarrow \quad \rho_f \overline{u}_x \overline{u}_z >> \rho_f \overline{u_x' u_z'} \end{aligned} \tag{10.30}$$

The advective momentum fluxes are much larger than the turbulent ones. Nevertheless, the momentum exchange with the bed is a contribution by the turbulent terms. A detailed quantitative description would require additional information, related to the aspect ratio of the cross sections of the flow, and the grain size distribution etc. The situation is very complex but can be understood in principle. Tsujimoto and Kitamura (1996) made a first attempt to use these concepts for numerical modeling.

10.5 Bed Forms Due to Sorting Effects

The bed form we considered consists of alternating stripes of smooth and rough texture, and they are due to secondary flows. A superposition of the main and secondary flow velocities produces a helical flow. As a first approximation, sediment transport may be thought to be the result of the transports in the flow and spanwise directions. This assumption is rather risky, because sorting occurs even in case the spanwise transport is subcritical, because the particles resuspended by the main flow direction can be transported sideways by the secondary flow.

10.5.1 Coarse Dunes

Dunes can also form on beds containing sediment mixtures but in a limited range of size distributions only. From Chap. 9 (Sect. 9.5), we can conclude that there must exist a limit in size, however its value is not known yet. One idea to estimate this value is to postulate that the transport must be due to a resuspension by the turbulent flow structures and that separation effects on single obstacles are irrelevant. In this case an estimate based on Eqs. 10.18–10.20 could be used. A criterion for wind transported sediment was given by Bagnold (1941)

$$u_{\tau c} = A\left(\rho' g d_{sc}\right)^{1/2} \quad \wedge \quad A = 0.09 \tag{10.31}$$

with $u_{\tau c}$ the critical wall shear velocity, $\rho' = (\rho_s - \rho_f)/\rho_f$, the relative density of the grains, g the gravitational acceleration, and d_c the critical grain diameter. A is an empirically evaluated coefficient. In other words, the critical diameter is given by

$$d_{sc} = \frac{u_\tau^2}{A^2 \rho' g} \tag{10.32}$$

In open channel flows with water, this value is about one tenth of the one in air due to the difference in fluid density and the absence of momentum transfer by grain collisions. For the formation of dunes, enough material of size d_c must be available to cover so that the bed and no rougher material should be exposed. But also no smoother grains in a fixed form.

Even when dunes develop on beds with coarse grain mixtures, the bed forms are not composed homogeneous anymore. They show a sorting in the vertical direction because the deposition in the separation zone is preferentially due to the free flight of the resuspended grains, that is, determined by their weight. The result is a stratified dune. In case these dunes are moving due to transport, the erosion process is confronted with a vertical variation of roughness. In the extreme case, the coarser grains are transported by a rolling downhill on a ramp on the lee side with a slope corresponding to the angle of repose. The grains then cover the bed in the trough, with the effect that the dunes migrate on a self produced armored bed (Zanke, 1982). A vertical sorting of grains in dunes was also described by Blom et al. (2003) based on Hirano's (1971) two-layer model. This model features an active transported layer on top and a layer beneath that is at rest and consists of coarser material, which does not need to have a uniform grain size height as in the armoring case. Similar mechanisms are mentioned by Crickmore and Lean (1962) and Ribberink (1987). In both of these papers, it was clearly shown that the classical transport equations fail when sorting occurs. Blom et al. (2003) and Blom (2003) tried to fill this gap by introducing a new transport law for the case where sorting occurs. She starts with an expanded continuity equation, making use of a concept by Parker et al. (2000) for the transport by the migration of dunes. For the motion of the grains, Einstein's (1950) concept of saltation was supplemented with a sorting function in the lee region of the dunes given by the variability of their shapes.

This sorting mechanism is not only important for sediment transport but also for the transport of contaminants as undesired chemicals often adhere preferentially to the small particles with their relatively large active surface per unit volume.

It is often mentioned that the transport in suspension is important for dune formation. It is evident that by a sorting process, the finer fraction will be moved away so that the bed and the dunes are formed from coarser material alone. A superposition of the two modes of transport, the one by migration and the one as suspension is, therefore, appropriate for calculating the total transport.

10.5.2 Bed Forms Resulting from Secondary Flows

As we saw in Chap. 10. (Sect. 10.4), the first self-organized bed form for grain mixtures are longitudinal stripes of a width h. For $h > d_{s,}$ one can observe the formation of bed-forms on the smooth stripes described in Chap. 9, as well as dunes described in Chap. 10 (Sect. 10.5.1), and in the thesis of Günther (1971).

Apart from the types of bed forms mentioned above one encounters so-called barchans. They are dunes formed when the transport capacity of fine material is larger than the supply due to erosion, when fine material moves on a solid nonerodible bed, or when the secondary flow interacts with the mean flow.

Normal dunes are more or less periodic and 2D bed-forms transversal to the orientation of the main flow. They consist of a rather long luff front of modest slope and a much shorter lee side with a slope equal to the natural angle of repose (Fig. 10.8.).

Fig. 10.8 Typical dune

By contrast, barchans are also spatially periodic but 3D features. The slopes are similar to those for dunes, but the bed forms are now similar to a half-moon, The so-called side horns are often longer than the width of the barchans (Fig 10.9), and they occur only on the smooth surfaces. The formation can be explained as follows. When a deposition of

Fig. 10.9 Typical barchanes

grains becomes large enough due to the intermittent motion of sediment in the turbulent flow regime, the deposition acts as an obstacle, which causes an accelerated flow over the luff side and a separation on the lee side.

For fine sand, the vorticity of the flow field due to these obstacles is concentrated in vortex skeletons, and is the cause for the formation of 2D ripples which act as disturbances necessary for the development of dunes (Gyr, 2003; Gyr and Kinzelbach, 2003). When secondary flows occur, we encounter a helical flow depositing the grains preferentially as shown in Fig. 10.10.

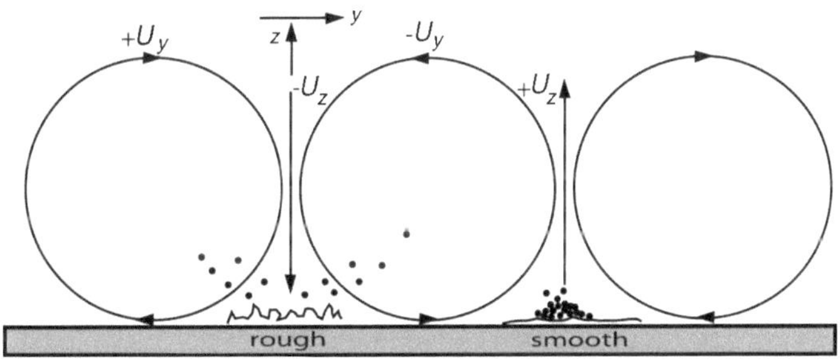

Fig. 10.10 Secondary flows over a bed with a spanwise variation in roughness. The finer sand is eroded on the rough area and deposited over the smooth areas producing a striped configuration

On the smooth areas barchans are formed instead of dunes.

The corresponding topology of the vortex skeleton is reproduced in Fig. 10.11.

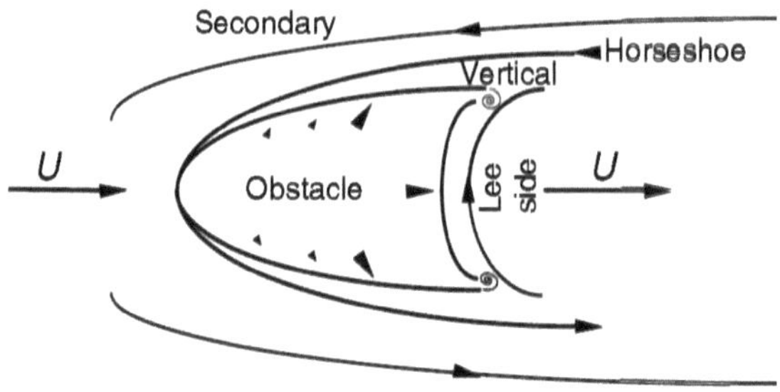

Fig. 10.11 Vortex skeleton of the separation on an obstacle with superimposed secondary flow produced by the spanwise variability of the roughness

For a quantitative estimate of the transport, several assumptions are required. As the longitudinal roughness stripes are developing, secondary flows are enhanced. In the mean, they are considered to be longitudinal vortices scaling with the water depth, so that, their radius is $H/2$, and the width of the stripes of different roughness is H in good approximation. The vortices are idealized as exhibiting a solid-body rotation with a circumferential velocity decreasing linearly toward their center at height $H/2$. The vorticity of these vortices is interacting with the one produced by bed forms, which is concentrated in a vortex skeleton around the corresponding feature. It is evident that as soon as the secondary flow becomes strong enough, 3D bed forms must develop because the deposition of the whirled up grains is controlled by these vortices, and they settle where the vorticity is small.

For a quantitative description of the sediment transport, the ratio of the vorticity of the secondary flow and the one due to the bed forms must have a characteristic value. The development of the stripes is very dynamical, but for an estimate one can use the final configuration. Such a configuration was investigated by Studerus (1982), but without sediment transport.

Assuming that the flow is a 2D open channel flow uniform in flow direction ($\partial/\partial x = 0$) and stationary ($\partial/\partial t = 0$), the mean momentum flux in y-direction from the smooth to the rough area, which is maximum at the transition between the two bed roughnesses, can be estimated using the continuity and the Reynolds equation as

$$\tau_{yxm}\Big|_{y=h/2} = \left(\frac{\lambda_{\mathrm{r}}}{\lambda_{\mathrm{m}}} - 1\right)\rho_{\mathrm{f}} g J \frac{h}{2} \tag{10.33}$$

with the subscripts r for rough and m for mean. λ is the roughness coefficient used in the Moody diagram, by using instead of the hydraulic radius R_{H}, $4R_{\mathrm{H}}$ as the correct length scales to determine Re and $k_s/4R_H$. J is the slope and is responsible for the rate energy and is fed into the system. The mean roughness coefficient is given by

$$\lambda_{\mathrm{m}} = \frac{1}{2}\left(\lambda_{\mathrm{rough}} + \lambda_{\mathrm{smooth}}\right) \tag{10.34}$$

and the hydraulic radius as

$$R_{\mathrm{H}} = \frac{hb}{2h+b} \tag{10.35}$$

The momentum flux in the spanwise direction is a linear function of y, being zero at the center of the rough element and given by Eq. 10.33 by matching the two functions for smooth and rough walls. For a mixture of the sediment, λ can be estimated since the nontransportable grains determine the roughness of the rougher surface, which is an armored bed.

The momentum transport in spanwise direction has two causes, the advective secondary flow and a turbulent momentum exchange, which will be neglected. To within this approximation, one may calculate the mean transversal velocity due to the secondary flow. With the above assumptions we can write

$$\rho_{\mathrm{f}} \int_0^h \bar{u}_x(z)\bar{u}_y(z)\mathrm{d}z = h\tau_{yxm}\Big|_{y=h/2}$$
$$\ni \bar{u}_x(z) = \frac{u_\tau}{\kappa}\ln(z) + C \quad \wedge \quad \int_0^h \bar{u}_y(z)\mathrm{d}z = 0 \tag{10.36}$$

with the von Karman constant κ. For a solid body rotation

$$\bar{u}_y = \bar{u}_{y0}\left(1 - \frac{2z}{h}\right) \tag{10.37}$$

Equation 10.36 can be integrated and get

$$\bar{u}_{y0} = -\frac{\kappa}{0.5\rho_{\mathrm{f}} u_\tau}\tau_{yxm}\Big|_{y=h/2} \tag{10.38}$$

A good approximation for the unknown u_τ, is $u_{\tau\mathrm{m}}$

$$u_{\tau\mathrm{m}} = \sqrt{\frac{\tau_{\mathrm{m}}}{\rho_{\mathrm{f}}}} \quad \wedge \quad \tau_{\mathrm{m}} = \rho_{\mathrm{f}} gJh \tag{10.39}$$

The center of the secondary vortices lies just above the roughness change, such that the vorticity ω_x has its maximum there as well, and defined by Eq. 10.37

$$\omega_x = \frac{\partial \bar{u}_y}{\partial z} = -\frac{2\bar{u}_{y0}}{h}$$

This vorticity is enhanced by the vorticity concentrated in the side vortices, because the vorticity of the lateral vortices is of the same sign as the one of the vortices of the secondary flow as shown in Fig. 10.11. The barchans remain 3D since no deposition occurs on the rough areas of the bed. As a consequence, the barchans have no direct neighbors which would create 2D forms as in case of the ripple formation.

The total vorticity ω_x has to be compared with the vorticity ω_y of the lateral separation vortices. The horns of the barchans grow with increasing ω_x/ω_y, for a low value of this ratio we encounter stripes of dunes as reported by Günther (1971).

Since the separation in the lee side of the obstacles is of mixing layer type, the main vortex skeleton in this area is a highly deformed spanwise structure, in the mean containing the vorticity created by the velocity difference over the forming shear layer. If the thickness of this layer is δ^*, the vorticity is given by the derivative of Eq. 10.36 in terms of a new coordinate ζ for the z-direction, with its origin at the crest of the obstacle we obtain

$$\omega_y(\zeta) = \frac{\partial \bar{u}_x(\zeta)}{\partial \zeta} \tag{10.40}$$

The average vorticity can be calculated by using the circulation theorem

$$\bar{\omega}_y = \frac{\Gamma}{A} = \frac{1}{A}\int_0^{\delta^*} \omega_y(\zeta)\bar{u}_x(\zeta)\mathrm{d}\zeta \quad \wedge \quad A = \int_0^{\delta^*} \bar{u}_x(\zeta)\mathrm{d}\zeta \tag{10.41}$$

As for the well-known case of a logarithmic velocity profile, we have to replace 0 by a small value to compute the important ratio ω_x/ω_y of the two vorticities.

The flow over the rough part of the bed is helical and has a ω_y which is even larger than over the smooth part. The topology shown in Fig. 10.11 is more complicated than a simple interaction of two vortices which are perpendicular to each other, so ω_z plays a role too, and the interaction of ω_x with ω_y is highly dependent on the geometry of the vortex skeletons. They contain a larger vorticity than is found by averaging and the process of how vorticity becomes focused is of importance as well. With these results, we can formulate criteria to distinguish various classes of barchans.

To raise grains from the bed requires an impact of vortices with the bed; they are produced by the turbulent flow field characterized by τ_{wm} Eq. 10.39. τ_{wm} has to be larger than the critical wall shear stress given by the Shields diagram for the fine material. τ_{xwm} is always higher than τ_{ywm}, and therefore the secondary flow determines where the grains will deposit, but they are not responsible for the main transport mechanism. To obtain a good estimate, we assume that the spanwise component in the wall shear is about one tenth of the longitudinal one. Since this ratio in wall shear stress is equivalent to the ratio ω_x/ω_y at the bed, this ratio in vorticity can be used also if one would like to evaluate the wall shear ratio more precisely.

The ratio ω_x/ω_y is smaller than one but together with the lateral vortices in Fig. 10.11 with their vorticity ω_{sx}, the relevant vorticity ratio

$$\left(\omega_x + \omega_{sx}\right)/\,\omega_y\big|_{z\to 0} \tag{10.42}$$

is of order 1. For an increasing ratio ω_x/ω_y the ratio Eq. 10.42 grows as well, and with it grows the side horns of the barchans. This provides a criterion, and not a quantitative assessment, since the mechanism of vortex interaction is much more complicated, as stated earlier, in addition, it should be noted that the criterion is only valid within a small range. For a high value of the ratio in Eq. 10.42, new forms of barchans could be created, however these cases are not very realistic since stronger secondary flows are rather impossible because the needed high roughness variation is unrealistic.

The width of the barchans is of order H, and their horns grow with increasing vorticity ratio ω_x/ω_y.

10.5.3 Additional Bed Forms

One has to be aware that the flows mentioned here are essentially 2D open channel flows, whereas a river can meander. Since the secondary flow can deepen the stripe structure, the channel can be thought as composed of local subchannels. Any disturbance of such a configuration produces an instability, which starts a meandering of the flow. Such an explanation for the meandering problem would require a theory for the development of banks, as well as quantitative description of the winding of the river.

The main new aspects of meanders is a disturbance of the secondary flow, which has its origin in the difference between the momentum transport close to the bed and the one near the surface. We see that these processes are driven by the momentum transport, and controlled by the wall roughness as well as the topography of the bed. To understand meandering, the main goal should be to investigate the role of the large topological flow structures associated with the bed forms, since they are the main features responsible for the sediment transport and also help to simplify the simulations

11 Gravel Beds

By examining gravel beds, we start looking at the real world since the complexity in this case is so large that there is no universal theory available nor is there any hope of finding one. The best we can expect to have is suitable empirical formulas available, which are primarily recipes for certain cases rather than theories in the sense of a mathematical description. Such a situation requires a classification of the various applications, and hopefully a physical concept that helps to understand transport processes.

In this chapter, the concepts used at present are briefly described, and a new type of roughness is proposed which supplements the one discussed in Chap. 7. Again we make use of the separation mechanisms described in Chap. 8, to extend the physical description and to stimulate further research based on the transport processes which occur in rivers with gravel beds.

The concepts used today can be found in the literature, e.g., in the book of Hey et al. (1982).

11.1 Transport Processes on Gravel Beds

The concept of sediment transport used in Chaps. 2 and 3 was so successful that it was also applied to describe the process for gravel beds, although it soon became clear that the scatter of the measurements was so large that they cannot be described by a single theory. As a consequence, a number of different approaches were developed in which the original concept was modified by introducing a variety of empirical constants and functions. These "knobs" could then be used to adjust the equations for the case under consideration. To predict the transport for a particular river, it is therefore necessary to determine which one of the descriptions should to be used. To some extent, this is also required for the cases discussed in Chaps. 9 and 10, but as we shall see, the assumptions made in those cases are much more of physical nature.

In these two chapters we assumed a straight open channel as the basic geometry of the river in which the bed form could adapt primarily to a change in discharge, with secondary currents adding some complexity. In rivers with gravel beds it is evident from field observations that not only the bed forms but also the channel-form is changing in space and time. In other words, the assumption of a steady flow is problematic. In addition, the sediment transport itself is highly time dependent because there is a hysteresis between the rising and falling limb. These critical arguments turned out to be true also for river with finer material. In the classical theories, all these influences are adjusted by a series of knobs.

The transport concept used is the one introduced by Du Bois Eq. 2.45. The transport forces acting on a particle are given by the wall shear stress. The idea behind this description is that the origin of these forces is given by the flow regime. A more interactive theory would need an explanation of the acting stresses on the grains based on physical mechanisms. In analogy to the shear stress distribution in a turbulent channel flow over rough beds (Fig. 2.2) it was assumed for gravel beds too that the wall shear stress is represented by the (x, z)-component of the Reynolds tensor, Eq. 2.14, and since gravels stand for an extremely rough bed, it was thought that this concept should be even better suited for this regime than for the ones discussed in Chaps. 9 and 10. However, with the same restriction as discussed in Chap. 3 that at a given location on the bed one needs a momentaneous load in addition to the mean wall shear stress, as shown in Eq. 3.1. For a very rough regime, the value of the load in that equation could be higher.

This extreme value of the relevant bed shear stress components was thought to be the result of a high-turbulent intensity even for gravel beds. This was not explicitly said but assumed in analogy to the theory of turbulence over rough walls. We will show that this assumption is wrong, but it has to be recognized that the mean wall shear stress is one of the rare quantities, which can be measured, and this quantity thus remains a candidate for building empirical theories. The wall shear stress is usually given by

$$\tau_{\mathrm{w}} = \rho_{\mathrm{f}} g R_{\mathrm{H}} S_{\mathrm{E}} \tag{11.1}$$

with S_{E} denoting the so-called energy slope, or it is based on velocity measurements at two different distance z_1 and z_2 from the bed

$$\tau_{\mathrm{w}} = \rho_{\mathrm{f}} \left[\frac{\overline{u}_{x2} - \overline{u}_{x1}}{2.3/\kappa\left(\ln\left(z_2/z_1\right)\right)} \right]^2 \tag{11.2}$$

It is common to relate the shear stress to the velocity by means of resistance coefficients. The most commonly used ones are

$$U^2 = \frac{8\tau_{\mathrm{w}}}{f\rho_{\mathrm{f}}} \qquad f : \text{Darcy} - \text{Weisbach coefficient} \tag{11.3}$$

or

$$\frac{U}{\left(gHS_{\mathrm{E}}\right)^{1/2}} = \left(\frac{8}{f}\right)^{1/2} = \frac{H^{1/6}}{ng^{1/2}} = \frac{C}{g^{1/2}} \quad \begin{array}{l} n: \text{Manning coefficient} \\ C: \text{Chezy coefficient} \end{array} \quad \begin{array}{c} \text{Limerinos} \\ (1970) \end{array} \tag{11.4}$$

Using Eqs. 11.3 and 11.4 one can derive the mean velocity profile, using the basic logarithmic profile derived in Chap. 2 (Sect. 2.1.3) as

$$\frac{U}{u_\tau} = \frac{2.3}{\kappa} \ln(z) + C_0 \tag{11.5}$$

C_0 is a constant which depending on the characteristics of the boundary, and the Eq. 11.5 is usable for $z/H < 0.15$.

Colebrook (1939) suggested a more general equation for the velocity distribution, which he later modified to the form

$$\frac{U}{u_\tau} = \frac{2.3}{\kappa} \ln\left(\frac{H}{k_s}\right) + C_1 - C_2 - \frac{\Delta}{H} \tag{11.6}$$

with a k_{s} which is not strictly defined, and C_1 an empirical constant for the fully developed flow over roughness beds, C_2 accounts for the deviation of the velocity profile from the semi-logarithmic law near the edge of the boundary layer, again an empirically defined parameter. Finally Δ is the Clauser (1954) defect thickness, which relates the mean and the free stream or surface velocity. This equation is thought to be applicable as long as the following Reynolds number restrictions are fulfilled

$$u_\tau H/\nu \geq 2000 \quad \vee \quad u_\tau k_s/\nu \geq 100 \tag{11.7}$$

Hey (1979) found that the mean velocity can be predicted with good accuracy by

$$\frac{U}{\left(gHS_f\right)^{1/2}} = A\ln\left(\frac{a' R_H}{mk}\right) \quad \wedge \quad A = \frac{2.3}{\kappa} \tag{11.8}$$

where a' is a so-called channel shape factor; k, a roughness; and m, a constant for a given k.

A number of authors proposed additional expressions for the mean velocity. Three are mentioned here because they can often be found in the literature

$$U = \frac{H^{2/3}S_w^{1/2}}{n} \quad \wedge \quad \begin{array}{l} n = 0.041 d_{s50}^{1/6} \ \text{Chow}(1959) \\ n = 0.038 d_{s90}^{1/6} \ \text{Henderson}(1966) \end{array} \tag{11.9}$$

or with

$$n = \frac{0.113 H^{1/6}}{1.16 + 2\ln\left(H/d_{s84}\right)} \quad \text{Limerinos } (1970) \tag{11.10}$$

The Keulegan equation (1938)

$$U = \left(gHS_w\right)^{1/2}\left[6.25 + 5.75\ln\left(H/k_s\right)\right] \tag{11.11}$$

with a not strictly defined k_s.and the Lacey equation (1946)

$$U = 10.8 H^{2/3} S_w^{1/3} \tag{11.12}$$

These equations show what is needed to adjust the equations to a particular case. Information is needed for the roughness parameter, the roughness size distribution, the roughness orientation and shape, the roughness spacing, and the roughness scale. Additional parameters are required for the channel geometry, the longitudinally nonuniform flows, nonuniform bed profile, 3D flows, etc. Often assuming that function instead of the von Karman constant, a function does this.

The variety is so large that for a given case, one has first to make a search for the most appropriate description by estimates. The resulting bed load is not reliably predictable and even harder to measure. The usual procedure is therefore to divide the sediment into fractions and to evaluate the transport of the different classes using formulas like those described in Chap. 3. In addition, it is assumed that the finest sediment fraction is transported as suspended load, and that the largest grain size is specified by the criteria in Chap. 3 (Sect. 3.3.3). Therefore most gravel bed streams are paved at low flow, such that the median size d_{sp50} of the pavement is one and a half to three times the subpavement grain size d_{s50}. This does not mean that no finer sediment can be entrained from the bed by the flow, since whenever a particle of the armored bed becomes unstable; a part of the subarmor gets exposed to the flow and can be transported. Helland-Hansen et al. (1974) showed that a disturbance blow out process might occur which allows large and small particles to move together at conditions near incipient motion. Milhous (1973) showed an analogue result, namely that particles subject to incipient motion are similar to most sizes present in the bed surface. In addition, he showed that the armor of a gravel bed stream acts as a reservoir for fine sand. However, for gravel beds, the hiding factor becomes extremely important, a value that in the classical description can only be taken care of by empirically evaluated data.

11.2 Separation Versus Turbulence

Below the critical value at which the momentum transport from the flow to the bed is achieved by roughness elements instead of the viscosity, it is due to separation at these elements. The flow structure of the separations are vortices scaling with the size of the roughness elements, and they are not part of the turbulent ambient flow but embedded in it. Separation vortices are large structures feeding the turbulent flow with energy and vorticity. To be part of the ambient turbulence they would have to be in an equilibrium with flow structures of all sizes. The most evident example are separations on dunes, which are not part of the turbulent flow but embedded in it. It is therefore not surprising that turbulent flows are usually described as a flow over a smooth wall.

One of the main properties of a turbulent flow is that the nonlinear interaction between the various structures is so intense and swift that the flow becomes independent of the feeding mechanism, i.e., after a short time turbulence forgets its origin.

For fine sand, the feeding elements are so small that they are comparable with the structures over a smooth wall due to the instability process producing coherent structures. For this reason, we can assume that turbulence theories can be used to describe the flow even at small wall distances. The situation is quite different for gravels since the separation vortices are large and need much more time until they are "incorporated" in the turbulent flow. This kind of roughness requires another description based on the separation processes.

For practical purposes, the flow structures could be thought as being of turbulent origin and that the fluctuations have very skew distributions, but this view seems inadequate for separation regions.

It is clear that a suitable description needs a high resolution in space and time, since it must be based on the separation regions of a single roughness elements, and it should include a statistical approach to cover a sufficient bed area. The separation on an element is characterized by the velocity distribution of the oncoming flow and the obstacle size. On the basis of this information we can estimate the Strouhal frequency of the vortices shed, Eq. 6.23. Their size is about the size of the obstacle and they contain the vorticity accumulated during their evolution process. The family of these vortices and their interaction is the source of the forces acting on the gravel bed, and it is evident that flow structures with potentially high-lift forces are much more abundant than in a purely turbulent flow. Such forces can be modeled according to Fig. 9.3, combined with a hiding effect for small particles in the lee of the obstacles in the spirit of the approach by Einstein.

In a first step, we have to estimate the mean local velocity profiles. For gravel beds dominated by large grains, sparsely distributed over the bed, for example, one can just use the mean velocity profile for a bed with a roughness determined by these grains, and describe the flow field as a turbulent one in which vortices of the size of the dominant grains are embedded and are generated at a given Strouhal frequency.

A further approach could be the construction of pseudo turbulent spectra based on families of vortices of given size and frequency. Gyr (1965) used this concept to construct turbulent spectra based on artificial distributions of given vortices.

Such models could also consider the spacing of the roughness elements and other parameters, which are now only taken care of by unspecified knobs.

11.3 Bed Forms in Gravel Beds

Separations arise due to the interaction of an obstacle with the approach flow. Since the approach flow is highly influenced by nearby roughness elements, models based on the exposition of single obstacles will lead to a crude first approximation only. On the other hand, investigations on flow fields over distributed obstacles are very rare and usually for the configurations that one does not encounter in rivers. An alternative approach is tentatively proposed below, and we hope that it is convincing enough to stimulate new research.

The flow around an obstacle attached to a wall was discussed in Chap. 8 (Sect. 8.3.3), and it was shown to be characterized by the vortex skeleton resulting from the enrolled vorticity into the separation, and the one produced by the displacement of the flow due to the obstacle. The solid transport is, therefore, due to the pressure distribution resulting from the interaction of these vortices, which in good approximation can be described by Biot-Savart's law, Eq. 8.20, and by using the interaction properties described in Chap. 8 (Sect. 8.4). Finally, we can calculate the pressure and its distribution by using Eq. 5.24 with a truncation criterion for the integration volume. This shows that the analytical tools for the local evaluation are available, but they need to be incorporated into a statistical description of the ambient flow for typical roughness distributions, unfortunately, such information is not available yet.

Nevertheless, one can explain certain bed configurations based on the properties of vortex skeletons. We made use of this idea in Chap. 9 (Sect. 9.4.3), to explain the mechanism responsible for forming 2D bed forms, such as ripples and dunes. This mechanism is sketched in Fig. 9.17. It was shown that the two-dimensionalization is achieved by a damping of the displacement lateral vortices on sand humps due to the annihilation of their vorticity by diffusion of vorticity of opposite sign from neighbored bed forms. The vortical skeletons are very similar for gravel (Fig. 8.9), with the difference that the luff side of the obstacle is much steeper. As a result, a stagnation point occurs on the front producing horseshoe vortices with tails along its sides. The mechanism is the same as for ripples. However, the large gravels, only rarely align two-dimensionally. One side of a horseshoe vortex interacting with the stagnation flow of a pebble in front of it produces an asymmetric flow field with a skew vortex skeleton as shown in Fig. 11.1, which results in a staggered configuration for the most stable neighbor.

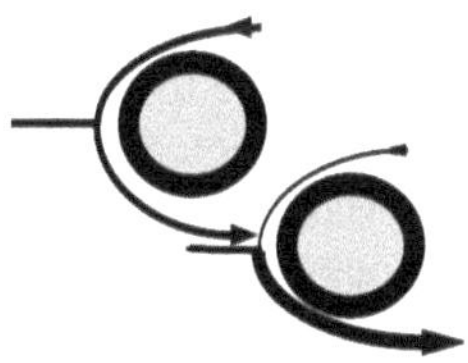

Fig. 11.1 The skeleton of the horseshoe vortices around two pebbles. The branches of the same sign pair, whereas the ones with opposite sign are damped by annihilation due to vorticity diffusion. This process favors a staggered configuration of the pebbles

The flow toward the obstacle has no vorticity in the flow direction; it is created by the lateral displacement of the flow. Lateral vorticity is turned into longitudinal one and so vorticity is fed into the horseshoe vortices, which start at the stagnation point. However, total circulation remains zero since the vorticity in the two branches is of opposite sign. In Fig. 11.1, the vortex skeleton at the outer front as shown in Fig. 11.2 too is growing, whereas the branches entering the formation of the pebbles are strongly damped. In other words, the large pebbles tend to align in a bank-form (Fig. 11.2). Because the ratio $h = k_s/H$ is large enough between the forming banks and the mean flow, a feedback system will develop. The flow becomes 3D in complete analogy to development of secondary flows described in Chap. 10 (Sect. 10.4).

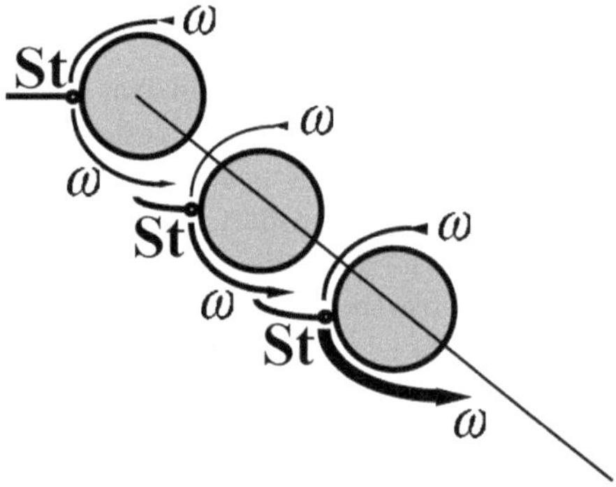

Fig. 11.2 Sketch of a bank-formation due to the asymmetric vortex skeletons of the horseshoe vortices. The flow toward the pebbles is thought to have no vorticity component in flow direction

This is an example of how the new concept can be used to predict bed configurations. Similar considerations apply for the prediction of hiding factors. A vortex skeleton defines a pressure field and a particle is hidden when the pressure forces on it are not strong enough to move it.

11.4 Complexity and Outlooks

The examples in Sect. 11.3 show that sediment transport is such a complex phenomenon that we can only hope to predict it by using a complex strategy. The main reason for this is that unsteady dynamical processes govern sediment transport, especially in case of gravel beds. We cannot hope to investigate a certain bed configuration and extract from it a transport theory because the configuration itself changes due to the transport and with it even the mean flow, so we are confronted with an unsteady condition. This means that gravel deposits have an influence even on channels that are normally classified as sand-bed rivers. During high flows, surface gravel deposits can become exposed or the bed can become armored due to the available gravel fraction. Armoring can limit the depth of bed scour and sediment transport rates, and thus control the morphological character of the river. Banks grow in low flow periods and are filled with fine sediment deposited under such conditions; they are therefore storage space for the finer fractions that are released during high-flow events. In other words, the strategy which can be used for finer sediment beds, an iterative procedure to estimate the load, will fail for gravel beds. This means that for this class of sediment our tools for dealing with the transport are very limited.

There are different strategies that can help in this situation. First one is to work with a more detailed classification of the sediment. Second, resort Einstein's comparison concept and third, one to start the very tedious work of building up a statistical description of the flow based on mechanistic elements as already postulated. We believe to know how the mechanisms of the sediment transport works locally for a specific configuration, this is a positive premise for further investigations, however, we must be aware that there is a lot of work to be done until a statistical description for such a complex system is achieved.

11.4.1 Subclasses of the Sediment

The grain size distribution was discussed in Chap. 1 (Sect. 1.2.2.1) and based thereon we can introduce a classification of the beds. One of it is its modality. An often-used modality is the uniform grain size in laboratory investigations or on fine-sand beds, others are bimodal distributions or multimodal ones. In gravel beds, these distinctions are essential since the largest stones are the ones which dominate the transport regime. When only a few large pebbles are embedded in fine sand material, the main transport can be described by the singular interaction of the two fractions with the scouring process around the gravels. The pebbles can engrave into the bed or can roll on the fine sand substrate due to the exposition function P evaluated by Fenton and Abbott (1977) (Chap. 10, Sect.10.1.1, and Fig. 10.1). On the other hand, a uniform gravel bed could be described by the flow field containing their separation-vortices, using the vortex skeleton model for evaluating such a flow field.

The presently used method of calculating the transport of different fractions, by using the wall shear stress as the only parameter is inadequate since such a description is neither capable of predicting the load nor can it explain the various bed forms.

However, if for any fraction, a special sediment transport law is used, the total of these equations with their knobs to adjust the load by calibration, this is a method for predictions and we can find in the literature several propositions of this kind (Hey et al., 1982).

11.4.2 The Morphological Concept

We return to the discussion with H. A. Einstein mentioned in Chap. 1(Sect. 1.1), and especially to his belief that complex systems can only be encountered by similarly complex system in their description, e.g., by making use of the manifold of the classified observations. The idea behind this strategy is trivial. If the boundary conditions are similar, the morphological result must be similar as well. The situation is analogous to that in long-term meteorological forecasting, where one of the best methods is to find the most similar weather map in the archives, bearing in mind that even if two maps would show the same pressure and temperature distribution, the result after a certain time will not be the same since the evolution remains chaotic. This concept should be appliable for the development of an alluvial channel and the sediment transport in it. By proceeding in this direction more and more morphologies as well as transport rate measurements can be compared and empirical formulas can be chosen which work best for a given class of channels. It is therefore better to provide a whole list of equations with an exact description under which circumstances they were obtained, than to insist on a specific equation.

Of some help in this respect are investigations of channel morphology. They provide empirical models that are often useful for management (Hey, 1982).

11.4.3 The Equal Mobility Approach

Parker developed a generalized relation for the transport on gravel beds based on a rather comprehensive study of field data sets for rivers with coarse bed sediment, Parker et al., 1982; Parker and Klingeman, 1982). His description is based on properties of the subarmor bed material and involves nondimensional analysis and curve fitting techniques. They subdivided the particle size distributions into ten classes and found what they called the "equal mobility" to describe the bed-load transport for gravel-bed streams. This hypothesis proposes that the bed armor controls the entrainment of particles by the stream, resulting in various sizes being approximately equal in mobility with particle sizes transported at rates proportional to their presence in bed materials. Bed load thus typically contains all sizes and tends to match the bed material in size gradation. Shih and Komar (1990) declared Parker's approach as a first order one, which needs to be improved by higher orders taking care of observed variations in bed-load particle size distributions at different flow rates. Bakke et al. (1999) proposed a refined P–K model, but it has to be calibrated for each river.

12 Data and Strategies to Calculate Sediment Transport

We have described many aspects of sediment Transport and given formulas or calculating the transport rate under certain circumstances. But we also made clear that there is no unique theory to calculate the transport and even the given equations hold only for steady-state conditions. This is generally not the case for the transport in a river, which requires an iterative procedure. In other words, any simulation starts with a problem analysis.

Many of the given results are not yet implemented in available software but need to be incorporated in future theories and programs. They were meant as stimulation to overcome the stagnation of progress observed in this field and are discussed further for those interested in this research.

In the past, predictions were often made using physical models of Froude-similarity laws, which were evaluated in specialized laboratories all around the world. With increasing computer capacity, these laboratories disappeared or combined computer simulations and physical models for their predictions, while the computer share became increasingly dominant. Physical modeling now is restricted to time-dependent processes because of their still exhaustive calculation requirements. Nevertheless, there is no doubt that the future belongs to computer simulations, which can only be as good as the algorithms we are using.

It is desirable to go along with the general philosophy for sediment transport calculations, which is best expressed as to use the simplest theory capable of predicting the transport.

It makes little sense to use a complicated form if the problem analysis allows a simpler approach to satisfy a first estimate. We therefore recommended using the classical representations as discussed in Chaps. 2 and 3 when the problem statement does agree with the assumptions applied in these theories. However if, e.g., the rheology changes, or turbulence becomes significant, one has to construct one's own algorithm that applies the knowledge relevant for these particular cases. This is all easier said than done since one needs decision criteria to direct the code.

We hope to have made clear that the sediment size-distribution function is the most important data set to be analyzed and classified first using recommendations given in Table 12.1. The simplest case is the fine-sand regime, where all the classical concepts are applicable. In Chap. 9, we discussed the evolution of bed forms using such modern elements as coherent structures, representative for the turbulent flow close to the bed, to explain ripple formation. Also the turbulent shear layer separations, showing evidence of Kelvin-Helmholtz instabilities, are key players to be used for predictions in the dune formation. The situation is quite acceptable also for sand mixtures (Chap. 10). Here, the classical formulas are to be used together with criteria for sorting and armoring of the sediment distribution also with secondary flows. Generally, an iterative procedure can handle these slow and homogeneous dynamics.

The real problems start with gravel beds, where we are only beginning to treat the sediment transport more in physical terms. Lacking better theories, the only help for a quick estimate is based on the old concepts. However, often one encounters a problem, which is local enough to already apply newer concepts, e.g., the scouring behind a pillar and similar configurations. For such problems, the vortex skeleton approach is already better than any other method.

Table 12.1: A classification for the size distributions of the sediment

The dominant grain size	Composition	Criteria
Fine sand	Uniform grains (1)	$d^+_s < 12.5$
	Mixtures (2)	$d^+_s < 12.5$
Sand mixture	Fine + medium (3)	$d^+_s < 12.5$; $d^+_s < 100$
	Mean mixtures (4)	$12.5 < d^+_s < 100$
	Fine + coarse (5)	$d^+_s < 12.5$; $d^+_s > 100$
	Mean + coarse (6)	$d^+_s < 100$; $d^+_s > 100$
	Fine+ mean + coarse (7)	
Gravel	Uniform gravels (8)	$d^+_s > 100$
	Gravel mixture (9)	$d^+_s > 100$

In Table 12.1, the fine sand is not defined as usual through hydraulically smooth flow conditions, e.g., $d^+_s < 5$, but through the size of those grains still capable of forming ripples $d^+_s \approx k^+_s$. The sediment transport is due to coherent structures, and therefore one should strive to use this new concept, although the classical theories give quite good results. The difference between classes (1) and (2) lies only in the sorting mechanism, which leads to bed armoring at low flow rates. For uniform grain size distributions the distribution of $\overline{\tau}_w$ and $\overline{\tau}_{wc}$ in the representation of Grass, Fig. 3.4 is much narrower than for mixtures. Very fine sand is the most common material and once suspended must not be forgotten in the estimate for the total load. At higher flow rates also dune formation must be considered.

Sediment mixtures are more difficult to calculate not only because of the sorting effects but also because the total load cannot be calculated from a superposition of the different fractions. Often, this approach must be chosen for the lack of a better representation.

The material of classes (3) and (4) is fine enough for the coexistence of coherent structures and larger turbulent structures. Here we can use the coherent structure concept with some modifications, as done by Grass, or use the turbulent transport concept as it is used by the classical theories. The suspended load is smaller for class (4), and often only the bed-load component is evaluated under these circumstances. Class (3) is capable of dune formation, which also needs to be considered here. Due to sorting effects, criteria for armoring and secondary flow occurrence should not be neglected. If class (4) is of uniform size distribution, the Grass-distribution shows a similar trend as in class (1).

Class (5) is typically defined by a bimodal size distribution, and it is important to know which part of the bimodal distribution is more important. It happens that the fine sand not only suspends but also settles in the wake of larger grains being thus protected from transport. In the classical representation, it is appropriate to use a hiding factor to apply Einstein's representation to this situation. But if the very coarse material is a small fraction of the size distribution, these grains will roll or engrave. Applying the vortex skeleton concept to the larger elements allows prediction of these cases. This class, however, will always tend to form an armored bed. Barchans and sand banks are possible and therefore several criteria have to be built into an algorithm.

Class (6) always needs an armoring criterion, which defines the point when the bed has been armored so that no transport will occur anymore, whereas before we still had a rather high suspension transport. The transport is very dynamic in that case, and we see that iterative procedures must be used. The same can be said for class (7) only that in that case again the suspension plays a much more significant role.

Class (8) should be treated using the new concept of local separations, however, for a uniform gravel size distribution a classical bed-load formula, e.g., Meyer-Peter, should give good results.

As mentioned in the previous chapter, Class (9) represents grain size distribution, which is the most difficult to calculate and needs further development through new theories based on local flow separations.

Since sediment transport is determined by the interaction of the flow and the sediment, it is evident that besides the grain size distribution the flow parameter must also be considered when analyzing a given problem. The discharge is given by the hydrology and the in situ sediment distribution, which are dependent quantities since the bed form and bed configuration change and react back on the flow. This interaction is best parameterized by the wall shear stress, which therefore remains the key parameter next to the sediment size distribution for a river classification. With this in mind, we characterize the flow by two sets of parameters. Both are Reynolds numbers, whereas one is based on the channel flow variables (U, H, ν) and the other one is based on the wall shear stress and the grain size (u_τ, k_s, ν). The bed load is characterized by the channel slope or by two analogous sets of parameters for the Froude number representation, the hydraulic radius R_H and the relative roughness k_s/H for a given channel geometry. All these new flow characteristics can occur in combination with all of the mentioned grain size distributions, and it becomes clear that a simulation must be organized in modular form, using the appropriate approach discussed in the theoretical part of this book. This classification is the main strategy tool to decide on the proper handling of these possible cases. However, nature is even richer and one must be aware of parameter extremes, such as flows in wadis and mountain rivers where d_s/H often exceeds unity. In this case, gravel protrudes through the water surface, and new program developments are unavoidable.

12.1 The Input Parameters

The analysis of the problem is only a first step, and it needs information that is often hard to obtain such as the grain size distribution. Other parameters cannot be measured properly, and not all parameters we would like to use in the calculation are available. Therefore, we have to first introduce those measurable parameters which can be used in defining the equations needed for the calculation.

12.1.1 The Description of a Channel

In a first step, we define the geometry of the channel. If simple enough, it can be represented by a functional for the boundary conditions; otherwise a common grid representation is needed.

The straight rectangular channel is the simplest case, which will be used for discussion, but we know that trapezoidal, triangular, and half pipe cross sections are often encountered as well as channels with bends. It is clear that not all cases can be discussed within the framework of such a book, and we have to cite the special literature where the reader can find the appropriate information.

Channel geometries can be measured using ultrasound but the effort is large, and it is only done if the channel geometry deviates from the usual simple configurations. If the geometry is really complex, it is appropriate to introduce a pseudo-cylindrical coordinate system, which should be used for half pipe cross sections. An example is shown in Fig. 12.1.

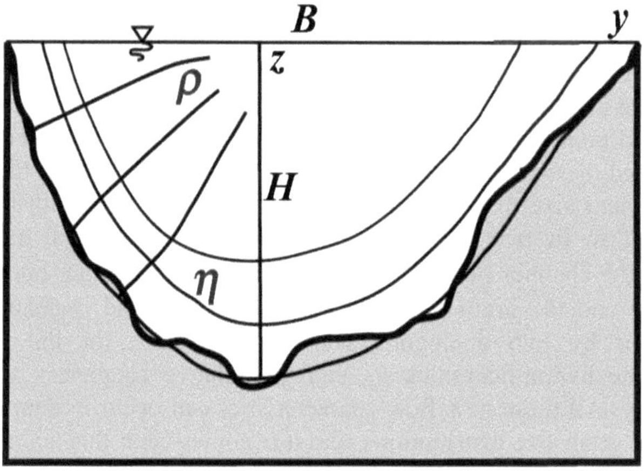

Fig. 12.1 Sketch of a pseudo-cylindrical non-Cartesian coordinate system (ξ, ρ, η), where ρ is the pseudo-radial and η the pseudo-circumferential coordinates defining the cross section and ξ the local cylinder axis. B is the width at the water surface at $z = H$. This procedure effectively smoothes the wall geometry in the cross section and the flow direction as indicated in the graph

We were already confronted with a similar problem when defining the proper coordinate origin for a flow over a rough bed. The generation of a discrete surface mesh is an art in itself and is discussed at length in applied mathematical textbooks. The specialty we encounter in modeling rivers is that lines of constant flow properties, e.g., iso-velocity lines become smooth at a certain distance, since the flow "forgets" its origin due to the turbulent redistribution. The distance from the wall necessary for this assumption is therefore given by the theory of turbulence.

12.1.2 The Flow Parameters

Mean flow: The amount of water flowing into the system is given mainly by the precipitation over the drainage area (Chap. 1, Sect. 1.3). This amount discharges to a large part through the river or channel. This quantity influx, $Q(t)$ is time dependent and the calculations need to incorporate this fact. Averaging over the cross section defines the mean velocity through the continuity equation.

$$Q(t) = A(t)\bar{U}(t) \tag{12.1}$$

with $A(t)$ being the local cross section which changes with time due to variations in the discharge $Q(t)$ and consequently also the water depth $H(t)$. The mean velocity $\bar{U}$ results from an area integration of the velocity profiles over the cross section divided by A.

Since the change in $Q(t)$ is usually slow, it is recommended to discretize the time and treat the flow as stationary between two time steps. Using this method, most of the mentioned results can be worked out. The flow can be treated separately from the sediment transport using a very similar argument since the ratio of the respective time scales is large. This statement although is not valid for suspension sediment transport. Applying all the mentioned restrictions leads to a flow representation of open channel hydraulics as it can be found in practically every textbook on this subject.

Equation 12.1 can usually be evaluated since A can be measured and Q gauged at special measuring sites. However, these values are more of hydrological interest than important for the sediment transport since the amount of transported sediment is usually not directly coupled to Q through an equation, although there exist also empirical descriptions of this kind. Especially for rivers with high suspension loads, Eq. 12.1 can be used well for estimations.

Introducing turbulence concepts needs adaptations even when one treats the flow in an averaged form. The simplest description is given by the Reynolds stress tensor, which shows clearly that there are very local effects that cannot be described by an averaged quantity over the entire height anymore. The Reynolds stress tensor is a locally defined quantity, which requires a variable mesh size for an efficient computation.

What does this mean with respect to an input quantity? It presumes that the flow field with its fluctuation is given at any location in the river. This is so unrealistic that this concept can obviously only be used with a series of assumptions. But even then it is very hard to measure the needed parameters. This is the main reason why the classical transport equations remained so popular, although their physical deficits were known. They are based on one single measurable value representative of the turbulent flow, namely the wall shear stress. It can be assumed to be local and time dependent but remains in a very treatable form since it changes smoothly as long as no separation occurs.

This representation is sufficient for the classical sediment transport evaluation. However, the moment the feedback mechanism becomes important, then the velocity profiles must also be evaluated. Here the calculation grids with a higher resolution as subsystems embedded in the mean flow model should be used, for example to treat only the near-wall region with a grid of higher resolution.

The flow state is given by the Navier–Stokes equation (Eq. 1.15) with its initial and boundary conditions. It describes the acceleration $\underline{a}$ of a fluid element $dV = dx\, dy\, dz$ in

Cartesian coordinates due to external forces acting on its surface and its internal inertia or other body forces. As long as the "fluid" remains Newtonian, the density of the sediment–water mixture is ρ_{fs} and the dynamic equilibrium is represented by

$$a_i = \frac{\mathrm{D}u_i}{\mathrm{D}t} = \frac{1}{\rho_{\mathrm{fs}}}\left(-p + \frac{\partial \tau_{ij}}{\partial x_j}\right) + g_i \quad \wedge \quad \sigma_i = \tau_{ii} \tag{12.2}$$

with σ_i representing the normal stresses.

The mean momentum equations: The Navier–Stokes equation is an integro-differential equation for the momentum transport, and one can integrate over the volume V by considering the forces on the surface ∂V to obtain a global representation. When multiplying momentum Eq. 12.2 with ρ_{fs} it becomes a force equation. With the application of the continuity equation this results in

$$\begin{gathered}\frac{\mathrm{d}}{\mathrm{d}t}\int_V \rho_{\mathrm{fs}} u_i \mathrm{d}V + \int_{\partial V} \rho_{\mathrm{fs}} u_i \left(u_j \frac{\partial x_j}{\partial n}\right) \mathrm{d}\partial V = \\ \int_V \rho_{\mathrm{fs}} g_i \mathrm{d}V - \int_{\partial V} p \frac{\partial x_i}{\partial n} \mathrm{d}\partial V + \int_{\partial V} \left(\tau_{ji} \frac{\partial x_j}{\partial n}\right) \mathrm{d}\partial V\end{gathered} \tag{12.3}$$

where $\underline{n}$ the unit-vector normal to the surface element pointing outward of ∂V.

In case of a straight channel, Eq. 12.3 reduces to

$$\begin{gathered}\int_{\partial V} \rho_{\mathrm{fs}} u_x^2 \frac{\partial x}{\partial n} \mathrm{d}\partial V + \int_{\partial V} p \frac{\partial x}{\partial n} \mathrm{d}\partial V = \\ \int_V \rho_{\mathrm{fs}} g_x \mathrm{d}V + \int_{\partial V} \tau_{yx} \frac{\partial y}{\partial n} \mathrm{d}\partial V + \int_{\partial V} \tau_{zx} \frac{\partial z}{\partial n} \mathrm{d}\partial V\end{gathered} \tag{12.4}$$

With additional assumptions, of which some will be worked out, this equation can be further simplified, and it is recommended to test the extent of the assumptions with respect to the actual problem. The simpler the equations can be formulated, the larger savings in computation time are achievable.

The first generally valid assumption is that the mixture is incompressible. This does not mean that the mixture cannot change its density, e.g., by suspensions. But this would have to be introduced as a source term depending on another timescale. The density ρ_{fs} of the mixture can therefore be taken in front of the integral. In the literature, the first integral in Eq. 12.4 is often replaced by mean flow quantities and the so-called momentum correction factor β.

$$\beta = \frac{1}{AU_x^2} \int_A u_x^2 \mathrm{d}A \tag{12.5}$$

The integration is most easily achieved when the control volume surfaces A are chosen in directions parallel and perpendicular to the mean flow velocity. The parallel boundaries show no flux, and the integration yields contributions on the areas of the inflow reproduced as $(..)_1$ and outflow $(..)_2$ cross sections of the river element only.

When the control volume is given by (x_c, B, H), Eq. 12.3 reads

$$\begin{aligned}&\beta\rho_{\mathrm{fs}}A_2U_{x2}^2+p_2A_2-\beta\rho_{\mathrm{fs}}A_1U_{x1}^2-p_1A_1-f\left(A_{\mathrm{r}}\right)=\\&\rho_{\mathrm{fs}}gV\sin\alpha-\tau_{\mathrm{w}}BX_{\mathrm{c}}-\tau_{\mathrm{s}}2HX_{\mathrm{c}}+f\left(\tau_{\mathrm{wind}}\right)\end{aligned}\tag{12.6}$$

where the influence of the rain $f(A_{\mathrm{r}})$ can generally be neglected within the observed timescale. The wind influence $f(\tau_{\mathrm{wind}})$ is also small and can be neglected as long as the river exhibits a moderate minimum slope.

$$\begin{aligned}&\wedge\quad f\left(A_{\mathrm{r}}\right)=\rho_{\mathrm{f}}A_{\mathrm{r}}U_{\mathrm{xr}}^2\sin\left(\alpha+\alpha_{\mathrm{r}}\right)\rightarrow 0\\&\text{and}\quad f\left(\tau_{\mathrm{wind}}\right)=\tau_{\mathrm{wind}}BX_{\mathrm{c}}\rightarrow 0\\&\text{and}\quad \sin\alpha\cong S_0\end{aligned}\tag{12.7}$$

We exclude mountain rivers and river deltas here, for which these assumptions are not valid. With these simplifications Eq. 12.6 reads

$$\begin{aligned}&p_2A_2+\beta\rho_{\mathrm{fs}}A_2U_{x2}^2+\tau_0\left(B+2H\right)X_{\mathrm{c}}=\\&p_1A_1+\beta\rho_{\mathrm{fs}}A_1U_{x1}^2+\rho_{\mathrm{fs}}g\left(\frac{A_1+A_2}{2}\right)X_{\mathrm{c}}S_{\mathrm{w}}\end{aligned}\tag{12.8}$$

Even more restrictive is the assumption of constant cross section $A = A_1 = A_2$

$$\begin{aligned}&\tau_0=\rho_{\mathrm{fs}}g\frac{A}{B+2H}S_{\mathrm{f}}\quad\wedge\quad S_{\mathrm{f}}=S_{\mathrm{w}}\\&\tau_0=\rho_{\mathrm{fs}}g\frac{A}{P}S_{\mathrm{f}}=\rho_{\mathrm{fs}}gR_{\mathrm{H}}S_{\mathrm{f}}\quad\wedge\quad R_{\mathrm{H}}=\frac{A}{P}\quad\text{and}\quad P=B+2H\end{aligned}\tag{12.9}$$

For B >> H, the hydraulic radius R_{H} in Eq. 12.9 reduces to the water depth H, which leads to the definition of the Darcy-Weisbach friction factor f

$$\begin{aligned}&\tau_{\mathrm{w}}=\rho_{\mathrm{fs}}gHS_{\mathrm{f}}=\rho_{\mathrm{fs}}gR_{\mathrm{H}}S_{\mathrm{w}}\\&\wedge\quad f=8\tau_{\mathrm{w}}/\rho_{\mathrm{fs}}U_x^2\quad\equiv\quad\lambda=8\frac{u_\tau^2}{U^2}\,(2.46)\wedge u_\tau=\sqrt{\frac{\tau_{\mathrm{w}}}{\rho_{\mathrm{f}}}}\,(2.12)\\&\therefore\quad\left(f/8\right)\rho_{\mathrm{fs}}gU_x^2=\rho_{\mathrm{fs}}gR_{\mathrm{H}}S_{\mathrm{f}}\quad\vee\quad U_x=\sqrt{\frac{8g}{f}}R_{\mathrm{H}}^{1/2}S_{\mathrm{f}}^{1/2}\end{aligned}\tag{12.10}$$

For a steady-state flow in a wide rectangular channel $R_{\mathrm{H}} \approx H$ and $q \approx U_xH$. Then the flow will be at equilibrium at the normal depth,

$$\begin{aligned}&H_{\mathrm{n}}=\left(fq^2/8gS_{\mathrm{f}}\right)^{1/3}=\left(f/8g\right)\left(U_x^2/S_{\mathrm{f}}\right)\\&\wedge S_{\mathrm{w}}=S_{\mathrm{f}}\quad\therefore\quad H_{\mathrm{n}}=\left(fq^2/8gS_{\mathrm{w}}\right)^{1/3}=\left(f/8g\right)\left(U_x^2/S_{\mathrm{w}}\right)\end{aligned}\tag{12.11}$$

The ratio H_{n}/H at constant discharge q and friction factor f can be related to the ratio of the friction and the surface slope

$$\frac{S_{\mathrm{f}}}{S_{\mathrm{w}}}=\left(\frac{H_{\mathrm{n}}}{H}\right)^3\tag{12.12}$$

A special problem arises with the use of the equivalent sand roughness, k_{s}, since it has to be evaluated experimentally. If it is not known, it is recommended to use d_{s} as it was introduced in Table 12.1. However, if k_{s} is known it makes sense to use it since this value represents a mean roughness. For this case, Prandtl and von Karman developed the logarithmic velocity profile using Eq. 2.35 with k_s instead of ε, and Eq. 2.41

$$\frac{U}{u_\tau} = 5.75 \ln \frac{y}{k_s} + 8.5 \tag{12.13}$$

With a known velocity profile, a series of new equivalent relations can be formulated instead of τ to describe the flow and its energy dissipation in a given river segment (Chap. 11, Sect. 11.1).

For example, using Eq. 12.10 Eq. 12.13 can be transformed into

$$\begin{aligned}
&\text{Eq. } 12.13 \rightarrow U = u_\tau \left(6.25 + 2.5 \ln \frac{R_H}{k_s} \right) \\
&\wedge U = \sqrt{8g/\lambda} S_w^{1/2} R_H^{\;1/2} \text{ Eq.} 12.17 \\
&\text{and } u_\tau = \sqrt{g} S_w^{1/2} R_H^{\;1/2} = U\sqrt{8/\lambda} \\
&\therefore \quad \frac{U}{u_\tau} = \frac{1}{\sqrt{8/\lambda}} \rightarrow \frac{u_\tau^2}{U^2} = \frac{\lambda}{8} \rightarrow \text{Eq. } 2.38 \wedge \text{Eq. } 11.3 \\
&\Delta H = \lambda \frac{U^2}{2g} \frac{1}{4R_H} \\
&\frac{1}{\sqrt{\lambda}} = 0.78 + 0.88 \ln \left(\frac{R_H U}{\nu} \sqrt{\lambda} \right) \rightarrow \lambda = \frac{1}{0.77 \left(\ln \frac{12.32 R_H}{k_s} \right)} \\
&\wedge B/H \geq 10 \rightarrow R_H \approx H \rightarrow \lambda_{\text{Moody}} \Rightarrow D = 4H
\end{aligned} \tag{12.14}$$

This yields $1/\sqrt{\lambda}$ as a common scaling parameter in the literature.

The same approach can be applied using the Chezy friction coefficient (Eq. 11.4)

$$\begin{aligned}
&U = C\sqrt{R_H S_w} \therefore u_\tau \left(5.75 \ln \frac{y}{k_s} + 8.5 \right) = C\sqrt{R_H S_w} \\
&C = 18 \frac{H}{k_s} + 19.5 \quad \wedge \quad y(u=0) \approx \frac{k_s}{30.2} \\
&\therefore \quad C = \sqrt{\frac{8g}{\lambda}}
\end{aligned} \tag{12.15}$$

Additionally, the Strickler formulation and its coefficient K shall be mentioned since it is used in several hydraulics books.

$$U = K\sqrt{S} R_H^{4/3} \rightarrow K = \frac{21}{\sqrt[6]{d_s}} \tag{12.16}$$

$$U = K \left(\frac{H}{k_s} \right)^{1/6} \sqrt{HS_0} = \frac{K}{k_s^{1/6}} H^{2/3} S_0^{1/2} = \frac{1}{n} H^{2/3} S_0^{1/2} \tag{12.17}$$

with n the Manning value depending on the particle diameter.

The flow and transport in special channels: Most representations for sediment transport are restricted to straight open channel flows of rectangular cross section. But engineers are frequently also interested in other specialized channel designs, for which we will give some analytical results.

Using topological arguments, we will look at the influence of the different geometries on the secondary flows to explain these additional mechanisms.

Simple trapezoidal channels can be treated as being rectangular, by adjusting the hydraulic radius R_H. However, when the sidewalls are inclined at a small angle, corrections need to be made as they can be developed from triangular profiles. A special case is the compound profile as it is encountered in rivers with floodplains. When the river inundates the floodplains, the different channel cross sections will show different streamwise velocities at different height. Between every two flows of different velocities, vertical shear produces vortices, which intermittently produce low pressure at the bed. This leads to higher sediment transport producing grooves in the primary channel, which is the starting point for a whole system of secondary flows.

Therefore, it is evident that even the straight channel geometry is not simply given by its initial condition. It changes during the process, which requires the geometry to be an adaptive parameter in the necessarily iterative calculation procedure. In a next step, we will discuss the flow equations applicable to these special channel geometries.

Wormleaton (1996) gives a solution for the 2D flow equation of compound trapezoidal channels. He starts with Eq. 2.8 in the form of Eq. 10.20, which with the use of the continuity equation reads

$$\begin{aligned} &\rho_f\left(\bar{u}_z\frac{\partial\bar{u}_x}{\partial z}+\bar{u}_y\frac{\partial\bar{u}_x}{\partial z}\right)=\rho_f gS_0+\frac{\partial\tau_{zx}}{\partial z}+\frac{\partial\tau_{yx}}{\partial y}\\ \therefore\quad &\rho_f\left(\frac{\partial\bar{u}_x\bar{u}_z}{\partial z}+\frac{\partial\bar{u}_x\bar{u}_y}{\partial y}\right)=\rho_f gS_0+\frac{\partial\tau_{zx}}{\partial z}+\frac{\partial\tau_{yx}}{\partial y} \end{aligned} \tag{12.18}$$

Integrating Eq. 12.8 over the water depth leads to the following equation for the averaged terms

$$\begin{aligned} &\frac{\partial\left(H\left\langle\rho_f\bar{u}_x\bar{u}_y\right\rangle\right)}{\partial y}-\frac{\partial z_s}{\partial y}\left(\rho_f\bar{u}_{xH}\bar{u}_{yH}-\left\langle\tau_{yx}\right\rangle_H\right)=\\ &\rho_f gHS_0+\frac{\partial\left(H\left\langle\tau_{yx}\right\rangle\right)}{\partial y}-\tau_w\left(1+\frac{1}{s^2}\right)^{1/2}=\Xi \end{aligned} \tag{12.19}$$

where the subscript $(..)_H$ refers to the values at the water surface and s is the bank slope of the channel. Ξ stands for the vertical forces acting on a fluid element of unit width, and it comprises the weight, the internal, and the wall shear stresses, respectively. With the apparent shear stress

$$\tau_a=\frac{1}{H}\int_0^B\left(\tau_w\left(1+\frac{1}{s^2}\right)^{1/2}-\rho_f gH(y)S_0\right)\mathrm{d}y \tag{12.20}$$

Eq. 12.19 reduces to

$$\Xi = \frac{\partial\left(H\left\langle\tau_{yx}\right\rangle\right)}{\partial y} - \frac{\partial\left(H\tau_a\right)}{\partial y} = \frac{\partial\left(H\Psi\right)}{\partial y}$$
$$\wedge \quad \frac{\partial\left(H\tau_a\right)}{\partial y} = \tau_{\mathrm{w}}\left(1+\frac{1}{s^2}\right)^{1/2} - \rho_{\mathrm{f}} gHS_0 \tag{12.21}$$
$$\text{and} \quad \Psi = \frac{1}{H}\int_0^y \Xi \mathrm{d}y = \left\langle\tau_{yx}\right\rangle - \tau_a$$

Equation (12.19) shows that Ξ is composed of two terms; the first represents forces from the secondary flow, whereas the second is due to the side slope and disappears if the water surface is normal to the gravity vector. If further the water surface is horizontal in spanwise direction, then it follows that also $\tau_{yx}=0$ and from Eqs. 12.19 and 12.21, we have

$$H\frac{\partial\left\langle\rho_{\mathrm{f}}\bar{u}_x\bar{u}_y\right\rangle}{\partial y} = \Xi = -H\frac{\partial\tau_a}{\partial y} \tag{12.22}$$

In such flows, a constant Ξ was observed, which has consequences for the flow structures in the floodplains, especially for the strength of the spanwise flow belonging to a secondary flow forced by the main stream. For the case where the floodplains are vegetated zones see Tsujimoto (1996).

For triangular channels, the sidewalls define the bed, and unless the whole sediment transport is in suspension, the cross section will fill in the trough to form a trapezoidal cross section.

In triangular channels, the flow always develops secondary cells into the edges as discussed in Chap. 10 (Sect. 10.4). In nature, triangular channels are usually sub-elements of the cross section but also grooves in the bed which themselves are the result of secondary flows in the main channel. This is relevant in river bends only where secondary flows always occur. For this case the lateral sediment transport needs also to be evaluated and can be superimposed to the streamwise transport in a first approximation, which leads to an additional boundary condition. The sidewall slope can never become larger than the natural angle of repose; see Chap. 1(Sect. 1. 2.2.2, a criterion, which has also to be applied when calculating the flow on the lee side of a dune. This criterion is also mostly relevant at the outer side of a bend. Here the bed is most vulnerable to erosion and if this happens, sediment will glide down the slope and be transported away by the flow (Thorne, 1982). For computer modeling of river bends the reader is referred to the excellent information and references found in Alabyan (1996).

The triangular cross section of rivers usually occurs under two different conditions, namely river bends and junctions. Similar to the compound channels, a vertical shear layer usually results at river junctions and scours the bed in longitudinal direction. McLelland et al. (1996) described this case and we would like to refer to the papers mentioned by these authors. Rhoads (1996) presents the confluence of a dominant main-stream and a side branch with the measurements made in a field investigation.

Such geometries are essential for the description of the flow in bends, although is often unknown why a bend appears. In the literature, one often finds that the rotation of the earth is one of the reasons. This is rather improbable. For an existing mean rotation, this concept allows formulating the main equations describing the flow around a curved section. In that case, two additional inertial forces appear; the Coriolis-force F_c and the centrifugal force F_{ce}. With the angular velocity Ω, the equation of motion can be written as

$$\frac{D\underline{u}}{Dt} = -\frac{1}{\rho}\nabla p - \left[\underline{\Omega},\left[\underline{\Omega},\underline{r}\right]\right] - 2\left[\underline{\Omega},\underline{u}\right] + \nu\nabla^2\underline{u} \tag{12.23}$$

The second and third term on the right correspond to the centrifugal and the Coriolis force, respectively. For the sediment transport both are rather irrelevant, however, not so for the river bends. In that case, one has to complement Eq. 12.18 with the centrifugal force as an outer force

$$\begin{aligned} &\underline{F_{ce}} = \frac{mu_\varphi^2}{r} = m\omega^2\underline{r} \quad \wedge \quad \omega = \frac{u_\varphi}{r} \quad \text{and} \quad u_\varphi = u_\varphi(r,z) \\ &\mathrm{d}\underline{F_{ce}}(r,z) = \rho_f\omega^2\underline{r}\mathrm{d}V \quad \wedge \quad \mathrm{d}V = r\mathrm{d}r\mathrm{d}\varphi\mathrm{d}z \end{aligned} \tag{12.24}$$

The topologies of such flows become rather complicated and are encountered in meandering rivers. Alabayan (1996) made an attempt to simulate the flow in such geometries. For further reading on the problem of meandering rivers we refer to following authors; Willetts and Rameshwaran (1996), Leopold (1982), and Ackers (1982), for so-called double meander, Naish and Sellin (1996) and for some speculations on the interactions of several parameters, Parker (1996).

The mean energy equation: Starting from the momentum Eq. 12.2, one can also calculate the energy balance over the control volume. The change in energy can be derived from the work done on the system, which can be found by integrating the force density ($\rho_{fs}\underline{a}$) multiplied by the velocity. If the integration is done over the same control volume as used before, we find

$$\begin{aligned} &\int_V \rho_{fs}(a_i u_i)\mathrm{d}V = \int_V \rho_{fs}(u_i g_i)\mathrm{d}V + \int_V u_i \frac{\partial\tau_{ji}}{\partial x_j}\mathrm{d}V \\ &\wedge \quad \tau_{ii} = \sigma_i \end{aligned} \tag{12.25}$$

Applying the Gauss-Theorem, Eq. 12.25 can be rewritten as

$$\begin{aligned} &\frac{\mathrm{d}}{\mathrm{d}t}\int_V \rho_{fs}\left(\frac{u^2}{2} - \Gamma\right)\mathrm{d}V + \int_{\partial V} \rho_{fs}\left(\frac{u^2}{2} - \Gamma\right)u_i\frac{\partial x_i}{\partial n}\mathrm{d}\partial V = \left(u_j\tau_{ij}\right)\frac{\partial x_i}{\partial n}\mathrm{d}\partial V - \\ &\int_V\left[\tau_{ij}\frac{\partial u_i}{\partial x_i} + \tau_{xy}\left(\frac{\partial u_x}{\partial y} + \frac{\partial u_y}{\partial x}\right) + \tau_{yz}\left(\frac{\partial u_z}{\partial y} + \frac{\partial u_y}{\partial z}\right) + \tau_{zx}\left(\frac{\partial u_x}{\partial z} + \frac{\partial u_x}{\partial x}\right)\right]dV \\ &\wedge \quad g_i = \frac{\partial\Gamma}{\partial x_i} \end{aligned} \tag{12.26}$$

The convective terms reduce to the net fluxes across the surface, and the integral on the right hand side of the equation stands for the work done by the outer stresses on the fluid element. The equation is conservative and the volume integral stands for the dissipation of kinetic energy into heat.

For the steady-state straight open channel flow, Eq. (12.26) reduces to

$$\int_{\partial V} \rho_{\mathrm{fs}} \left(\frac{u^2}{2} - \Gamma \right) u_x \frac{\partial x}{\partial n} \mathrm{d}\partial V = \int_{\partial V} -u_x p \frac{\partial x}{\partial n} \mathrm{d}\partial V + \int_{\partial V} u_x \tau_{zx} \frac{\partial y}{\partial n} \mathrm{d}\partial V - \int_V \tau_{xz} \frac{\partial}{}$$

$$\wedge \quad u_y = u_z = 0 \quad \& \quad \tau_{ii} = \tau_{yz} = \tau_{zy} = \tau_{xy} = \tau_{yx} = 0 \quad \& \quad \tau_{xz} = \tau_0 = \tau_w \tag{12.27}$$

$$\therefore \quad \int_{\partial V} \rho_{\mathrm{fs}} g \left(\frac{u^2}{2g} + \hat{z} + \frac{p}{\rho_{\mathrm{fs}} g} \right) u_x \frac{\partial x}{\partial n} \mathrm{d}\partial V = \int_V \tau_{xz} \frac{\partial u_x}{\partial z} \mathrm{d}V \quad \wedge \quad \Gamma = -g\hat{z}$$

and the second surface integral on the right hand side disappears because of $(u_x)_w \equiv 0$.

As for the momentum equation, it is common practice to introduce an energy correction factor α for the 1D flow model

$$\alpha = \frac{1}{U_x^3 A} \int_{\partial V} u_x^3 d\partial V \approx 1 \tag{12.28}$$

The volume integral in Eq. 12.28 represents the dissipation head loss ΔH_{L}, which can be given by Bernoulli-sums H_{L}:

$$\Delta H_{\mathrm{L}} = \frac{1}{\rho_{\mathrm{fs}} g Q} \int_V \tau_{xz} \frac{\partial u_x}{\partial z}$$

$$\wedge \quad \rho_{\mathrm{fs}} g Q \left[\frac{p_2}{\rho_{\mathrm{fs}} g} + \hat{z}_2 + \alpha_2 \frac{U_{x2}^2}{2g} - \frac{p_1}{\rho_{\mathrm{fs}} g} - \hat{z}_2 - \alpha_1 \frac{U_{x1}^2}{2g} \right] = \rho_{\mathrm{fs}} g Q \Delta H_{\mathrm{L}} \tag{12.29}$$

$$\wedge \quad \underbrace{\phantom{\frac{p_2}{\rho_{\mathrm{fs}} g} + \hat{z}_2 + \alpha_2 \frac{U_{x2}^2}{2g}}}_{\hat{H}_2} \quad \underbrace{\phantom{\frac{p_1}{\rho_{\mathrm{fs}} g} - \hat{z}_2 - \alpha_1 \frac{U_{x1}^2}{2g}}}_{\hat{H}_1}$$

$$\text{and} \quad \Delta H_{\mathrm{L}} = \hat{H}_1 - \hat{H}_2$$

In this representation, the integral form of the specific energy $\tilde{E}$ reads,

$$\tilde{E} = \frac{p}{\rho_{\mathrm{fs}} g} + \alpha \frac{U_x^2}{2g}$$

$$\frac{\Delta \hat{H}}{\Delta x} = \frac{\Delta}{\Delta x} \left(\frac{p}{\rho_{\mathrm{fs}} g} + \hat{z} + \alpha \frac{U_x^2}{2g} \right) = -S_{\mathrm{f}} \tag{12.30}$$

$$\frac{\Delta \tilde{E}}{\Delta x} = \frac{\Delta}{\Delta x} \left(\frac{p}{\rho_{\mathrm{fs}} g} + \alpha \frac{U_x^2}{2g} \right) = -\frac{\Delta \hat{z}}{\Delta x} - S_{\mathrm{f}} = S_w - S_{\mathrm{f}}$$

This representation is significant for the sediment transport because it allows calculating the backwater curves and the flow regime. The backwater curve describes the elevation of the local water surface, here only shown for the steady-state 1D case,

$$
\begin{aligned}
&\frac{\mathrm{d}\tilde{E}}{\mathrm{d}x} = \frac{\mathrm{d}\tilde{E}}{\mathrm{d}H}\frac{\mathrm{d}H}{\mathrm{d}x} = S_{\mathrm{w}} - S_{\mathrm{f}} \\
&\wedge \quad p = \rho_{\mathrm{fs}} g H \quad \text{and} \quad q = U_x H \\
&\therefore \left(1 - \alpha \frac{q^2}{gH^3}\right)\frac{\mathrm{d}H}{\mathrm{d}x} = S_{\mathrm{w}} - S_{\mathrm{f}} \\
&\wedge \quad \mathrm{Fr} = \frac{U_x}{\sqrt{gH}} = \frac{q}{H\sqrt{gH}} = \left(\frac{H_{\mathrm{c}}}{H_{\mathrm{n}}}\right)^{3/2} \qquad (12.31) \\
&\wedge \quad H_{\mathrm{c}} = \left(\frac{q^2}{g}\right)^{1/3} \quad \text{and} \quad H_{\mathrm{n}} = \left(\frac{fq^2}{8gS_{\mathrm{w}}}\right)^{1/3} \\
&\therefore \frac{\mathrm{d}H}{\mathrm{d}x} = S_0\left[1 - \left(\frac{H_{\mathrm{n}}}{H}\right)^3\right] \Big/ \left[1 - \left(\frac{H_{\mathrm{c}}}{H}\right)^3\right]
\end{aligned}
$$

where H_{n} is the normal depth that appears at equilibrium, when the system is uniform. The critical water depth H_{c} defines two flow regimes where one is named subcritical, where the velocity of the shallow gravity wave is larger than the flow velocity and thus the wave also propagates upstream. The other flow regime is labeled supercritical, where the flow velocity has become larger than the speed of the wave, which is no longer able to propagate toward the upstream direction. Changing from supercritical to subcritical, the flow generally develops a so-called hydraulic jump, where the stress on the bed may become extremely high. Scouring will occur due to the strong vorticity developing at that location. In such areas, it is recommended writing a submodule for the transport by consulting the literature on hydraulic jumps and to use transport models to be developed for strong vortex tubes (vortex skeletons). The separations on obstacles are also different for the two types of regimes, which is important to know when using separation models for the roughness. Extreme values of the gradient of the water depth in flow direction characterize the situation:

$$
\begin{aligned}
&\frac{\mathrm{d}H}{\mathrm{d}x} \to 0 \quad \ni \quad H \to H_{\mathrm{n}} \\
&\frac{\mathrm{d}H}{\mathrm{d}x} \to \infty \quad \ni \quad H \to H_{\mathrm{c}} \quad \mathrm{Fr} \to 1
\end{aligned} \qquad (12.32)
$$

Using this concept, we can define the five backwater profiles presented in Table 12.2

Table 12.2. The flow characterized by its depth in comparison to H_{n} and H_{c}.

Criterion	**Name of the profile**	**Behavior of the water surface**
$H_{\mathrm{n}} \to 0$	H	Horizontal water surface
$H_{\mathrm{n}} > H_{\mathrm{c}}$	M	Mild slope
$H_{\mathrm{n}} = H_{\mathrm{c}}$	C	Critical slope or hydraulic jump
$H_{\mathrm{n}} < H_{\mathrm{c}}$	S	Steep slope
$S_0 > 0$	A	Increasing water surface

The numerical calculation can be initiated at any location with a given inflow boundary condition H_1 and by choosing an increment Δx. With these initial conditions, we have for the water surface at H_2 $(x + \Delta x) = H_1(x) + \Delta H(\Delta x)$ at can be calculated by using Eq. 12.31

$$\Delta \boldsymbol{x} = \Delta \boldsymbol{H}\left[1-\left(\frac{\boldsymbol{H}_n}{\boldsymbol{H}_1}\right)^3\right] / \boldsymbol{S}_0\left[1-\left(\frac{\boldsymbol{H}_c}{\boldsymbol{H}_1}\right)^3\right] \vee \Delta \boldsymbol{H} = \Delta \boldsymbol{x} \boldsymbol{S}_0\left[1-\left(\frac{\boldsymbol{H}_c}{\boldsymbol{H}_1}\right)^3\right] / \left[1-\left(\frac{\boldsymbol{H}_n}{\boldsymbol{H}_1}\right)^3\right] \qquad (12.33)$$

Another so-called shooting method is to increase ΔH until a predetermined Δx is achieved.

Assuming that q and f are constant and using Eq. 12.12, one can analyze the differences in the friction or energy slope with respect to the slope of the bed

$$\begin{aligned} S_\mathrm{f} &< S_\mathrm{w} \quad \Leftrightarrow \quad H_\mathrm{n} > H \\ S_\mathrm{f} &= S_\mathrm{w} \quad \Leftrightarrow \quad H_\mathrm{n} = H \\ S_\mathrm{f} &> S_\mathrm{w} \quad \Leftrightarrow \quad H_\mathrm{n} < H \end{aligned} \qquad (12.34)$$

index w = wall, index f = friction

If we have a gradually varying water depth, due to a variation in profile at constant q and f, the wall shear stress can be expressed relative to the wall shear stress under normal conditions. It is important to account for this variation since it generally is evaluated at equilibrium. For conditions other than normal conditions, we find:

$$\frac{\tau_\mathrm{w}}{\tau_\mathrm{wn}} \cong \frac{\rho_\mathrm{fs} g H S_\mathrm{f}}{\rho_\mathrm{fs} g H_\mathrm{n} S_\mathrm{w}} = \frac{H}{H_\mathrm{n}}\left(\frac{H}{H_\mathrm{n}}\right)^3 = \left(\frac{H}{H_\mathrm{n}}\right)^2 \qquad (12.35)$$

$$\therefore \tau_\mathrm{w} > \tau_\mathrm{wn} \quad \wedge \quad H < H_\mathrm{n}$$

For converging and diverging flows, the wall shear stress increases and decreases in downstream direction, respectively;

$$\tau_\mathrm{w} \uparrow \text{conv.} \quad \text{and} \quad \tau_\mathrm{w} \downarrow \text{div.}$$

When the flow can no more be characterized as steady-state, the variation in q results in changing velocity profiles. Here τ_w must be known since this quantity contains the information of the turbulent regime, which enters into the transport equations in the classical form Eq. 2.44, as for a four-zone model Eq. 2.34 with the roughness laws (Eq. 2.42), where λ is given by Eq. 2.37 replacing D by R_H.

12.2 Coherent Structures

When the bed sediment is fine sand it is transported by coherent structures as shown in Chap. 9. Coherent structures are quasi-periodic flow events due to an instability process of the near-wall fluid, when the bed can be considered smooth or nearly smooth (Chap. 3, Sect. 3.2.4). These structures are embedded within the mean flow and can be treated as superimposed structures. It is assumed that they are homogeneously distributed laterally and represent the most intense velocity fluctuation of a turbulent flow. The significant improvement is that we know the "periodicity" and the strength of the events now. All this information enters into the distribution functions as introduced by Grass, see Chap. 3. (Sect. 3.2.3). It has been shown (Chap. 6, Sect. 6.2.5) that these values can be estimated analytically or calculated using a direct numerical integration method of the

Navier–Stokes equation. First attempts in this direction were published by Moin and Kim (1982, 1985, 1986), Kline and Robinson (1990) and references mentioned therein. Later calculations enhanced the resolution; the method however remained the same. Leonard (1980) investigated the development of disturbances on spanwise vorticity lines near the wall and found by numerical simulations that they develop into turbulent spots of Λ-type shape. These results were very informative because they showed the physical evolution, and it is well possible that in the near future these results can be incorporated into a simulation program for the sediment transport of this kind of sediment.

Further detailed information and stimuli on how this can be achieved are compiled with an extended list of references in Holmes et al. (1996), Lumley et al. (1999), Farrell and Ioannou (1999), Vassilicos (1999), and Gyr et al. (1999). But today it is still unrealistic to believe that the sediment transport can be calculated using these direct simulation methods. The alternatives are to evaluate empirically or calculate such instabilities and use the found structures or structure distributions in an abstract form as a module.

The main deficiency of such a procedure is that the feedback system with the bed configuration gets lost since the coherent structures depend on the wall configuration too. Numerical simulations need to prove their ability to handle the formation of longitudinal stripes and channeling effects, assuming that the bed contains these forms. All these investigations cannot be found in the literature yet. However, before the occurrence of these longitudinal stripes, one can calculate the transport or the velocity· the velocity profile using an approximate flow structure as for example used by Beljaars and Prasad et al. (1981). These authors assumed that the outer flow includes 2D large eddy disturbances in lateral direction, which are in phase with the burst cycle. The near-wall zone, where the viscous shear stress is large, can be treated independently of the outer flow where these stresses are much weaker and the flow can be treated as friction free. The coherent structures are coupling these two zones. The measured velocity profiles are in good agreement with the one calculated by the above-mentioned methods. For a complete overview the reader should consult the original literature.

For the transport, one needs the pressure distribution on the grains. It is therefore only a first step to evaluate the velocity fluctuations, for the description of the sediment transport via coherent structures. The long-time correlation of the velocity fluctuations will not properly represent the instantaneous value of the Reynolds stress tensor at a given place or time. When using the velocity fluctuations one would have to calculate the pressure distribution tediously through Eq. 5.24, which is too time consuming to be adequate for simulations. However, since we know more about coherent structures and especially about the corresponding vorticity distribution we can reduce the computational effort drastically, if we assume a pressure field as induced by the concentrated vortex cores. The pressure fluctuations on the bed can be evaluated (Métais et al., (1999). The main processes involved are described in Chap. 9 (Sect. 9.1) and shown in Fig. 9.3.

The main remaining problem lies in the description of the feedback mechanisms, which cannot be formulated without incorporating flow separation mechanisms and alignments of the perturbations due to special bed configurations as they were described heuristically in Chap. 9. There we also showed that the sediment transport calculation by coherent structures needs an iterative method since any change in the bed configuration alters the coherent structures. Every bed configuration provokes a different instability process and with it a different coherent structure.

The first attempt to apply coherent structure mechanisms was made by introducing a linearized wave interacting model for simulating the coherent structures, Landahl (1990) and the influence of 3D surface elements by Gustavsson and Wallin (1990). Since it is not yet possible to incorporate these feedback mechanisms it is prudent to use the statistical description by Grass (Chap. 3, Sect. 3.2.3) or to assume a certain bed configuration as an initial condition, e.g., a streaky bed.

12.3 Turbulent Flows

The dominance of the coherent structures diminishes with increasing roughness of the sand mixture bed until they finally disappear as discussed in Chap. 10. Then the flow can be assumed similar to the turbulent flow field as it exists over rough walls. For bed-load transport from turbulent flow, it is sufficient to use the wall shear stress τ_w with a given distribution, due to the fluctuating instantaneous Reynolds stress $u'v'$. It deviates significantly from the mean value representing the mean Reynolds stress component which cannot be applied to the transport of suspended particles, for the turbulent structures are responsible for keeping the particles suspended in this regime during the entire transport time. Ideally one would like to have the full representation of the turbulent flow structures in a statistical sense. We all know that this can only be achieved approximately with special numerical models, which themselves constitute a whole research field.

A good strategy lies in consulting the textbook of Tennekes and Lumley (1972 and following new editions) or the following review articles and choosing the method most adequate for the problem; Lumley (1996), Kraichnan (1991), Moin (1993, 1996), Germano (1999), Meneveau et al. (1999), Oberlack (1999), and Leonard (1999). Depending on the chosen method, one has to define the initial and boundary condition of the system and it is essential that the simulation resolves the relevant details.

The approach will be different when local interactions are important, e.g., for a flow around an obstacle, which is to be used as flow module to find a statistical representation. In this case, it is suggested to use the rapid distortion theory (RDT) as discussed by Hunt et al. (1991).

The turbulent flow has been tackled by a series of numerical methods, but not all seem appropriate for the discussed problem of sediment transport. The most sophisticated is the direct numerical simulation (DNS) of the Navier–Stokes equation through integration. The so-called large eddy simulation (LES) is a simplification from DNS, which results when the grid spacing will not allow resolving the smallest scales. Instead, the contributions from the small scales enter as a disturbance, which are estimated using a subgrid model. We therefore would have to introduce a threshold for what grid size of the LES is compatible with the problem. However, the biggest deficiency of these methods is that they need periodic boundary conditions when the size of the observation volume is larger than a calculation box, which is very problematic for a natural boundary and practically not applicable for flows interacting with obstacles initiating a flow separation.

A more modest approach is the kinematical representation of the turbulent flow introduced by Wray and Hunt (1990). In their representation, the turbulent flow field has four statistically significant active regions, which were labeled and defined by a given structure: (1) eddies, (2) convergence zones, (3) shear zones, and (4) stream zones.

With this classification one can calculate the interaction of the particles in suspension, and using the statistical distribution of these flow elements the total interaction can be evaluated.

Until now none of these methods has been used to calculate the sediment transport, which shows how little attention was paid to the mechanics of turbulent flow in the modeling of sediment transport.

12.4 Flow with Separations

Also the different types of flow separation must be classified in the context of sediment transport. The flow separation on a 3D obstacle like coarse gravel differs significantly from a flow separation on a 2D bed form like a dune.

12.4.1 Flow Separations at a Roughness or a 3D Obstacle

We distinguished the different roughness types by their local flow separation effect (Chap. 7). For the formulation of numerical modules we refer to Chap. 8, especially Chap. 8 (Sect. 8.3.3) treating the separation on exposed obstacles.

Again we restrict our discussion to the two examples as they were treated in Chap. 8. Sect. 8.3.3), which should help to develop one's own representations if confronted with other configurations. The two obstacles mentioned have a length L in flow direction and a half width $b = B/2$. Assuming symmetry about these two directions yields the main dimensionless parameters h/A and A/B. If the obstacles are not symmetric by orientation or shape, one has to introduce correction factors.

Many details of the flow past obstacles at high Reynolds numbers are still subject to current research, and we advise consulting the literature before introducing such flows into sediment transport simulations. The simplest flow configuration is the so-called potential flow, which is related to an inviscid flow. The flow field $\underline{u}$ can then be derived from a velocity potential ϕ. The upstream flow velocity U is assumed uniform and

$$\left[\nabla,\underline{\boldsymbol{u}}\right]-0 \quad \therefore \quad \underline{\boldsymbol{u}}=\nabla\boldsymbol{\phi}\rightarrow\boldsymbol{\phi}(x,y,z) \tag{12.36}$$

With the continuity condition this implies

$$\left(\nabla,\underline{\boldsymbol{u}}\right)=0 \quad \therefore \quad \Delta\boldsymbol{\phi}=0 \tag{12.37}$$

The boundary conditions at the surfaces imply that the normal component of the velocity u_n must vanish. This system of equations specifies the potential flow, and the flow pattern adjusts to $U = U(t)$.

For viscous fluids however, the boundary condition reads $\underline{\boldsymbol{u}} = 0$ on the surface which cannot be solved with potential flow theory and the solution departs from a potential flow. A boundary layer forms, which can separate at the surface. Such separation may occur in various places depending on the flow, the shape and the expositions of the sidewalls of the obstacle and they are mainly influenced by the inclination angle of the upstream face. To say, there exist substantial volumes of fluid which are directly affected by the boundary processes and characterized by near singular behavior, e.g., large gradients and the topology of their near flow field, whereas other regions can be approximated well with a potential flow field. The flow around an obstacle is given by a large variety of topological flow patterns, which are shape- and Reynolds number dependent.

Separation on a cube-like obstacle: The topology of this first example was discussed shortly in Chap. 8. (Sect. 8.3.3.1), where it was also shown that the main sediment transport is achieved through the side- or horseshoe-vortices, shown as 2a and 2b in Fig. 8.9. They possess high vorticity since the entire boundary layer fluid is concentrated and rolled up into the separation eddies. For a single horseshoe vortex we can estimate the circulation Γ_s

$$\Gamma_s \approx \frac{b}{2}\int_0^{\delta_\omega} \omega_y \mathrm{d}z \quad \wedge \quad \delta_\omega^+ \approx 12 \tag{12.38}$$

The thickness of the original boundary layer, δ_w must be evaluated empirically since the theoretical evaluation is too complicated to realize in a simulation. For sand mixtures $\delta_w^+ \approx 12$ is a good estimate, which can also be used as the characteristic size of the vortex core diameter.

Often one can observe that two and more flow separations occur in front of the obstacle (cf. Fig. 8.9). If this is the case, one has to account for each individual Γ_s the rolled up from the vorticity sheet. The two side branches have opposite vorticity and annihilation occurs. The overlapping area of the diffusing vorticity can be used to estimate the dissipated vorticity. The radius of the vortex grows with $\sqrt{\nu t}$, where $t = x_s/u(\delta_w/2)$ an x is a vertical distance from the stagnation point and the advection-velocity given at the location of the core. This is not strictly correct as the helical flow in the core lags in x-velocity compared to the mean flow. However, this error is too small for requiring a higher order approximation.

To evaluate the other contribution, one needs the topology of the flow field or at least the pattern of its wall shear lines, which can be described in this case by an "owl face" footprint. The flow separates at the rear edges of the obstacle and produces two vertical vortices with strength ω_z and of a horizontal vortex branch at the top crest of strength ω_y. The ω_z is of tornado-like shape, and as we know, this produces a concentric flow field at the wall together with a low pressure core. This likely resuspends the grains that get dragged up into the core and therefore contributes to the scouring and transport. The estimation of these combined effects is hard since it is already problematic to measure the vorticity rolling up from the boundary layer. When the obstacle is rather smooth (Chap. 9, Sect. 9.4.3), the vertical vortices have the least vorticity and can be neglected in a first approximation. The lee-vortex is generally much stronger, but its centerline is away from the bed and therefore also not important for the transport of suspended particles in the wake. Contrary to the expectation, the load on the particles in the rear of such an obstacle is so small that they remain on the bed, or if suspended even can settle, although they should move due to the forces generated by the mean flow. This is exactly what Einstein described by the hiding factor. On the other hand, this vortex system creates the dead-water zone behind the obstacle resulting in a low pressure wake, which is often strong enough to drag the obstacle. Therefore, one has to calculate the flow field around the obstacle and especially the pressure distribution over its surface. If the obstacles and their distribution are representative for the bed configuration, it makes sense calculating these local interactions and using them as statistical modules in the simulation. Are the grains distributed such that too many modules need to be introduced? This description may still be beyond our computer capacities.

As discussed in Chap. 8, the topology of the flow is important because it provides us with a picture where to find the concentrated vorticity skeletons. Having those, we can calculate the flow field using Biot-Savarts equation

$$\underline{u}(\underline{x},t)=\frac{\mathrm{d}\underline{x}}{\mathrm{d}t}=-\frac{\Gamma}{4\pi}\int_{C}\frac{\left[\left(\underline{x}-\underline{r}(s)\right),\left(\partial\underline{r}/\partial s\right)\right]}{\left(\left|\underline{x}-\underline{r}(s)\right|^{2}+\alpha a^{2}\right)^{3/2}} \tag{12.39}$$

In Eq. 12.39, $\underline{u}$ is the induced velocity whereas $\underline{x}$ is the spatial coordinate in a co-transported system, $\underline{r}$ is the distance vector from the vortex center which has s as its line coordinate. The core radius a together with its vorticity distribution α and $\partial\underline{r}/\partial s$ is the unit tangent vector along ds. For a vortex ring α is given by the value 0.22.

In a first approximation, the core vorticity was thought to be well represented by a Rankine-vortex model as given in Eqs. 10.12 and 10.13 for viscous fluids. It is however more appropriate to use

$$\omega(r)=\begin{matrix}\exp\left[f(r)\right] & \in & r\le 1\\ 0 & \in & r>1\end{matrix}$$
$$f(r)=-\frac{r^{2}}{1-r^{2}}+r^{2}\left(1+r^{2}+r^{4}\right)\quad\wedge\quad r=R/a \tag{12.40}$$

where R is the radial distance from the vortex core which has a core radius a.

Separations on polynomial hills: The topology of such an obstacle was described in Chap. 8. (Sect. 8.3.3.2) and shown in Fig. 8.10. The main purpose for constructing the topology of the flow field is to locate the vortex tubes generated after separation and use this information to evaluate the corresponding pressure field responsible for the transport. In the case discussed, three vortex elements are dominant; the central leevortex and the two side vortices, which differ from the ones discussed in this section because they are not connected via the front part of a horseshoe vortex. The method for evaluating their strength, however, remains the same since one has to evaluate the amount of the boundary layer entering the open separation bubble and the entrained vorticity. The total vorticity fed into the bubble is given by the integral Eq. 8.3, defining the total circulation that remains conserved.

$$\Gamma=\oint\underline{u}\mathrm{d}\underline{l}=\iint\left[\nabla,\underline{u}\right]\mathrm{d}\underline{A}=\text{const}\approx\left(\delta\underline{A},\left[\nabla,\underline{u}\right]\right) \tag{12.41}$$

If one can evaluate the diameter of the vortex tube after it has formed, one can attribute to it the circulation calculated with Eq. 8.3 by using a measured width of the boundary layer entering the separation bubble. This is insofar more complicated as the vorticity concentrates in the vortex tubes as shown in Fig. 8.10. and Eq. 12.41 therefore reads

$$\Gamma_{0}=\Gamma_{\text{lee}}+2\Gamma_{\text{s}}+\Gamma_{\text{dis}} \tag{12.42}$$

The dissipated circulation Γ_{dis} is counted within the separation due to annihilation of vorticity produced and recirculated at the bed. This contribution is small and can be neglected. It can be used for a check of a closed separation bubble, which is the equilibrium state when

$$\Gamma_{0}\approx\Gamma_{\text{dis}} \tag{12.43}$$

The circulation of the three vortex tubes is of the same order of magnitude, although Γ_{lee} is less important for the sediment transport since its vortex core is at a higher distance from the bed. In addition the vorticity of the open branches of this vortex have, compared to the side vortices, opposite signs and they therefore damp each other, but together with the mirror images they stabilize the pattern. Of course, one has to estimate the strength of the three vortex tubes, their core diameter together with their wall distances as well as possible, but for an estimate one can roughly assume $\Gamma_0/3$ for each of them.

It has to be pointed out that this description and its numerical evaluation is a rough idealization. We deal with an averaged structure and with increasing Reynolds number also the complexity of the flow topology increases since the wake vortex starts to shed intermittently. These vortices are characterized by inclined loops which are transported toward the wall and scour the bed a short distance downstream. These brief statements give an impression of the difficulties one faces in the realization of an algorithm accounting for such separation events. Nevertheless, statistical implementations of such processes will become necessary when we want to model the feedback systems responsible for the bed-form configuration.

With known vortical tubes, the Biot-Savart law allows estimating the velocity at any point in the region of interest. If we only know the vorticity distribution on the surface and in the near flow field, numerical methods are available that achieve a similar result. A rather large variety of methods are available, one of which is the "Vortex in Cell" Algorithm (Cottet and Koumoutsakos, 2000). We would also or proceed classically by applying DNS for the interaction with the bed (Chang et al., 1997).

12.4.2 Flow Separation at 2D Bed Forms

The smallest 2D bed forms are ripples, which have been discussed in Chap. 9. Their formation mechanism was shown together with their feedback influence on the sediment transport. Ripples are interesting from a conceptual point of view, but they are irrelevant for the transport itself since their appearance is rather limited and restricted by a low flow discharge over fine sand. Therefore, it is sufficient to add an additional flow resistance in case they appear.

The situation changes completely, the instant dunes appear Chap. 9, Sect. 9.5) because they cause intermittent conditions for the sediment transport on the boundary. There is an obvious need to evaluate their forms, which can be achieved using more or less sophisticated theories as described in Chap. 9 (Sect 9.5). Once the bed forms are known, they could be introduced in the formulation of new channel geometries to proceed as shown in Chap. 12 (Sect. 121.1). However, this helps only if we are able to measure dome relevant parameter. This is not always the case because in defining the flow separation one also needs flow field criteria, which are often not available. We have to introduce the flow separation in some form and the simplest case is assuming a mixing layer separation at the dune crest which characterizes the flow by the scales defined in Fig. 9.23. A vortex shedding theory based on such parameters is discussed in Chap. 9 (Sect. 9.5.4). However, it would be better to use an alternative sediment transport model based on the two zones composing a dune wavelength

$$\Lambda = \Lambda_A + \Lambda_{trans} \tag{12.44}$$

The more relevant contribution for the sediment transport is Λ_{trans} since in the region Λ_A mostly sedimentation occurs, which nevertheless is important for the change in dune shape.

Consequently, the transport is given by a slow contribution given by the dune progression speed and a fast contribution of the suspended matter which becomes significant in the nonequilibrium flow states. Both can be accounted for using a continuity description for the sediment as was shown by Engelund and Hansen (1966), Eq. 9.56 resulting in Eqs. 9.60 and 9.61. If the erosion in Λ_{trans} increases, the sediment body must react, either by changing the wavelength or adjusting the amplitude. The contributions given by Eq. 12.44 remain constant only in the equilibrium state. If not, either the length will change with time, $\Lambda(t)$, or the subdivision in the two length scales will be altered.

At first one has to evaluate the reattachment point using Eq. 9.24. The transport varies considerably along Λ_{trans}, and the main erosion will occur near the attachment zone where the shed shear layer vortices hit the bed. The scouring and suspension of material can be estimated using models shown in Chap. 8 (Sect. 8.4) and Fig. 9.3. Immediately downstream an accelerating turbulent boundary layer with a small momentum thickness establishes, which increasingly saturates the fluid with suspended sediment. The transport should scale with the existing wall shear stress, which increases due to the acceleration on the luff slope.

In a first approximation, one can estimate the transport as given by the contribution of the attachment area and assume that no additional erosion will occur and the transport later on remains constant. Here it is important to know the strength of the interacting vortices, which will depend on the pairing that occurred before touching the bed.

Another special problem related to such a splitting is the transport by suspension. In the eroding area, the amount of suspended sediment is large and not all of this sediment will settle again in zone Λ_A. We are confronted with a vortical flow field as Meiburg and coworkers investigated it (Chap. 9, Sect. 9.5.1). Through the motion of the whole sediment body, all material gets repeatedly exposed to the fluid shear and thus the fine material washes out with time. Thus the newly suspending load will diminish in the equilibrium state and can be neglected, which however must not be done for the suspension load arriving from an upstream region.

12.5 Suspended Load

The calculations for the suspended load transport were presented in Chap. 3. While the models discussed therein were 1D and restricted to granular material of a certain size being homogeneously distributed over the bed. These assumptions are not always valid and we have therefore to supplement Chap. 3 with cases, which occasionally arise in a complex calculation of sediment transport.

12.5.1 Washload

In nature, one often encounters a situation where the suspended material is composed of fine grains for which the entrainment conditions are valid and even finer material, e.g., silt, which is permanently in suspension. Whereas for the first group, the relevant flow structures are close to the bed, turbulence in its statistical form can be used for permanently suspended material. The very fine part of the suspended material is called washload, however there is no rigorous criterion for its definition and only empirical classifications are used. For overall fine material, one often finds as definitions $d_s < d_{10}$ or an absolute value like $d_{sw} \approx 0.0625$ mm.

Using a 1D mixing concept for an amount of nonsettling fine sediment released at time zero at $y = 0$, Eq. 3.63 reduces to the transversal mixing as encountered if the washload is released locally and

$$\frac{\partial c_V}{\partial y} = \frac{\varepsilon_y \partial^2 c_V}{\partial y^2} \tag{12.45}$$

The solution in the cross-sectional area A or plane (x, z) is

$$c_V(y,t) = \frac{m}{A\sqrt{4\pi\varepsilon_y t}} \mathrm{e}^{-z^2/4\varepsilon_y t} \tag{12.46}$$

where the mass m in V is given by

$$m = \int_V c_{vwl}\, dV \tag{12.47}$$

When the concentration is distributed normally as given by Eq. 12.46, the variance will increase linearly with time

$$\sigma_y^2 = 2\varepsilon_y t \tag{12.48}$$

If the material is released locally, the width of the mixing cloud can be estimated from

$$4\sigma_y = 4\sqrt{2\varepsilon_y t} \tag{12.49}$$

The relevant mixing coefficients as well as the length- and timescales are given in Table 12.3 for a channel flow width B and depth H and characterized by the wall shear velocity u_τ. The timescales are obtained from the normal distribution Eq. 12.46 via the following relations

$$H^2 = 2\varepsilon_z t_v \quad \wedge \quad B^2 = 2\varepsilon_y t_y \quad \wedge \quad L_x^2 = 2\varepsilon_x t_x \tag{12.50}$$

The timescales follow from combining the mean flow velocity U and the length scales.

Table 12.3 Mixing coefficients, time- and length scales for the dispersion of washload

	Mixing coefficient	Timescales	Length scales
x	$\varepsilon_x \approx 250\, Hu_\tau$ Or $0.11 U^2 B^2 / Hu_\tau$	$t_x = L_x/U \approx L_x^2/500 Hu_\tau$.	$L_x = 500 Hu_\tau/U$
y_{norm}	$\varepsilon_y \approx 0.15\, Hu_\tau$	$t_y = L_y/U \approx B^2/Hu_\tau$.	$L_y = UB^2/Hu_\tau$.
y_{nat}	$\varepsilon_y \approx 0.6\, Hu_\tau$		
z	$\varepsilon_z \approx 0.067\, Hu_\tau$	$t_z = L_z/U \approx H/0.1u_\tau$.	$L_z = HU/0.1u_\tau$.

From Table 12.3 one can conclude that the ratio of the vertical to the transversal time- and length scales is given by $H^2/0.1B^2$ stating that the vertical mixing is faster than the transversal unless B is small compared to H.

This kind of mixing and transport is mainly applicable to a local supply like a plume from a steady point source or mixing at stream confluence; for these cases see Fischer et al. (1979) and Rutherford (1994), respectively.

12.6 The Significance of Experiments for the Simulations

Many of the encountered theories are of empirical or half-empirical nature, meaning their reliance on experimentally evaluated data and we have to discuss two data categories related to the problem.

It is clear that no simulation is at all possible without a quantitative characterization of the system under investigation, which requires measurements taken from the river. Those measurements may be very difficult to carry out and we do not want to discuss here all the problems related to this issue. It is thus evident that practically every measurement needs adaptations of the measurement devices and methods due to the individual circumstances of the field site. Furthermore, one needs special constructions anywhere in the cross section in order to obtain, e.g., $Q(t)$ or $U(A)$. Case studies which normally cannot be found in textbooks but mainly in conference proceedings or specialized journals are important sources for relevant datasets.

It became fashionable to think that further experiments are not necessary once we have good datasets, since the experimental prediction can be acquired numerically with a computer. The problem is in this case that we need the knowledge of the involved physical laws occurring in the complex interaction process called sediment transport. These laws are generally not known but even if they were, their complexity would still exceed today's computer capacities for obtaining a good resolution on the small scale together with the developments on the larger scales. Turbulence, flow separations, transport capacity, rheology, and feedback mechanisms are only some keywords to underscore the significance of this statement. We need simplifications, which need to pass confirmation or face rejection while numerically we have to work with modules. For the development of such subprograms, we need information which can only be found by performing experiments, and the experimental approach has changed drastically therefore. The simulation of the whole system by physical models belongs to the past. The experimental work returned to classical laboratory experiments of much smaller size by which parameters and interactions relevant for the module construction are studied.

This development goes hand in hand with an enormous evolution in the measurement techniques for water flows for which a good overview can be found in Eckelmann (1997). Large progress was made integrating optical systems to evaluate the velocity at certain points or even the whole velocity field of the fluid and of the particle load as well. The methods used are laser-doppler-anemometry (LDA), particle image velocimetry (PIV), particle tracking velocimetry (PTV), and laser induced fluorescence (LIF) combined with gradient field tracking. Supplemented by very sophisticated hot-wire systems to evaluate the vorticity in situ (not in a sediment laden flow). A good overview can be found in the ERCOFTAC Series Dracos (1996), or see also Rösgen and Totaro (2003) for multiphase flow. However, they are methods only practicable for laboratory experiments.

13 References

Ackers, P.: Meandering channels and the influence of bed material. In: Hey, R.D., Bathurst J.C., Thorne, C.R. (eds.) Gravel-Bed Rivers, pp. 389–421. John Wiley and Sons, Chichester (1982)

Ackers, P., White, W.R.: Sediment transport: New approach and analysis. J. Hydr. Div. ASCE **99**(HY11), 2041–2060 (1973)

Adrian, R.J.: Particle-imaging techniques for experimental fluid mechanics. Ann. Rev. Fluid Mech. **23**, 261–304 (1991)

Adrian, R.S.: Vortex packets and the structure of wall turbulence-Extended abstract. In: Gyr, A. et al. (eds.) Science and Art Symposium 2000 77 + Plates (2000)

Alabyan, A.: A computer model of bank erosion based on secondary flow simulation. In: Ashworth, P.J., et al. (eds.) Coherent Flow Structures in Open Channels, pp. 567–580. John Wiley and Sons, Chichester, New York, Brisbane, Toronto, Singapore (1996)

Alam, A.M.Z., Kennedy, J.F.: Friction factors for flow in sand bed channels. Proc. ASCE **95**(HY6), 1109–1127 (1969)

Alam, A.M.Z., Cheyer, T.F., Kennedy, J.F.: Friction factors for flow in sand bed channels. Hydrodynamics Laboratory Report No. 78. MIT, Cambridge (1966)

Albering, W.: Elementarvorgänge fluider Wirbelbewegungen. Akademie-Verlag, Berlin (1981)

Alfredsson, H., Johansson, A.V.: On the detection of turbulence-generating events. J. Fluid Mech. **139**, 325–345 (1984)

Aliseda, A., Cartellier, A., Hainaux, F., Lasheras, J.C.: Effect of preferential concentration on the settling velocity of heavy particles in homogeneous isotropic turbulence. J. Fluid Mech. **468**, 77–105 (2002)

Allen, J.: Bed forms due to mass transfer in turbulent flows: A kaleidoscope of phenomena. J. Fluid Mech. **49**, 49–63 (1971)

Allen, J.R.L.: The nature and origin of bed-form hierarchies. Sedimentology **10**, 161-82 (1968)

Anderson, R.S.: A theoretical model for impact ripples. Sedimentology **34**, 943–956 (1987)

Anderson, R.S.: Eolian ripples as example of self-organization in geomorphological system. Earth Sci. **29**, 77–96 (1990)

Andreotti, B., Claudin, P., Douady, S.: Eur. Phys. J. **B 28**, 321-29 (2002)

Aref, H.: Application of dynamical systems theory to fluid mechanics. In: Lumley et al. (eds.) Research Trends in Fluid Dynamics, pp. 15–30. American Institute of Physics, Woodbury, New York, 15-30 (1996)

Arndt, R.E.A., Ippen, A.T.: Rough surface effects on cavitation inception. J. Basic Eng. **90**, 249–261 (1968)

Arnold, V.I.: Small denominators and the problem of stability of motion in classical and celestal mechanics. Russ. Matn. Surv. **18**, 85–191 (1963)

Arnold, V.I.: Mathematical Methods of Classical Mechanics. Springer-Verlag, New York (1978)

Arnold, V.I., Avez, A.: Ergodic Problems of Classical Mechanics. Benjamin, NY (1968)

Ashworth, P.J., Bennet, S.J., Best, J.L., McLelland, S.J.: Coherent Flow Structures in Open Channels. John Wiley and Sons, Chichester (1996)

Asmolov, E.S[R5].: Numerical simulation of the coherent structures in a homogeneous sedimenting suspension. In: Gyr, A., Kinzelbach, W. (eds.) Sedimentation and Sediment Transport, Kluwer Academic Publ. Dordrecht, Boston, London pp. 159–164 (2003)

Baas, J.H., Best, J.L.: Turbulence modulation in clay-rich sediment-laden flows and some implications for sediment deposition. J. Sediment Res. **72**, 336–340 (2002)

Bagnold, R.A.: The Physics of Blown Sand and Dunes. Methuen, London (1941)

Bagnold, R.A.: Experiments on gravity-free dispersion of large solid spheres in a Newtonian fluid under shear. Proc. R. Soc. **A225**, 49–63 (1954)

Bagnold, R.A.: Auto-suspension of transported sediment; turbidity currents. Proc. R. Soc. **A265** (Nr. 1322) (1962)
Bagnold, R.A.: An approach to the sediment transport problem for general physics. US Geological Survey Professional Paper 422-I. Washington DC (1966)
Bakke, P.D., Basdekas, P.O., Dawdy, D.R., Klingeman, P.C.: Calibrated Parker-Klingeman model for gravel transport. J. Hydr. Eng. ASCE **125**, 657–660 (1999)
Bakker, P.G., de Winkel, M.E.M.: On the topology of three-dimensional separated flow structures and local solutions of the Navier-Stokes equation. In: Moffatt, H.K., Tsinober, A. (eds.) Topological Fluid Mechanics, pp. 384–394. Cambridge University Press, Cambridge, New York, Port Chester, Melbourne, Sydney (1990)
Ball, R., Richmond, P.: Dynamics of colloidal dispersions. J. Phys. Chem. Liq. **9**, 99–116 (1980)
Barenblatt, G.I.: Similarity, Self-Similarity, and Intermediate Asymptotics. Consultants Bureau, NY (1979)
Barnes, H.A.: Dispersion Rheology: 1980. Royal Society of Chemistry, Industrial Division, London (1981)
Barnes, H.A., Walters, K.: The yeld stress myth? Rheol. Acta **24**, 323–326 (1985)
Barnes, H.A., Hutton, J.F., Walters, K.: An Introduction to Rheology. Rheology Series, 3. Elsevier Amsterdam, Oxford, New York, Tokyo (1989)
Batchelor, G.K.: Theory of Homogeneous Turbulence. Cambridge University Press, Cambridge (1953)
Batchelor, G.K.: An Introduction to Fluid Dynamics. Cambridge University Press, London, New York (1967)
Batchelor, G.K.: Sedimentation in a dilute dispersion of spheres. J. Fluid Mech. **52**, 245–268 (1972)
Batchelor, G.K.: The effect of Brownian motion on the bulk stress in a suspension of spherical particles. J. Fluid Mech. **83**, 97–117 (1977)
Batchelor, G.K., Green, J.T.: The hydrodynamic interactions of two small freely-moving spheres in a linear flow field. J. Fluid Mech. **56**, 375–400 (1972)
Batchelor, G.K., Townsend, A.A.: The nature of turbulent motion at large wave numbers. Proc. R. Soc. **A199**, 238–255 (1949)
Bechert, D.W., Bruse, M., Hage, W., van der Hoeven, J.G.T., Hoppe, G.: Experiments on drag reducing surfaces and their optimisation with adjustable geometry. J. Fluid Mech. **338**, 59–87 (1997)
Beljaars, A.C.M., Prasad, K.K.: A module for periodic structures in turbulent boundary layers. Lect. Notes Phys. **136**, 93–118 (1981)
Beljaars, A.C.M., Prasad, K.K., de Vries, D.A.: A structural model for turbulent exchange in boundary layers. J. Fluid Mech. **112**, 33–70 (1981)
Benjamin, T.B.: Shearing flow over a wavy boundary. J. Fluid Mech. **6**, 161–205 (1959)
Bennet, S.J., Best, J.L.: Mean flow and turbulence structure over fixed ripples and the ripple-dune transition. In: Ashworth, P.J., et al. (eds.) Coherent Flow Structures in Open Channels, pp. 281–304. John Wiley and Sons, Chichester, New York, Brisbane, Toronto, Singapore (1996)
Bennet, S.J., Best, J.L.: Mean flow and turbulent structure over fixed, two-dimensional dunes: Implication for sediment transport and bedform stability. Sedimentology **42**, 491–514 (1995)
Benney, D.J.: A non-linear theory for oscillations in a parallel flow. J. Fluid Mech. **10**, 209–236 (1961)
Benney, D.J., Lin, C.C.: On the secondary motion induced by oscillations in a shear flow. Phys. Fluids **3**, 656–657 (1960)
Bernal, L.P.: The coherent structure in turbulent mixing layers. Ph.D. Thesis CALTECH (1981)
Bernard-Michel, G., Monavon, A., Lhuiller, D., Abdo, D., Simon, H.: Particle velocity fluctuations and correlation length in dilute sedimenting suspensions. Phys. Fluids **14**, 2339–2349 (2002)
Best, J., Bennet, S., Bridge, J., Leeder, M.: Turbulent modulation and particle velocities over flat sand beds at low transport rate. J. Hydr. Eng. ASCE **123**, 1118–1129 (1997)

Binding, D.M.: An approximate analysis for concentration and converging flows. J. Non-Newtonian Fluid Mech. **27**, 173–89 (1988)
Blackwelder, R.F., Eckelmann, H.: Streamwise vortices associated with the bursting phenomenon. J. Fluid Mech. **94**, 577–594 (1979)
Blackwelder, R.F., Kaplan, R.E.: On the wall structure of the turbulent boundary layer. J. Fluid Mech. **76**, 89–112 (1976)
Blom, A.: A sediment continuity model for rivers with non-uniform sediment and bed forms. Ph.D. Thesis, University of Twente, Enschede (2003)
Blom, A., Ribberink, J.S., Parker, G.: Sediment continuity for rivers with non-uniform sediment, dunes, and bed load transport. In: Gyr, A., Kinzelbach, W. (eds.) Sedimentation and Sediment Transport, pp. 179–182. Kluwer Academic Publ. Dordrecht, Boston, London
Bogardi, J.L.: European concepts of sediment transportation. Proc. ASCE **91**(HY1), 29–54 (1965)
Bogardi, J.L.: Sediment Transport in Alluvial Streams. Akademiai Kiado, Budapest (1974)
Bonnefille, R.: Essais de synthese des lois de debut d'entrainement des sediments sous l'äctiond'un courant en regime continu. Essais de synthese des lois de debut d'entrainement des sediments sous l'äctiond'un courant en regime uniforme. Bull. Du CREC, Nr. 5, Chatou, (1963)
Brady, J.F., Bossis, G.: Stokesian dynamics. Ann. Rev. Fluid Mech. **20**, 111–157 (1988)
Brebner, A., Wilson, K.C.: Determination of the regime equation from relationships for pressurized flow by use of the principle of minimum energy degradation. Proc. ICE **36**, 47–62 (1967)
Brenner, H.: Rheology of two-phase systems. Ann. Rev. Fluid Mech. **2**, 137–176 (1970)
Brenner, H.: Rheology of a dilute suspension of axisymmetric Brownian particles. Int. J. Multiphase Flow **1**, 195–341 (1974)
Brooks, N.H.: [R10]Loose Boundry Hydraulics. Pergamon Press, New York (1956)
Brown, C.B.: Sediment transport In: Engineering Hydraulics. Edit. H. Rouse p. 796. Wiley New York (1950)
Bruse, M., Bechert, D.W., Hage, W.: The flow over riblets; Velocity measurements with hot film probes. In: Meier, G.E.A., Viswanath, P.R. (eds.) IUTAM Symposium on Mechanics of Passive and Active Flow Control, pp. 115–120. Kluwer Academic Publishers (1999)
Callander, R.A.: River meandering. Ann. Rev. Fluid Mech. **10**, 129–158 (1978)
Cantwell, B.J.: Organized motion in turbulent flow. Ann. Rev. Fluid Mech. **13**, 457–515 (1981)
Cantwell, B.J., Coles, D., Dimotakis, P.: Structure and entrainment in the plane of symmetry of a turbulent spot. J. Fluid Mech. **87**, 641–672 (1978)
Cartwright, J.H.E., Magnosco, M.O., Piro, O. and Tuval, I.: Bailout Embeddings and Neutrally Buoyant Particles in Three-Dimensional Flows. Phys. Rev. Lett. **89**, 264–501 (2002)
Cauchy, A.-L.: Ex de Math. **3**, Oeuvre 8, 195–226 (1828)
Cebeci, T., Smith, A.M.O.: Analysis of Turbulent Boundary Layers. Academic Press, New York (1974)
Cellino, M., Graf, W.H.: Sediment-laden flow in open channels under noncapacity and capacity conditions. J. Hydr. Eng. ASCE **125**, 455–462 (1999)
Chabert, J., Chauvin, J.L.: Formation de dunes et des rides dans les modèles fluviaux. Bull. Cen. Rech. Ess. Chatou, Nr. 4 (1963)
Chang, T.Y., Hertzberg, J.R., Kerr, R.M.: Three-dimensional vortex/wall interaction: Entrainment in numerical simulation and experiment. Phys. Fluids **9**, 57–66 (1997)
Chang, H.H.: Geometry of gravel streams. J. Hydraul. Div., ASCE **106** (HY9), 1443-56 (1980)
Chapman, G.T., Yates, L.A.: Topology of flow separation on three-dimensional bodies. Appl. Mech. Rev. **44**, 329–345 (1991)
Chen, C.-H.P., Blackwelder, R.F.: Large-scale motion in a turbulent boundary layer: A study using temperature contamination. J. Fluid Mech. **89**, 1–31 (1978)
Chen, Y.-C., Chung, J.N.: The linear stability of an oscillatory two-phase channel flow in the limit of small Stokes number. Phys. Fluids **7**, 1510 (1995)
Cherry, N.J., Hiller, R., Latour, M.P.: Unsteady measurements in a separated and reattaching flow. J. Fluid Mech. **144**, 13–46 (1984)

Chien, N.: The present status of research on sediment transport. Trans. ASCE **121**, 833–868 (1956)
Chong, M.S., Perry, A.E., Cantwell, B.J.: A general classification of three-dimensional flow patterns. Report No. SUDAAR 572, Dept. Aeron. & Stron., Stanford University 1988. A general classification of three-dimensional flow fields. In: Moffatt, H.K., Tsinober, A (eds.) Topological Fluid Mechanics, pp. 408–420. Cambridge (1990)
Chuna, L.V.: About the roughness in alluvial channels with comparative coarse bed material. Proc. 14th Congr. IAHR **1**, 76–84 (1967)
Cimbala, J.M., Nagib, H.M., Roshko, A.: Large structures in the wakes of two-dimensional bluff bodies. J. Fluid Mech. **190**, 265–298 (1988)
Clauser, F.H.: Turbulent boundary layers in adverse pressure gradients. J. Aero. Sci. **21**, 91–108 (1954)
Coffman, D.M., Keller, E.A., Melhorn, W.N.: New topological relationship as an indicator of drainage network evolution. Water Resour. Res. **8**, 1497–1505 (1972)
Colby, J.W.: Discharge of sand and mean velocity relationship in sand-bed stream. US Geological Survey Professional Paper 462-A. Washington DC (1964a)
Colby, J.W.: Practical computations of bed –material discharge. J. Hydr. Eng. **90**(HY2), 217–246 (1964b)
Colby, J.W., Hembree, C.H.: Computation of total sediment discharge: Niobrara river near Cody, Nebraska. US Geological Survey Water-Supply Paper **1357** (1955)
Colebrook, C.F.: Turbulent flow in pipes with particular reference to the transition region between the smooth and rough pipe laws. J. Inst. Civil Engineers **11**, 133–156 (1939)
Coleman, N.L.: A theoretical and experimental study of drag and lift forces acting on a hypothetical streambead. In: Proceedings of the 12th Congress IAHR, vol. 3, pp. 185–192. Fort Collins (1967)
Coleman, N.L.: Flume studies of the sediment transfer coefficient. Water Resour. Res. **6**, 801–809 (1970)
Coleman, N.L.: Effect of suspended sediment on the open channel velocity distribution. Water Resour. Res. **22**, 1377–1384 (1986)
Coles, D.: The law of the wake in the turbulent boundary layer. J. Fluid Mech. **1**, 191–226 (1956)
Coles, D., Barker, S.J.: Turbulent Mixing in Nonreactive and Reactive Flows. Plenum Press, NY (1975)
Coles, D., Savas, O.: Interactions for regular patterns of turbulent spots in laminar boundary layer. In: Laminar Turbulent Transition, Eds. R. Eppler and H. Fasel. Springer-Verl. Berlin, Heidelberg, New York, 277-87 (1980)
Cottet, G.H., Koumoutsakos, P.: Vortex Methods: Theory and Practice. Cambridge (2000)
Crickmore, M.J., Lean, H.G.: The measurement of sand transport by means of radioactive tracers. Proc. R. Soc. Lond. **A266**, 402–421 (1962)
Dallmann, U., Schulte Werning, B.: Topological changes of axisymmetric and non-asymmetric vortex flows. In: Moffatt, H.K., Tsinober, A. (eds.) Topological Fluid Mechanics, pp. 373–383. Cambridge (1990)
Davis, R.H.: Particulate flows and sedimentation. In: Lumley et al. (eds.) Research Trends in Fluid Dynamics, pp. 60–68. American Institute of Physics, NY (1996)
Davis, R.H., Hassen, M.H.: Spreading of the interface at the top of a slightly polydisperse sedimenting suspension. J. Fluid Mech. **196**, 107–134 (1988)
De Bruin, H.A.R., Moore, C.J.: Zero-plane displacement and rough length for tall vegetation, derived from a simple mass conservation hypothesis. Bound. -Layer Meteorol. **31**, 39–49 (1985)
Denn, M.M.: Issues in non-Newtonian fluid mechanics and rheology. In: Lumley et al. (eds.) Research Trends in Fluid Dynamics, pp. 69–76. American Institute of Physics, Woodbury, New York (1996)
Dimas, A.A., Kiger, K.T.: Linear stability of a particle-laden mixing layer with a dynamic dispersed phase. Phys. Fluids **10**, 2539–2557 (1998)

Dinkelacker, A., Hessel, M., Meier, G.E.A., Schewe, G.: Investigation of pressure fluctuations beneath a turbulent boundary layer by means of optical method. Phys. Fluids **20**, S216–S224 (1977)

Doligalski, T.L., Smith, C.R., Walker, J.D.A.: Vortex interactions with walls. Ann. Rev. Fluid Mech. **26**, 573–616 (1994)

Dracos, T. (ed.): Three-Dimensional Velocity and Vorticity Measuring and Image Analysis Techniques. Kluwer Academic Publishers, Dordrecht, Boston, London (1996)

Druzhinin, O.A.: On the two-way interaction in a two-dimensional particle laden flows: The accumulation of particles and flow modification. J. Fluid Mech. **297**, 49–76 (1995)

Druzhinin, O.A.: The dynamics of concentration interface in a dilute suspension of solid heavy particles. Phys. Fluids **9**, 315–324 (1997)

Druzhinin, O.A.: The influence of inertia on the two-way coupling and modification of isotropic turbulence by microparticles. Phys. Fluids **13**, 3738–3755 (2001)

Du Bois, M.P.: Le Rhone et les rivieres a lit affoulable, Mem. Doc. Ann. Pont et chaussees, Ser.5, **18** (1879)

Eckelmann, H.: Einführung in die Strömungsmesstechnik. Teubner Studienbücher; Mechanik, Stuttgart (1997)

Eckelmann, H., Nychas, S.G., Brodkey, R.S., Wallace, J.M.: Vorticity and turbulence production in pattern recognized turbulent flow structures. Phys. Fluids **20**, S225–S231 (1977)

Eichelbrenner, E.A., Preston, J.H.: On the role of secondary flow in turbulent boundary layers in corners (and salient). J. de Mécanique **10**, 91–112 (1971)

Einstein, A.: Eine neue Bestimmung der Moleküldimension. Ann. Physik **19**, 289–306 (1906)

Einstein, A.: Berichtigung zu meiner Arbeit: Bestimmung der Moleküldimension. Ann. Physik **34**, 591–592 (1911)

Einstein, A.H.: Formulae for transportation of bed-load. Trans. ASCE, **107**, 561–577 (1942)

Einstein, H.A.: The bed-load function for sediment transportation in open channel flows. US Dept. Agri. Techn. Bull. **1026** (1950)

Einstein, H.A. Barbarossa: River channel roghness. Proc. ASCE **92**(HY2), 315–326 (1952)

Einstein, H.A., Chien, N.: Effects of heavy sediment concentration near the bed on velocity and sediment distribution. MRD Ser. 8. Univ. Calif., Inst. Eng. Res. & US Army Eng. Div. Miss. Riv. Corps of Eng. Omaha, Nebraska (1955)

Einstein, H.A., El Samni, E.S.A.: Hydrodynamic forces on rough wall. Rev. Mod. Phys. **21**, 520–524 (1949)

Einstein, H.A., Li, H.: The viscous sublayer along a smooth boundary. ASCE Trans. **123**, 293–317 (1958a)

Einstein, H.A., Li, H.: Secondary currents in straight channels. Am. Geophys. Union **39**, 1085–1088 (1958b)

Elghobashi, S., Truesdell, G.C.: On the two-way interaction between homogeneous turbulence and dispersed solid particles I: Turbulence modification. Phys. Fluids **A5**, 1790–1801 (1993)

Emmerling, R.: The instantaneous structure of the wall pressure under a turbulent boundary layer flow. Max-Planck Inst. F. Strömungsforschung Ber. **9**, (1973)

Engelund, F.: A criterion of the occurrence of suspended load. La Houille Blanche Nr. 6 (1965)

Engelund, F.: Hydraulic resistance of alluvial streams. J. Hydr. Div. ASCE **92**(HY2), 315–326 (1966a)

Engelund, F.: Closure and discussion of hydraulic resistance of alluvial streams. Proc. ASCE **93**(HY4), 287–296 (1966b)

Engelund, F., Fredsoe, J.: Sediment ripples and dunes. Ann. Rev. Fluid Mech. **14**, 13–37 (1982)

Engelund, F., Hansen, E.: A Monograph on Sediment Transport to Alluvial Streams. Copenhagen:Teknik Vorlag (1967)

Engelund, F., Hansen, E.A.: Investigations of flow in alluvial streams. Tech. Univ. Denmark Hydr. Lab. Bull. **9** (1966)

Faber, T.E.: Fluid Dynamics for Physicists. Cambridge University Press, Cambridge, New York, Melbourne (1995)

Fallon, T., Rogers, C.B.: Turbulence induced preferentional concentration of solid particles in microgravity conditions. Exp. Fluids **33**, 233–241 (2002)
Farrell, B.F., Ioannou, P.J.: Origin and growth of structures in boundary layer flows. Fundamental problematic issues in turbulence. In: Gyr et al. (eds.) Trend in Mathematics, pp. 75–82. Birkhäuuser (1999)
Farris, R.J.: Prediction of the viscosity of multimodal suspensions from unimodal viscosity data. Trans. Soc. Rheol. **12**, 281–301 (1968)
Favre, A., Gaviglio, J., Dumas, J.: Structure of velocity space-time correlation in a boundary layer. Phys. Fluids **10**, 138–145 (1967)
Fenton, J.D., Abbott, J.E.: Initial movements of grains on a stream bed: The effect of relative protrusion. Proc. R. Soc. Lond. **A352**, 532–537 (1977)
Ferrante, A., Elgobashi, S.: On the physical mechanisms of two way coupling in particle-laden isotropic turbulence. Phys. Fluids **15**(2), 315–329 (2003)
Fischer, H.B., Lit, E.J., Koh, R.C.Y., Imberger, J., Brooks, N.H.: Mixing in Inland and Coastal Waters. Academic Press, NY (1979)
Francis, J.R.D.: Experiments on the motion of solitary grains along the bed of a water stream. Proc. R. Soc. **322**, 443–471 (1973)
Frisch, U.: Turbulence. Cambridge University Press, Cambridge, New York, Melbourne (1995)
Frost, W., Bitte, J.: Statistical concept of turbulence. In: Frost, Moulden (eds.) Handbook of Turbulence, chap. 3, pp. 53–83. Plenum Press, New York, London (1977)
Führböter, A.: Zur Mechanik der Strömungsriffel. Mitt. Franzius-Inst., TH Hannover, **29** (1967)
Führböter, A.: Strombänke (Grossriffel) und Dünen als Stabilisierungsformen. Mitt. Leichtweiss-Inst., TH Braunschweig, **67** (1980)
Fuller, W.B., Thompson, S.E.: The laws of proportioning concrete. Trans. Am. Soc. Civ. Eng. **59** (1907)
Galland, J.-C.: Transport de sediments en suspension et turbulence. Rapport HE-42/96/007/A. Laboratoire National Hydraulique Environnement, EDF, 88ff (1996)
Garde, R.J., Ranga Raju, K.G. Mechanics of sediment transportation and alluvial stream problems, 2nd ed. Wiley, New York (1985)
Garg, R.P., Ferzinger, J.H., Monismith, S.G., Koseff, J.R.: Stably stratified turbulent channel flow. Phys. Fluids **12**, 2569–2594 (2000)
Garner, F.H., Jenson, V.G., Keey, R.B.: Flow pattern around spheres and the Reynolds analogy. Trans. Inst. Chem. Eng. **37**, 191–197 (1959)
Germano, M.: Basic issues of turbulence modeling. Fundamental problematic issues in turbulence. In: Gyr et al. (eds.) Trend in Mathematics, pp. 213–220. Birkhäuuser (1999)
Gessler, J.: Der Geschiebetrieb bei Mischungen untersucht an natürlichen Abpflästerung-serscheinungen in Kanälen, Mitt. Der VAW, ETHZ **69** (1965)
Gessler, J.: Critical shear stress of sediment mixtures. In: Proceedings of the 14th Congress IAHR, vol. 3(C1). Paris (1971)
Gessner, F.B.: The origin of secondary flow in turbulent flow along a corner. J. Fluid Mech. **58**, 1–25 (1973)
Gilbert, G.K.: Transportation of debris by running water. US Geological Survey Professional Paper Nr. 86 (1914)
Gill, M.A.: Height of sand dunes in open channel flows. Proc. ASCE **97**(HY12) (1971)
Goldstein, S.: The steady flow of viscous fluidpast a fixed spherical obstacle at small Reynolds numbers. Proc. R. Soc. **A123**, 225–235 (1929)
Graf, W.H.: Hydraukics of Sediment Transport. McGraw-Hill Book Company (1971)
Grass, A.: Initial instability of fine bed sand. J. Hydr. Div. ASCE **96**(HY2), 619–632 (1970)
Grass, A.: Structural features of turbulent flow over smooth and rough boundaries. J. Fluid Mech. **50**, 233–255 (1971)
Grass, A.: The influence of boundary layer turbulence on the mechanics of sediment transport. In: Sumer, B.M., Müller, A. (eds.) Proceedings of Euromech 156, Mechanics of sediment transport, pp. 3–17. Balkema (1983)

Grass, A., Stuart, R.J., Mansour-Tehrani, M.: Vortical structure and coherent motion in turbulent flow over smooth and rough boundaries. Phil. Trans. R. Soc. Ser. A **A336**, 35–65 (1991)

Grass, A., Mansour-Tehrani, M.: Generalized scaling of coherent bursting structures in near-wall region on turbulent flow over smooth and rough boundaries. In: Ashworth, P.J., Bennett, S.J., Best, J.L., McLelland, S.J. (eds.) Coherent Flow Structures in Open Channels, pp. 41–61. John Wiley and Sons, Chichester, New York, Brisbane, Toronto, Singapore (1996)

Grave, B.: Bewegung kugelförmiger Partikel in strömenden Medien. Studienarbeit Thermodynamik und Verfahrenstechnik, TU Berlin (1967)

Green, S.I. (ed.): Fluid Vortices (Fluid Mech. and its Appl. 30). Kluwer Academic Publishers, Dordrecht, Boston, London (1995)

Greimann, B.P.: Two-phase flow analysis of sediment velocity. In: Gyr, A., Kinzelbach, W. (eds.) Sedimentation and Sediment Transport, pp. 83–86. Kluwer Academic Publishers, Dordrecht, Boston, London (2003)

Greimann, B.P., Holly, F.M.: Two-phase flow analysis of concentration profiles. J. Hydr. Engr. ASCE **127**, 753–762. Dordrecht, Boston, London (2001)

Greimann, P.B., Muste, M., Holly, F.M.: Two-phase formulation of suspended sediment transport. J. Hydr. Res. **37**, 479–500. Dordrecht, Boston, London (1999)

Günther, A.: Die mittlere kritische Sohlenschubspannung bei Mischungen unter Berücksichtigung der Deckschichtbildung und der turbulenten Sohlenschubspannung. Ph.D. Thesis ETHZ (1971)

Gustavsson, L.H., Wallin, S.: Effect of three-dimensional surface elements on boundary layer flow. In: Gyr, A (ed.) Structure of Turbulence and Drag Reduction, pp. 399–406. Springer-Verlag, Berlin, Heidelberg, New York (1990)

Gyr, A.: Ein Tropfenakkreszenzmodell in Atmosphäre von homogen isotroper Turbulenz. ZAMP **16**, 721–739 (1965)

Gyr, A.: The vorticity diffusion of Λ-vortices in drag reducing solutions. In: Gampert, B. (ed.) The Influence of Polymer Additives on Velocity and Temperature Fields, pp. 233–247. Springer-Verlag (1980)

Gyr, A.: The vorticity diffusion of Λ-vortices in drag reducing solutions. In: The Influence of Polymer Additives on Velocity and Temperature Fields, ed. B. Gampert. (IUTAM Symp. 1984). Springer Verl. Berlin, Heidelberg, New York Tokyo 1985 233-47 (1985)

Gyr, A.: Natural riblets. In: Meier, G.E.A., Viswanath, P.R. (eds.) IUTAM Symposium on Mechanics of Passive and Active Flow Control, pp. 109–114. Kluwer Academic Publishers Dordrecht, Boston, London (1999)

Gyr, A.: The self-organization of ripples towards two-dimensional forms. In: Gyr, A., Kinzelbach, W. (eds.) Sedimentation and Sediment Transport, pp. 183–186. Kluwer Academic Publishers Dordrecht, Boston, London (2003)

Gyr, A., Bewersdorff, H.-W.: Drag Reduction of Turbulent Flows by Additives. Kluwer Academic Publishers, Dordrecht, Boston, London (1995)

Gyr, A., Kinzelbach, W.: Bed forms in turbulent channel flows. Appl. Mech. Rev. **57**, 77–93 (2004)

Gyr, A., Müller, A.: Alteration of structures of sublayer flow in dilute polymer solutions. Nature **253**, 185–187 (1975)

Gyr, A., Müller, A.: The role of coherent structures in developing bedforms during sediment transport. In: Ashworth, P.J., et al. (eds.) Coherent Flow Structures in Open Channels, pp. 227–235. John Wiley and Sons, Chichester, New York, Brisbane, Toronto, Singapore (1996)

Gyr, A., Tsinober, A.: On some local aspects of turbulent drag reducing flows of dilute polymers and surfactants. In: Gavrilakis et al. (eds.) Advances in Turbulence VI, pp. 449–452. Kluwer Academic Publishers, Dordrecht, Boston, London (1996)

Gyr, A., Kinzelbach, W., Tsinober, A.: Fundamental problematic issues in turbulence. In: Gyr et al. (eds.) Trend in Mathematics, pp. 213–263. Birkhäuuser, Basel, Boston Berlin (1999)

Ha, H.K., Chough, S.K.: Intermittent turbulent events over sandy current ripples: A motion-picture analysis of flume experiments. Sediment Geol. **161**, 295–308 (2003)

Hack, J.T.: Studies of longitudinal streams in Virginia and Maryland. US Geological Survey Professional Papers 294B (1957)

Hage, W., Bechert, D.W., Bruse, M.: Artificial shark skin on its way to technical application. In: Gyr, A. (ed.) Science and Art Symposium 2000, pp. 169–175. Kluwer Academic Publishers Dordrecht, Boston, London (2000)
Ham, J.M. and Homsy, G.M.: Hindered settling and hydrodynamic dispersion in quiescent sedimenting suspensions. Int. J. Multiphase Flow **14**, 533–46 (1988)
Hamilton, J.M., Kim, J., Waleffe, F.: Regeneration mechanisms of near-wall turbulence structures. J. Fluid Mech. **287**, 317–248 (1995)
Happel, J., Brenner, H.: Low Reynolds Number Hydrodynamics. Prentice-Hall (1965)
Hardtke, P.: Turbulenzerzeugte Sedimentriffeln. Mitt. Inst. Wasserbau III, Univ. Karlsruhe, Heft **47**, (1979)
Haritonidis, J.H., Kaplan, R.K., Wygnanski, I.: Interaction of a turbulent spot with a turbulent boundary layer. In: Lecture Notes in Physics, vol. 75, pp. 234–247. Springer-Verlag, Berlin, Heidelberg, New York (1978)
Head, H.R., Bandyopadhyay, P. (eds.): New aspects of turbulent boundary layer structure. J. Fluid Mech. **107**, 297–338 (1981)
Helland-Hansen, E., Milhous, R.T., Klingeman, P.C.: Sediment transport at low Shields-parameter values. J. Hydr. Div. ASCE **100**(HY1), 261–265 (1974)
Helman, J., Hesselink, L.: Analysis and visualization of flow topology in numerical data sets. In: Moffatt, H.K., Tsinober, A (eds.) Topological Fluid Mechanics, pp. 361–371. Cambridge University Press, Cambridge, New York, Melbourne (1990)
Hénon, M.: A two dimensional mapping with a strange attractor. Commun. Math. Phys. **50**, 69 (1976)
Hey, R.D.: Flow resistance in gravel bed rivers. J. ASCE HY, **105**(HY4), 365–379 (1979)
Hey, R.D.: Gravel-bed rivers: Form and processes. In: Hey et al. (eds.) Gravel –bed rivers, pp. 5–13. John Wiley and Sons, Chichester, New York, Brisbane, Toronto, Singapore (1982)
Hey, R.D., Bathurst, J.C., Thorne, C.R.: Gravel –Bed Rivers. John Wiley and Sons (1982)
Hill, H.M., Srinivasan, V.S., Unny, T.E.: Instability of flat bed in alluvial channels. ASCE Ann. Meeting & Nat. Mmee. of Water Resources Engineering. New Orleans, LA (1967)
Hinze, J.O.: Secondary currents in wall turbulence. Phys. Fluids Suppl. **10**, 122–125 (1967)
Hinze, J.O.: Experimental investigation on secondary currents in the turbulent flow. In: Hydraulic Problems Solved by Stochastic Methods, pp. 453–465. Water Resources Publications, Fort Collins (1973)
Hirano, M.: River bed degradation with armoring. Trans. JSCE **3**, 194–195 (1971)
Ho, C.-M., Huang, L.-S.: Subharmonics and vortex merging in mixing layers. J. Fluid Mech. **119**, 443–473 (1982)
Holmes, P., Lumley, J.L., Berkooz, G.: Turbulence, Coherent Structures, Dynamical Systems and Symmetry. Cambridge University Press, Cambridge, New York, Melbourne (1996)
Hopfinger, E.J., Linden, P.F.: Formation of thermoclines in zero mean shear turbulence subjected to a stabilizing buoyancy flux. J. Fluid Mech. **114**, 157–173 (1982)
Hornung, H.G., Perry, A.E.: Some aspects of three-dimensional separation. Part I: Streamsurface bifurcations. Z. Flugwiss. Weltraumforschung **8**, 77–87 (1984)
Horton, R.e.: Erosional development of streams and their drainage basin: Hydrophysical approach to quantitative morphology. Geol. Soc. Amer. Bull. **56**, 275–330 (1945)
Hsu, T.J., Jenkins, J.T., Liu, P.L.J.: On two phase sediment transport: Dilute flow. J. Geophys. Res. **108**, 3057 (2003)
Hu, C., Hui, Y.: Bed load transport. I: Mechanical characteristics. J. Hydr. Eng. ASCE, **122**, 245–261 (1996)
Hudy, L.M., Naguib, A.M., Humphreys, W.M., Bartram, S.M.: Wall-pressure-array measurements beneath a separating/reattaching flow. AIAA J. **40**, 1026–1036 (2002)
Hunt, J.C.R., Abell, C.J., Peterka, J.A., Woo, H.: Kinematical studies of the flows around free or surface mounted obstacles; applying topology of flow visualization. J. Fluid Mech. **86**, 179–2000 (1978); Corrigendum **95**, 796 (1979)

Hunt, J.C.R., Carruthers, D.J., Fung, J.C.H.: Rapid distortion theory as a means of exploring the structure of turbulence. In: Sirovich, L (ed.) New Perspective in Turbulence, pp. 55–103. Springer-Verlag, Berlin, Heidelberg, New York (1991)
Hunter, R.J.: Foundation of Colloid Science. 1, Oxford University Press, Oxford (1987)
Huppert, H.E., Turner, J.S., Hallworth, M.A.: Sedimentation and entrainment in dense layers of suspended particles stirred by an oscillating grid. J. Fluid Mech. **289**, 463–493 (1995)
Israelachvili, J.: Intermolecular and Surface Forces, 4th edn. Academic Press, London (1994)
Iwasa, Y., Kennedy, J.F.: Free surface shear flow over a wavy bed. Proc. ASCE J. Hydr. Div. **94**, 431–454 (1968)
Jackson, P.S.: On the displacement height in the logarithmic velocity profile. J. Fluid Mech. **111**, 15–25 (1981)
Jackson, R.: Sedimentological and fluid dynamic implications of the turbulent bursting phenomenon in geophysical flows. J. Fluid Mech. **77**, 531–566 (1995)
Jimenez, J.: A spanwise structure in the plane shear layer. J. Fluid Mech. **132**, 319–336 (1983)
Jimenez, J., Cogollos, M., Bernal, L.P.: A perspective view of the plane mixing layer. J. Fluid Mech. **152**, 125–143 (1985)
Jimenéz, J., Moin, P.: The minimal flow unit in a near-wall turbulence. J. Fluid Mech. **225**, 213–240 (1991)
Johansson, A., Alfredsson, P.: On the structure of turbulent channel flow. J. Fluid Mech. **122**, 295–314 (1982)
Julien, P.Y.: Erosion and Sedimentation. Cambridge University Press, Cambridge, New York, Melbourne (1995)
Julien, P.Y., Lan, Y.Q.: Rheology of hyperconcentration. J. Hydr. Eng. ASCE **115**, 346–353 (1991)
Julien, P.Y., Raslan, Y.: Upper regime plane bed. J. Hydr Eng. ASCE **124**, 1086–1096 (1988)
Julien, P.Y., Lan, Y.Q., Berthault, G.: Experiment of stratification of heterogeneous sand mixtures. Bull. Soc. Gööl. France **164**, 649–660 (1993)
Kalinske, A.A.: Movement of sediment as bed-load in rivers. Trans. Am. Geophys. Union **28**, 615–620 (1947)
Karim, F.: Bed configuration and hydraulic resistance in alluvial-channel flow. J. Hydr. Eng. **121**, 15–25 (1995)
Karim, F., Kennedy, J.F[R25].: Computer-based predictors for sediment discharge and friction factor of alluvial streams. IIHR Report, No. 242. University of Iowa, Iowa city, Iowa (1983)
Kaskas, A.: Berechnung der stationären und instationären Bewegung von Kugeln in ruhenden und strömenden Medien. Dipl. Thermodynamik und Verfahrenstechnik TU Berlin (1964)
Kennedy, J.F.: Mechanics of dunes and antidunes in errodible-bed channels. J. Fluid Mech. **16**, 521–544 (1963)
Kennedy, J.F.: The formation of sediment ripples, dunes and antidunes. Ann. Rev. Fluid Mech. **1**, 147–168 (1969)
Kennedy, J.F.: General report: Changes in alluvial beds composed of non-uniform material. In: Proceedings of the 24th Congress IAHR, vol. 6, pp. 241–252. Paris (1971)
Keulegan, G.H.: Laws of turbulent flows in open channels. J. Res. US NBS **21**, RP 1151, 707–741 (1938)
Kiger, K., Pan, C., Rivero, A.: Experimental measurement of sediment suspension and particle kinetic stress transport within a horizontal channel flow. In: Gyr, A., Kinzelbach, W. (eds.) Sedimentation and Sediment Transport, pp. 87–90. Kluwer Academic Publishers, Dordrecht, Boston, London (2003)
Kiger, K.T., Pan, C.: PIV technique for the simultaneous measurement of dilute two-phase flows. J. Fluids Eng. **122**, 811–818 (2000)
Kiya, M., Shimizu, M., Mochizuki, O.: Sinusoidal forcing of a turbulent separation bubble. J. Fluid Mech. **342**, 119–139 (1997)
Klaasen, G.J.: Report R657-XII Delft Hydraulics Laboratory, Delft (1980)
Klebanoff, P.S., Tidstrom, K.D., Sargent, L.M.: The three-dimensional nature of boundary layer instability. J. Fluid Mech. **12**, 1–34 (1962)

Kline, S.J., Robinson, S.K.: Turbulent boundary layer structure: Progress status, and chalange. In: Gyr, A. (ed.) Structure of Turbulence and Drag Reduction, pp. 3–22. Springer-Verlag, Berlin, Heidelberg, New York (1990)
Knorez, W.S.: The influence of macro-rugosity of the channel on the hydraulic resistance. Istwestia WNIIG, Nr. 62, (1959)
Koch, D.L., Shaqfeh, E.S.G.: The instability of a dispersion of sedimenting spheroids. J. Fluid Mech. **209**, 521–542 (1989)
Kolmogorov, A.N.: Dissipation of energy in local isotropic turbulence. Dokl. Akad. Nauk SSSR **32**, 16–18 (1941). English translation see selected works of A.N. Kolmogorov I (ed. Tikhomirov) Kluwer, 318–321 (1991)
Kraichnan, R.H.: Stochastic modeling of isotropic turbulence. In: Sirovich, L. (ed.) New Perspectives in Turbulence, pp. 1–54. Springer-Verlag, Berlin, Heidelberg, New York (1991)
Krieger, I.M.: Rheology of monodisperse latices. Adv. Colloid Interface Sci. **3**, 111–136 (1972)
Krieger, I.M., Dougherty, T.J.: A mechanism of non-Newtonian flow in suspensions of rigid spheres. Trans. Soc. Rheol. **3**, 82–90 (1959)
Kulik, J.D., Fessler, J.R., Eaton, J.K.: Particle response and turbulence modification in fully developed channel flow. J. Fluid Mech. **277**, 109–134 (1994)
Lacey, G.: A theory of flow in alluvium. J. Inst. Civil Eng. **27**, 16–47 (1946)
Lacey, G.: A theory of flow in alluvium. J. Inst. Civil Eng. **27**, 16–47 (1947)
Lamb, H.: Hydrodynamics (1879)–(1932) Dover Publication, New York (1945)
Landahl, M.T.: A wave guide model for turbulent shear flow. J. Fluid Mech. **29**, 305–322 (1967)
Landahl, M.T.: Hydrodynamic instability and coherent structures in turbulence. In: Gyr, A. (ed.) Structure of Turbulence and Drag Reduction, pp. 371–397. Springer-Verlag, Berlin, Heidelberg, New York (1990)
Landau, L.D.: Dokl. Akad. Nauk U.S.S.R. **44**, 339 ff, or Landau & Lifschitz § 26 (1944)
Landau, L.D., Lifschitz, E.M.: Lehrbuch der theoretischen Physik, 4. Hydrodynamik, Akademie Verlag, Berlin (1991)
Langbein, W.B., Leopold, L.B.: River meanders-Theory of minimum variance. US Geological Survey Paper 422-H, 15 pp. (1966)
Leighton, D.T., Acrivos, A.: Shear induced migration of particles in concentrated suspension. J. Fluid Mech. **181**, 415–439 (1987a)
Leighton, D.T., Acrivos, A.: Measurement of shear-induced self-diffusion in concentrated suspensions of spheres. J. Fluid Mech. **177**, 109–131 (1987b)
Leonard, A.: Turbulent structures in wall-bounded shear flows, observed via three-dimensional numerical simulations. Lect. Notes Phys. **136**, 119–146 (1980)
Leonard, A.: Subgrid modeling for the filtered scalar transport equation. Fundamental problematic issues in turbulence. In: Gyr et al. (eds.) Trend in Mathematics, pp. 257–263. Birkhäuuser Basel, Boston Berlin (1999)
Lesieur, M.: Turbulence in Fluids. Kluwer Academic Publishers, Dordrecht, Boston, London (1987)
Leopold, L.B.: Water surface topography in river channels and implications for meander development. In: Hey et al. (eds.) Gravel-Bed Rivers, pp. 359–388. John Wiley and Sons Chichester, New York, Brisbane, Toronto, Singapore (1982)
Leopold, L.B., Langbein, W.B.: River meanders. Sci. Am. **217**, 60–70 (1966)
Lighthill, M.J.: In: Rosenhead (ed.) Laminar Boundary Layers, pp. 48–88. Oxford University Press, Claredon Press, Oxford (1963)
Lilley, G.M., Hodgons, T.H.: On surface pressure fluctuations in turbulent boundary layers. AGARD Report **276**, (NATO) (1960)
Lim, T.T., Chong, M.S. and Perry, A.E.: The viscous tornado. Proc. 7th Austr. Conf. On Hydraulics and Fluid Mechanics, Brisbane, 250-3 (1980)
Limerinos, J.T.: Determination of the Manning coefficient from measured bed roughness in natural channels. US Geological Survey Water-Supply Paper 1898-B (1970)
Liu, C.K., Kline, S.J., Johnston, J.P[R30].: An experimental study of turbulent boundary layers on rough walls. Thermosci. Div. Mech. Eng. Eng. Dept. Stanford Univ. Rep. MD-15 (1957)

Liu, H.: The mechanics of sediment ripple formation. J. Hydr. Div. ASCE **183**(HY2), 1–23 (1957)
Ljus, C., Johansson, B., Almstedt, A.-E.: Turbulence modification by particles in a horizontal pipe flow. Int. J. Multiphase Flow, **28**, 1075–1090 (2002)
Lovera, F., Kennedy, J.F.: Friction factors for flat bed flows in sand. Proc. ASCE **95**(HY4), 1227–1234 (1969)
Luchini, P., Manzo, F., Pozzi, A.: Resistance of grooved surfaces to parallel flow and cross flow. J. Fluid Mech. **228**, 87–109 (1991)
Lüthi, B., Tsinober, A., Kinzelbach, W.: Lagrangian measurement of vorcity dynamics in turbulent flows. J. Fluid Mech. **528**, 87-118 (2005)
Lumley, J.L.: Drag reduction by additives. Ann. Rev. Fluid Mech. **1**, 367–384 (1969)
Lumley, J.L.: Stochastic Tools in Turbulence. Academic Press, NY (1970)
Lumley, J.L.: Drag reduction in two-phase and polymer flows. Phys. Fluids **20**, S64–S71 (1977)
Lumley, J.L.: Coherent structures in turbulence. In: Meyer, R. (ed.) Transition and Turbulence, pp. 215–242. Academic Press, London (1981)
Lumley, J.L.: Turbulence and turbulence modeling. In: Lumley et al. (eds.) Research Trends in Fluid Dynamics, pp. 167–177. American Institute of Physics. Woodbury, New York (1996)
Lumley, J.L., Blossey, P.N., Podvin-Delarue, B.: Low dimensional models, the minimal flow unit and control. Fundamental problematic issues in turbulence. In: Gyr et al. (eds.) Trend in Mathematics, pp. 57–66. Birkhäuuser (1999)
Lundbladh, A., Schmidt, P., Berlin, S., Hennigson, D.: Simulations of bypass transition for spatially evolving disturbances. In: Cantwell, B., et al. (eds.) Application of Direct and Large Eddy Simulation to Transition and Turbulence, AGARD Conf. Proc. AGARD-CP **551**, pp. 18.1–18.3 (1994)
Lyn, D.A.: Turbulence and turbulent transport in sediment-laden open-channel flows. Report No. KH-R-49 CALTECH Pasadena, California (1986)
Maas, H.-G.: Contributions of digital photogrammetry to 3-D PTV. In: Dracos, Th. (eds.) Three-Dimensional Velocity and Vorticity Measuring and Image Analysis Techniques, pp. 191–207. Kluwer Academic Publishers, Dordrecht, Boston, London (1996)
Magnus, G.: Poggendorf's Annalen der Pysik u. Chemie **88**, 1 (1853)
Mandelbrot, B.B.: The Fractial Geometry of Nature. W.H. Freeman and Company, NY (1977)
Martin, J.E., Meiburg, E.: The accumulation and dispersion of heavy particles in forced two-dimensional mixing layers. I. The fundamental and subharmonic case. Phys. Fluids **6**, 1116–1132 (1994)
Maxey, M.R.: The gravitational settling of aerosol particles in homogeneous turbulence and random flow fields. J. Fluid Mech. **174**, 441–465 (1987)
Maxey, M, Chang, E.J., Wang, L.-P.: Interaction of particles and microbubbles with turbulence. Exp. Thermal Fluid Sci. **12**, 417–425 (1996)
Maxey, M, Patel, B.K., Chang, E.J., Wang, L.-P.: Simulations of dispersed turbulent multiphase flow. Fluid Dyn. Res. **20**, 143–156 (1997)
McLean, S.R., Smith, J.D.: Turbulence measurement in the boundary layer over a sand wave field. J. Geophys. Res. Oc. Atm. **84**(NC12), 7791–7808 (1979)
McLean, S.R., Nelson, J.M., Shreve, R.L.: Flow-sediment interactions in separating flows over bedforms. In: Ashworth, P.J., et al. (eds.) Coherent Flow Structures in Open Channels, pp. 203–226. John Wiley and Sons, Chichester, New York, Brisbane, Toronto, Singapore (1996)
McLelland, S.J., Ashworth, P.J., Best, J.L.: The origin and downstream development of coherent structures at channel junctions. In: Ashworth, P.J., et al. (eds.) Coherent Flow Structures in Open Channels, pp. 459–490. John Wiley and Sons, Chichester, New York, Brisbane, Toronto, Singapore (1996)
Meiburg, E.: Numerical investigation of two-way coupling mechanisms in dilute, particle laden flows. In: Gyr, A., Kinzelbach, K. (eds.) Sedimentation and Sediment Transport, pp. 149–154. Kluwer Academic Publishers, Dordrecht, Boston, London (2003)
Meiburg, E., Wallner, E., Pagella, A., Riaz, A., Haertel, C., Necker, F.: Vorticity dynamics of dilute two-way-coupled particle-laden mixing layers. J. Fluid Mech. **421**, 185–227 (2000)
Meneveau, G., O'Neil, J.O., Port-Agel, F., Cerutti, S., Parlange, M.B.: Physics and modeling of small scale turbulence for large eddy simulation. Fundamental problematic issues in turbulence. In: Gyr et al. (eds.) Trend in Mathematics, pp. 221–232. Birkhäuuser (1999)

Métais, O., Lamballais, E., Lesieur, M.: Pressure fluctuations in a turbulent channel. Fundamental problematic issues in turbulence. In: Gyr et al. (eds.) Trend in Mathematics, pp. 329–335. Birkhäuuser (1999)
Meyer-Peter, E., Müller, R.: Formulas for bed-load transport. Proceedings of IHAR, Stockholm (1948)
Meyer-Peter, E., Müller, R.: Eine Formel zur Berechnung des Geschiebetriebes, Schw. Bauzeitung **67**, Nr. 3, (1949)
Milhous, R.T.: Sediment transport in a gravel-bottomed stream. Ph.D. Thesis, Oregon State University, Corvallis (1973)
Milliken, W.F., et al.: Effect of diameter of falling balls on the apparent viscosity of suspensions of spheres and rods. PCH **11**, 341–355 (1989)
Moin, P.: A new approach for large eddy simulation of turbulence and scalar transport. In: Dracos, Th., Tsinober, A. (eds.) New Approaches and Concepts in Turbulence, pp. 331–340. Bitkhäuser Verlag, Basel, Boston Berlin (1993)
Moin, P.: Direct and large eddy simulation of turbulence. In: Lumley et al. (eds.) Research Trends in Fluid Dynamics, pp. 188–193. American Institute of Physics (1996)
Moin, P., Kim, J.: Numerical investigation of turbulent channel flow. J. Fluid Mech. **118**, 341–377 (1982a)
Moin, P., Kim, J.: The structure of the vorticity field in turbulent channel flow. Part 1: Analysis of the instantaneous field in statistical correlations. J. Fluid Mech. **155**, 441–464 (1985). Part 2: Study of ensemble-averaged fields. J. Fluid Mech. **162**, 339–363 (1982b)
Moin, P., Kim, J.: The structure of the vorticity field in turbulent channel flow. Part 1: Analysis of the instantaneous field in statistical correlations. J. Fluid Mech. **155**, 441–464 (1985). Part 2: Study of ensemble-averaged fields. J. Fluid Mech. **162**, 339–363 (1986)
Moody, L.F.: Friction factor for pipe flow. Trans. ASME **66**, 8 (1944)
Morrison, W.R.B., Bullock, J.K., Kronauer, R.E.: Experimental evidence of waves in the sublayer. J. Fluid Mech. **47**, 639–656 (1971)
Müller, A.: Sediment transport: Gaps between phenomena, concepts, and the need for predicting tools.-Discussion. In: Nakato, Ettema (eds.) Issues and Directions in Hydraulics, pp. 93–95. Balkema, Rotterdam (1996)
Müller, A., Gyr, A.: Visualisation of the mixing layer behind dunes. In: Sumer, B.M., Müller, A. (eds.) Mechanics of Sediment Transport, pp. 41–45. A.A. Balkema, Rotterdam (1983)
Müller, A., Gyr, A.: On the vortex formation in the mixing layer behind dunes. J. Hydr. Res. **24**, 359–375 (1986)
Müller, A., Gyr, A.: Geometrical analysis of the feedback between flow, bedforms and sediment transport. In: Ashworth, P.J., et al. (eds.) Coherent Flow Structures in Open Channels, pp. 237–247. John Wiley and Sons, Chichester, New York, Brisbane, Toronto, Singapore (1996)
Müller, A., Studerus, X.: Secondary flow in an open channel. In: Proceedings of XVIII IAHR Congress, vol. 3, pp. 19–24. Cagliari (1979)
Müller, A., Studerus, X.: A three component velocity measurement in open channel by a combination of LDA and hot film anemometry. Proceedings of 7th Biennial Symposium on Turbulence, Rolla, Missouri (1981)
Müller, A., Wiggert, D.C.: Analysis of turbulent shear flows using photochromic visualization. In: Adrian et al. (eds.) Laser Anemometry in Fluid Mechanics, vol. **4**, pp. 518–534. Springer-Verlag, Berlin, Heidelberg, New York (1989)
Müller, A., Gyr, A., Dracos, T.: Interaction of rotating elements of the boundary layer with grains of a bed: A contribution of the problem of the threshold of sediment transportation. J. Hydr. Res. **9**, 373–411 (1971)
Munro, R.J., Dalziel, S.B.: Particle resuspension by an impacting vortex ring. In: Gyr, A., Kinzelbach, W. (eds.) Sedimentation and Sediment Transport, pp. 105–108. Kluwer Academic Publishers, Dordrecht, Boston, London (2003)

Naish, C., Sellin, R.H.J.: Flow structure in a large-scale model of a doubly meandering compound river channel. In: Ashworth, P.J., et al. (eds.) Coherent Flow Structures in Open Channels, pp. 631–654. John Wiley and Sons, Chichester, New York, Brisbane, Toronto, Singapore (1996)
Nakagawa, H., Nezu, I., Tominaga, A.: Spanwise streaky structure and macroturbulence in open channel flows. Mem. Fac. Eng. Kyoto Univ. **43–41**, 34–67, Kyoto (1981)
Nasner, H.: Über das Verhalten von Transportkörper im Tidegebiet. Mit. Franzius Inst. TU Hannover **40** (1974)
Nelson, J.M., McLean, S.R., Wolfe, S.R.: Mean flow and turbulence fields over 2-dimensional bed forms. Water Resour. Res. **29**, 3935–3953 (1993)
Nezu, I., Nakagawa, H.: Turbulence in open channel flows. IAHR Monograph, Balkema (1993)
Nezu, I., Rodi, W.: Experimental study on secondary currents in open channel flows. In: Proceedings of 21st IAHR Congress, vol. 2 (1985)
Nezu, I., Rodi, W.: Open-channel flow measurements with a laser Doppler anenometer. J. Hydr. Eng. ASCE **112**, 335–355 (1986)
Nikora, V, Koll, K., McLean, S., Dittrich, A., Aberle, J.: Zero-plane displacement for rough-bed open-channel flows. In: Bousmar, D., Zech, Y. (eds.) Proceedings of International Conference on Fluvial Hydraulics, River Flow, vol. 1, pp. 83–91. Balkema Publishers, Belgium (2002)
Nikuradse, J.: Verl. Deutsch. Ing. Forschungsheft **361** (1933)
Nomicos, G.N.: Effect of sediment load on the velocity field and friction factor of turbulent flow in an open-channel. Ph.D. Thesis, CALTECH (1956)
Nordin, C.F., Algert, J.A.: Geometrical properties of sand waves: A discussion. Proc. ASCE **81**(HY5) (1965)
Nordin, C.F., Dempster, G.R.: Vertical distribution of velocity and suspended sediment. US Geological Survey Professional Paper 462-B, Middle Rio Grande, New Mexico (1963)
Nowell, A.R.M., Church: Turbulent flow in depth limited boundary layer. J. Geophys. Res. **84**(C8), 4816–4824 (1979)
Oberlack, M.: Symmetries of the Navier-Stokes equation and their implications for subgrid-models in large eddy simulation of turbulence. In: Gyr et al. (eds.) Fundamental Problematic Issues in Turbulence: Trend in Mathematics, pp. 247–256. Birkhäuuser (1999)
O'Brien, J.S., Julien, P.Y.: Physical properties and mechanics of hyperconcentrated sediment flows. In: Proceedings of ASCE Conference on Delineation of Landslide, Flashflood and Debris Flow Hazards, Ser.UWRL/G-85–03, pp. 260–79 (1985)
Onishi, Y., Jain, S.C., Kennedy, J.F.: Effects of meandering in alluvial streams. J. Hydr. Div. Proc. ASCE **102**, 899–917 (1976)
Oseen, C.W.: Über die Stokes'sche Formel und über eine angewandte Aufgabe in der Hydrodynamik. Ark. Math. Astr. Fys. **6**(Nr. 29) (1910)
Pan, Y., Banerjee, S.: Numerical simulation of particles interactions with wall turbulence. Phys. Fluids **8**, 2733–2755 (1996)
Panchev, S.: Random Functions and Turbulence. Pergamon Press, Oxford (1971)
Parker, G.: Some speculations on the relation between channel morphology and channel-scale flow structures. In: Ashworth, P.J., et al. (eds.) Coherent Flow Structures in Open Channels, pp. 423–458. John Wiley and Sons, Chichester, New York, Brisbane, Toronto, Singapore (1996)
Parker, G., Klingeman, P.C.: On why gravel bed streams are paved. Water Resour. Res. **18**, 1409–1423 (1982)
Parker, G., Klingeman, P.C., McLean, D.G.: Bedload and size distribution in paved gravel- bed streams. J. Hydr. Div. ASCE **108**(HY4), 544–571 (1982)
Parker, G., Paola, C., Leclair, S.: Probabilistic Exner sediment continuity equation for mixtures with no active layer. J. Hydr. Eng. ASCE **126**, 818–826 (2000)
Perry, A.E., Abell, C.J.: Asymptotic similarity of turbulence structures in smooth and rough-walled pipes. J. Fluid Mech. **79**, 785–799 (1977)
Perry, A.E., Chong, M.S.: On the mechanism of wall turbulence. J. Fluid Mech. **119**, 173–217 (1982)
Perry, A.E., Chong, M.S.: A description of edding motions and flow patterns using critical point concepts. Ann. Rev. Fluid Mech. **19**, 125–156 (1987)

Perry, A.E., Hornung, H.G.: Some aspects of three-dimensional separation. Part II: Vortex skeletons. Z. Flugwiss. Weltraumforschung **8**, 155–160 (1984)
Perry, A.E., Schofield, W.H., Joubert, N.: Rough wall turbulent boundary layers. J. Fluid Mech. **37**, 383–413 (1969)
Perry, A.E., Lim, T.T., Te, E.W.: A visual study of turbulent spots. J. Fluid Mech. **104**, 387–405 (1981)
Peschke, G.: Zur Anwendbarkeit statistischer Modelle für die Untersuchung des Meanderproblems. Acta Hydrophysica **17**, 235–247 (1973)
Phillips, R.J., et al.: Constitutive equation for concentrated suspensions that accounts for shear-induced particle migration. Phys. Fluids **A4**, 30 (1992)
Pointcaré, H.: Les Methodes Nouvelles de la Mechanique Celeste. Gauthier-Villars, Paris (1892)
Prandtl, L.: Über die au8sgebildete Turbulenz. Verh. D. 2. intern. Kongr. F. Techn. Mechanik, Zürich (1926)
Qian, N., Wan, Z.: A critical review of the research on the hyperconcentrated flow in China, Beijing. International Research and Training Centre on Erosion and Sedimentation (1986)
Ranga Raju, K.G., Soni, J.P.: Geometry of ripples and dunes in alluvial channels. J. Hydr. Res. **14**(Nr. 3) (1976)
Ranga Raju, K.G., Garde, R.J., Bhardwaj, R.C.: Total load transport in alluvial channels. J. Hydr. Div. ASCE **107**(HY2), 179–192 (1981)
Rao, K.N., Narasimha, R., Badri Narayanan, M.A.: The 'bursting' phenomenon in a turbulent boundary layer. J. Fluid Mech. **48**, 339–352 (1971)
Raudkivi, A.J.: Study of sediment ripple formation. J. Hydr. Div. ASCE Proc. 3**89**(HY6), 15–33 (1963)
Raudkivi, A.J.: Loose Boundry Hydraulics. Pergamon Press, Oxford (1976)
Raudkivi, A.J.: Grundlagen des sedimenttransportes. Springer-Verlag, Berlin, Heidelberg, New York (1982)
Raupach, M.R.: Conditional statistics of Reynolds stress in rough-wall and smooth-wall turbulent boundary layers. J. Fluid Mech. **108**, 363–382 (1981)
Rempfer, D., Parson, L., Xu, S., Lumley, J.: Turbulent boundary layers over compliant walls: Low-dimensional models and direct simulations. In: Gyr, A., Kinzelbach, W. (eds.) Sedimentation and Sediment Transport, pp. 43–50. Kluwer Academic Publishers, Dordrecht, Boston, London (2003)
Reynolds, W.C. and Cebeci, T.: Calculation of turbulent flows. In: Toppics in applied Physics; Turbulence ed. P. Bradshaw. Springer Ver. Berlin, Heidelberg, New York, 193–229 (1976)
Rhoads, B.L.: Mean structure of transport-effective flows at an asymmetrical confluence when the main stream is dominant. In: Ashworth, P.J., et al. (eds.) Coherent Flow Structures in Open Channels, pp. 491–517. John Wiley and Sons, Chichester, New York, Brisbane, Toronto, Singapore (1996)
Ribberink, J.S.: Mathematical modeling of one-dimensional morphological changes in rivers with non-uniform sediment. Ph.D. Thesis, Delft University (1987)
Richards, F.: The formation of ripples and dunes on an erodible bed. J. Fluid Mech. **99**, 597–618 (1980)
Robinson, K.R.: The kinematics of turbulent boundary layer structure. NASA Tech. Memo. 103859, Ames Res.Center (1991a)
Robinson, S.K.: A review of vortex structures and associated coherent motions in turbulent boundary layers. In: Gyr, A. (ed.) Structure of Turbulence and Drag Reduction, pp. 23–50. Springer-Verlag, Berlin, Heidelberg, New York (1990)
Robinson, S.K.: Coherent motion in the turbulent boundary layer. Ann. Rev. Fluid Mech. **23**, 601–639 (1992)
Rösgen, T., Totaro, R.: Low coherence techniques for imaging in multiphase flows. In: Gyr, A., Kinzelbach, W. (eds.) Sedimentation and Sediment Transport, pp. 255–267. Kluwer Academic Publishers, Dordrecht, Boston, London (2003)
Rutherford, J.C.: River Mixing. John Wiley and Sons (1994)
Saffman, P.G.: On the stability of laminar flow of a dusty gas. J. Fluid Mech. **22**, 120–128 (1961)

Saffman, P.G.: Vortex Dynamics. Cambridge Monographs on Mechanics and Applied Mathematics. Cambridge University Press, Cambridge, New York, Melbourne (1992)

Saint-Venant, de B.: Comptes Rendus **17**, 1240 (1843)

Scheidegger, A.E.: A thermodynamic analogy for meander systems. Water Resour. Res. **3**, 1041–1046 (1967)

Schlichting, H.: Experimentelle Untersuchungen zum Rauhigkeitsproblem. Ing.-Arch. **7**, 1–34 (1936)

Schlichting, H.: Grenzschicht-Theorie, Verlag G.Braun Karlsruhe (1958)

Schmid, A.: Wandnahe turbulente Bewegungsabläufe und ihre Bedeutung für die Riffelbildung. Diss. ETHZ Nr. 7697 (1985)

Schmidt, W., Gyr, A.: Stability of an erodible bed of lead granulates at low wall shear stress. In: Rys, F.S., Gyr, A. (eds.) Physical Processes and Chemical Reactions in Liquid Flows, pp. 93–208. A.A. Balkema1, Rotterdam (1998)

Schofield, W.H., Logan, E.: Viscous flow around wall mounted obstacles. ARL-Aero-Prop-Rept-172. Deptartment of Defence, Melbourne (1986)

Schoklitsch, A.: Handbuch des Wasserbaus. Springer-Verlag (1950)

Sechet, P., Le Guennec, B.: Bursting phenomenon and incipient motion of solid particles in bed-load transport. Int. J. Hydr. Res. **37**, 683–696 (1999)

Segre, P.N., Herbolzheimer, E., Chaikin, P.M.: Long ranged correlations in sedimentation. Phys. Rev. Lett. **79**, 2574–2577 (1997)

Serrin, J.: Mathematical principles of classical fluid mechanics. In: Flügge, S., Truesdell, C. (eds.) Fluid Dynamics I, Encyclopedia of Physics VIII/I, pp. 125–263. Springer-Verlag, Berlin, Heidelberg, New York (1959)

Shaqfeh, E.S.G., Koch, D.L.: Orientational dispersion of fibers in extensional flows. Phys. Fluids **2**, 1077–1093 (1990)

Shen, H.W., Hung, C.S.: An engineering approach to total bed-material load by regression analysis. In: Shen Berkeley, H.W. (ed.) Proceedings of Sedimentation Symposium, chap. 14. Water Resources Publications, California (1972)

Shields, A.: Anwendung der Ähnlichkeitsmechanik und der Turbulenzforschung auf die Geschiebebewegung, Mitt der Preussischen Versuchsanstalt für Wasser- Erd- und Schiffbau, **26** (1936)

Shih, S.M., Komar, P.D.: Hydralic controls of grain-size distribution of bedload gravels in Oak Creek, Oregon USA. Sedimentology **37**, 367–376 (1990)

Shih, S.M., Komar, P.D.: Differential bedload transport rates in a gravel-bed stream, a grain-size distribution approach. Earth Surf. Processes Landforms **15**, 539–552, (1990)

Simons, D.B.: Theory and design of stable channels in alluvial material. Ph.D. Dissertation, Colorado State University (1957)

Simons, D.B., Li, R.M., Fullerton, W.: Theoretical derived sediment transport equations for Pima county, Arizona. Prepared for Prma county DOT and Flood control district, Tucscon, Ariz. Ft. Collins, Colo.:Simons, Li and Assoc (1981)

Simons, D.B., Richardson, F.V., Nordin, C.F. Jr.: Sedimentation structures generated by flow in alluvial channels. Rep. CER 64, DB S-EVR-CNF 15, Colorado State University, Fort Collins (1964)

Simons, D.B., Senturk, F.: Sediment Transport Technology. Water Resources Publications, Fort Collins (1977)

Singh, B.: Bed load transport in channels, irrigation and power. J. Central Board of Irrigation and Power **18**, 411–430 (1961)

Smart, J.S.: Quantitative characterization of channel network structure. Water Resour. Res. **8**, 1487–1496 (1972)

Smith, C.R.: Coherent flow structures in smooth-wall turbulent boundary layers: Facts mechanisms and speculation. In: Ashworth, P.J., et al. (eds.) Coherent Flow Structures in Open Channels, pp. 1–40. John Wiley and Sons, Chichester, New York, Brisbane, Toronto, Singapore (1996)

Squires, K.D., Eaton, J.K.: Particles response and turbulence modification in isotropic turbulence. Phys. Fluids **A2**, 1191–1203 (1990)
Squires, K.D., Eaton, J.K.: Preferential concentration of particles by turbulence. Phys. Fluids **A3**(5), 1169–1178 (1991)
Sreenivasan, K.R.: Fractal geometry and multifractal measures in fluid mechanics. In: Lumley et al. (eds.) Research Trends in Fluid Dynamics, pp. 263–285. American Institute of Physics Woodbury, New York (1996)
Stehr, E.: Grenzschicht-theoretische Studie über die Gesetze der Strombank und Riffelbildung. Hamburger Küstenforschung Heft **34** (1975)
Stimson, M., Jefferey, G.B.: The motion of two spheres in a viscous fluid. Proc. R. Soc. **A111**, 110–116 (1926)
Stokes, G.G.: On the theories of the internal friction of fluids in motion, and of the equilibrium and motion of elastic solids. Trans. Camb. Phil. Soc. **8**, 287 (1845)
Strahler, A.N.: Hypsometric (area-altitude) analysis of erosional topography. *Geological Society of America Bulletin* **64**, 165–176 (1952)
Studerus, X.: Sekundärströmungen im offenen Gerinne über rauhen Längsstreifen. Dissertation ETHZ 7035 (1982)
Sundaram, S., Collins, L.: A numerical study of the modulation of isotropic turbulence by suspended particles. J. Fluid Mech. **379**, 105–143 (1999)
Surkan, A.J., van Kan, J.: Constrained random walk meander generation. Water Resour. Res. **5**, 1343–1352 (1969)
Taylor, G.I.: Diffusion by continuous movements. Proc. Lond. Math. Soc. Ser 2 **20**, 196–212 (1921)
Taylor, G.I.: The statistical theory of turbulence. Proc. R. Soc. **A151**, 421–478 (1935)
Tennekes, H., Lumley, J.L.: A First Course in Turbulence. The MIT Press (1972) and following new editions. Cambridge Mass., London
Thakur, T.R., Scheidegger, A.E.: A test of the statistical theory of meander formation. Water Resour. Res. **4**, 317–329 (1968)
Thorne, C.R.: Processes and mechanisms of river bank erosion. In: Hey et al. (eds.) Gravel-Bed Rivers, pp. 227–271. John Wiley and Sons, Chichester, New York, Brisbane, Toronto, Singapore (1982)
Tobak, M., Peake, D.J.: Topology of three-dimensional separated flows. Ann. Rev. Fluid Mech. **14**, 61–85 (1982)
Tollmien, W.: Über Kräfte und Momente in schwach gekrümmten oder konvergenten Strömungen. Ing-Arch. **9**, 308–326 (1938)
Torobin, L.B., Gauvin, W.H.: Fundamental aspects of solids-gas flow. Part I-V. Can. J. Chem. Eng. **37**, 129-41, 167-76, 224-36 (1959)
Torobin, L.B., Gauvin, W.H.: Fundamental aspects of solids-gas flow. Part I-V. Can. J. Chem. Eng. **38**, 142-53, 189-200 (1960)
Torobin, L.B., Gauvin, W.H.: Fundamental aspects of solids-gas flow. Part I-V. Can. J. Chem. Eng. **39**, 113-20 (1961)
Trenberth, K.E. (ed.): Climate System Modelling. Cambridge University Press, Cambridge (1992)
Truesdell, G.C., Elghobashi, S.: On the two-way interaction between homogeneous turbulence and dispersed solid particles. II: Particle dispersion. Phys. Fluids **6**, 1405–1407 (1994)
Tsinober, A.: An Informal Introduction to Turbulence. Kluwer Academic Publishers, Dordrecht, Boston, London (2001)
Tsinober, A.: Nonlocality in turbulence. In: Gyr, A., Kinzelbach, W. (eds.) Sedimentation and Sediment Transport, pp. 11–22. Kluwer Academic Publishers, Dordrecht, Boston, London (2003)
Tsujimoto, T.: Fractional transport rateand fluvial sorting. In: Proceedings of Grain Sorting Seminar, vol. 117, pp. 227–249. VAW-ETHZ, Mitteilungen (1992)

Tsujimoto, T.: Coherent fluctuations in a vegetated zone of 0pen-channel flow: Causes of bedload lateral transport and sorting. In: Ashworth, P.J., et al. (eds.) Coherent Flow Structures in Open Channels, pp. 375–396. John Wiley and Sons, Chichester, New York, Brisbane, Toronto, Singapore (1996)

Tsujimoto, T., Kitamura, T.: Interaction between cellular secondary currents and lateral alternate sorting. In: Ashworth, P.J., et al. (eds.) Coherent Flow Structures in Open Channels, pp. 359–374. John Wiley and Sons, Chichester, New York, Brisbane, Toronto, Singapore (1996)

Ungarish, M.: Hydrodynamics of Suspensions. Springer-Verlag (1993)

Vanoni, V.A.: Transportation of suspended sediment by water. Trans. ASCE **1**(2267), 67–133 (1960)

Vanoni, V.A.: Sedimentation engineering. ASCE Manuel and Reports on Engineering Practice **54** (1975)

Vanoni, V.A.: Sedimentation engineering. ASCE Manuel and Reports on Engineering Practice **54** (1977)

van Rjin, L.C.: Bedform and alluvial roughness. J. Hydr. Div. ASCE **110**, 1733–1754 (1984)

Vassilicos, J.C.: Near singular flow structure in small-scale turbulence. Fundamental problematic issues in turbulence. In: Gyr et al. (eds.) Trend in Mathematics, pp. 107–116. Birkhäuuser Basel, Boston Berlin (1999)

Verbanck, M.A.: Sediment-laden flows over fully-developed bed forms: First and second harmonics in a shallow, pseudo-2D turbulence environment. In: Jirka, Uijttewaal (eds.) ShallowFlows. Balkema, Rotterdam (2004)

Verbanck, M.A.: [R49]How fast can a river flow over alluvium? J. Hydr. Res. in press (2005)

Vlugter, H.: Sediment transportation by running water and the design of stable channels in alluvial soils. Bouwn-en Waterbouwn-kunde. De Ingenieur **74**(36), 227–231 (1962)

Vollmers, H., Pernecker, L.: Neue Betrachtungsmöglichkeiten des Feststofftransportes in offenen Gerinnen, Die Wasserwirtschaft **55** (1965)

Vollmers, H., Pernecker, L.: Der Beginn des Feststofftransportes für feinkörnige Materialien in einer richtungskonstanten Strömung. Die Wasserwirtschaft Heft **6** (1967)

von Schelling, H.: Most frequent particle path in a plane. Trans. Am. Geophys. Union **32**, 222–226 (1951)

von Schelling, H.: Plane most frequent particle path in a straight channel. Contr. to "Stud. In onore di Corrado Gini. Univ. Studi Roma **1**, 357–372 (1961)

von Schelling, H.: Most frequent random walks. G.E. Advanced Tech. Lab. Rept. Nr. 64GL92, 60 pp. (1964)

Wakeman, R.: Packing densities of particles with log-normal size distribution. Powder Tech. **11**, 297–299 (1975)

Walker, J.D.A., Smith, C.R., Cerra, A.W., Doligalski, T.L.: The impact of a vortex ring on a wall. J. Fluid Mech. **181**, 99–140 (1987)

Wallace, J.M., Eckelmann, H., Brodkey, R.S.: The wall region in a turbulent shear flow. J. Fluid Mech. **54**, 39–48 (1972)

Wallner, E., Meiburg, E.: Vortex pairing in two-way coupled, particle-laden mixing layers. Int. J. Multiphase Flow **28**, 325–346 (2002)

Wang, H., Nickerson, E.C.: Response of a turbulent boundary layer to lateral roughness discontinuities. Fluid Dyn. & Diff. Lab. Coll. of Eng. Colorado State University (1972)

White, W.R., Bettes, R., Paris, E.: Analytical approach to river regime. J. Hydr. Div. ASCE **108**(HY10), 1179–1193 (1982)

Williams, D.T. and Julien, P.Y.: On the selection of sediment transport equations. J. Hyd. Eng. ASCE 115, 1578–81 (1989)

Willetts, B.B., Rameshwaran, P.: Meandering overbank flow structures. In: Ashworth, P.J., et al. (eds.) Coherent Flow Structures in Open Channels, pp. 609–692. John Wiley and Sons, Chichester, New York, Brisbane, Toronto, Singapore (1996)

Willmarth, W.W., Lu, S.S.: Structure of the Reynolds stress near the wall. J. Fluid Mech. **55**, 65–92 (1972)

Woodcock, L.V.: Molecular dynamics and relaxation phenomena in glasses. In: Bielefeld, Z.I.P. (ed.) Proceedings of a Workshop on Glassforming Liquids. Springer Lecture Series in Physics, vol. 277, pp. 113–124 (1985)

Wormleaton, P.R.: Floodplain secondary circulation as a mechanism for flow and shear stress redistribution in straight compound channels. In: Ashworth, P.J., et al. (eds.) Coherent Flow Structures in Open Channels, pp. 581–608. John Wiley and Sons, Chichester, New York, Brisbane, Toronto, Singapore (1996)

Wray, A.A., Hunt, J.C.R.: Algorithms for classification of turbulent structures, In: Moffatt, Tsinober (eds.) Topological Fluid Dynamics, pp. 95–104. Cambridge, New York, Melbourne (1990)

Wygnanski, I.: The effect of Reynolds number and pressure gradient on the transitional spot in laminar boundary layer. Lecture Notes in Physics **136**, Springer-Verlag 304–332 (1981)

Wygnanski, I., Haritonidis, J.H., Kaplan, R.E.: On a Tollmien-wave packet produced in a turbulent spot. J. Fluid Mech. **92**, 505–528 (1979)

Wygnanski, I., Zilberman, M., Haritonidis, J.H.: On the spreading of a turbulent spot in the absence of a pressure gradient. J. Fluid Mech. **123**, 69–90 (1982)

Yalin, M.S.: Geometrical properties of sand-waves. Proc. ASCE **90**(HY5) (1964)

Yalin, M.S.: Mechanics of sediment transport. Pergamon Press ASCE **105**(HY11), 1433–1443 Braunschweig (1972)

Yang, C.T.: Incipient motion of sediment transport. Proc. ASCE J. Hydr. Div. **99**, 10 (1973)

Yang, C.T., Sonh, C.C.S., Woldenberg, M.J.: Hydraulic geometry and minimum rate of energy dissipation. Water Resour. Res. **17**, 1014–1018 (1981)

Yizhaq, H., Balmforth, N.J., Provenzale, A.: An integro-differential model for the dynamics of Aeolian sand ripples. In: Gyr, A., Kinzelbach, W. (eds.) Sedimentation and Sediment Transport, pp. 187–194. Dordrecht, Boston, London (2003)

Zanke, U.: Über den Einfluss von Kornmaterial, Strömung und Wasseständen auf die Kenngrössen von Transportkörpern in offenen Gerinnen. Mitt. Franzius-Inst. TU Hannover **44** (1976)

Zanke, U.: Über die Abhängigkeit der Grösse des turbulenten Diffusionsaustauschkoeffizienten von suspendierten Sedimenten. Mitt. Franzius-Inst. TU Hannover, Heft **49** (1979)

Zanke, U.: Grundlagen der Sedimentbewegung. Springer-Verlag (1982)

Zilberman, M., Wygnanski, I., and Kaplan, R.E.: Transitional boundary layerspot in a fully turbulent environment. Phys. Fluids **20**, 258–271 (1977)

Znamenskaya, N.S.: Calculation of dimensions and speed of shifting of channel formations. Soviet Hydrology (American Geophysics Union) Nr. 5 (1962)

Znamenskaya, N.S.: Morphological principle of modeling of river bed processes. In: Proceedings of the 13th Congress 1AHR, Vol. 5 p.1. Kyoto (1969)

14. Appendix

14.1 Albert Einstein's Letter of Recommendation for His Son

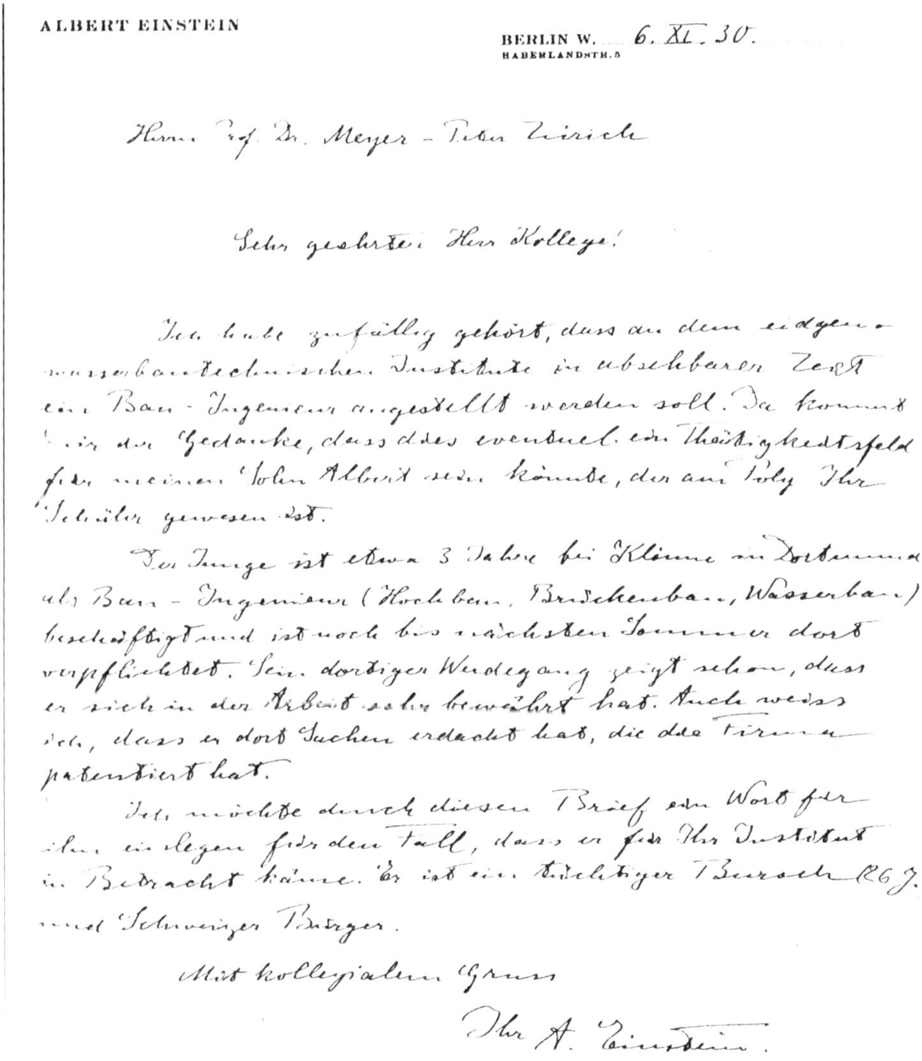

ALBERT EINSTEIN

BERLIN W. 6. XI. 30.
HABERLANDSTR. 5

Herrn Prof. Dr. Meyer-Peter Zürich

Sehr geehrter Herr Kollege!

Ich habe zufällig gehört, dass an dem eidgen. wasserbautechnischen Institute in absehbarer Zeit ein Bau-Ingenieur angestellt werden soll. Da kommt mir der Gedanke, dass dies eventuell ein Thätigkeitsfeld für meinen Sohn Albert sein könnte, der am Poly Ihr Schüler gewesen ist.

Der Junge ist etwa 3 Jahre bei Klönne in Dortmund als Bau-Ingenieur (Hochbau, Brückenbau, Wasserbau) beschäftigt und ist noch bis nächsten Sommer dort verpflichtet. Sein dortiger Werdegang zeigt schon, dass er sich in der Arbeit sehr bewährt hat. Auch weiss ich, dass er dort Sachen erdacht hat, die die Firma patentiert hat.

Ich möchte durch diesen Brief ein Wort für ihn einlegen für den Fall, dass er für Ihr Institut in Betracht käme. Er ist ein tüchtiger Bursch (26 J.) und Schweizer Bürger.

Mit kollegialem Gruss

Ihr A. Einstein.

Translation:

ALBERT EINSTEIN

BERLIN W. 6.XI.30.
HABERLANDSTR. 5

Prof. Dr. Meyer – Peter Zurich

Dear Colleague!

I heard by chance that in the foreseeable future a Civil Engineer is to be hired by the Federal Institute of Hydraulic Structures. It then occurred to me that this field might offer suitable career prospects for my son Albert*, who was one of your students at the ETH.

The fellow is employed by Klinne in Dortmund for about 3 years as a Civil Engineer (Buildings, bridges, hydraulic structures), and he is obliged to stay there until next summer. His formation already shows that he has availed himself in his work.
I also know that some of his ideas have led to patents for the company.

With this letter I would like to put in a word of recommendation for him, should he be considered by your Institute. He is an industrious fellow (26 y.) and a Swiss citizen.

With collegial greetings

Yours A. Einstein.

*Hans Albert is Albert Einstein's first son he had with his first wife Mileva Einstein Maric was baptized as Albert Einstein. When his father became more and more famous, son Albert changed his name to Hans Albert for the rest of his life.

14.2 Tables

14.2.1 Table of Physical Values

Basic conversion constants

Quantity	To convert from	To	Multiply by
Length	Foot	Meter = m	0.3048
	Mile (statute)	Kilometer = km	1.609344
Force	Pound	Newton = N	4.448221
	Dyne	N	10^{-5}
Density	lbm/ft^3	kg/m^3	16.01846
	gm/cm^3	kg/m^3	1000
Mass	Lbm	Kilogram = kg	0.453592
Pressure	lb/ft^2	N/m^2	47.88026
	Lb/inc^2	N/m^2	6894.757
	N/m^2	Pascal = Pa	1
	Bar	N/m^2	100
	Atmosphere=atm	N/m^2	101.325
	10mm Hg (15°)	N/m^2	13.56
Temperature	Celsius = C	Kelvin = K	+ 273.15
	Fahrenheit = F	C	(T-32)•(5/9)
Kinematic viscosity	ft^2/s	m^2/s	0.09203
	Stoke = St	m^2/s	10^{-4}
Viscosity	Poise = po	N-s/m^2	0.1
Power	Btu/s	kW	1.0543503
	calorie/s	Watt/s = W	4.184
	Joule/s	W	1
	N-s/s	W	1
	Horsepower	kw	0.7456999

Universal constants

Acceleration of gravity	9.80665 m/s^2
Avogadro constant, N_A	6.0222×10^{26} kmole^{-1}
Boltzmann constant, k	1.3806×10^{-23} Jdeg^{-1}
Gas constant, R_0	8.3143 J deg^{-1}mole^{-1}
Planck constant, h	6.6262×10^{-34} Js
Speed of light (vacuum)	2.997925×10^8 m/s

Temperature dependent physical properties of pure water

Temperature, T (°C)	Density ρ_f (kg/m^3)	Viscosity (N-s/m^2 × 10^3)	Kinematic viscosity (m^2/s × 10^{-6})	Surface tension (N/m × 10^2)
0	999.9	1.787	1.787	7.56
5	1000	1.514	1.514	7.49
10	999.7	1.304	1.304	7.42
15	999.1	1.137	1.138	7.35
20	998.2	1.002	1.004	7.28
25	997.1	0.891	0.894	7.20
30	995.7	0.798	0.802	7.12

Temperature, T (°C)	Vapor pressure (kN/m^2)	Latent heat vap. (J/g × 10^3)	Specific heat [c_p] (J/g°C)	Specific heat [c_p–c_V] (J/g°C)
0	0.61	6.1	4.217	0.002
5	0.87	8.7	4.202	0
10	1.23	12.3	4.192	0.005
15	1.70	17.0	4.186	0.013
20	2.33	23.3	4.182	0.024
25	3.16	31.6	4.179	0.041
30	4.23	42.3	4.178	0.06

Temperature, T (°C)	Thermal cond. [k_H] (J/ms°C × 10^{-1})	Thermal diff. [κ_H] (mm^2/s × 10)	Prandtl Nr ν/κ_H	Coef. therm. ex. [β] °C^{-1}×10^{-4}
0	5.6	1.33	13.4	–0.6
10	5.8	1.38	9.5	+0.9
15	5.9	1.40	8.1	1.5
20	5.9	1.42	7.1	2.1
30	6.1	1.46	5.5	3.0

Temperature, T (°C)	Vol. of air in 1m^3 of saturated water (reduced to 0°C) m^3	Percentage of NaCl (in saturated solution) (%)	Velocity of sound waves (m/s × 10^3)
0	0.0292	26.4	1.407
5	0.0257		
10	0.0228		1.445
15	0.0205		
20	0.0187	26.5	1.484
25	0.0171		
30	0.0157		1.510

Surface tension between two fluids at 20°C. (for laboratory purposes)

	Water–Air	Benzol–Water	Alcohol–water	Olive oil–water	Parafinoil–water
N/m	0.073	0.033	<0	0.018	0.051

14.2.2 Tables of sediment characterization

Size definitions in (mm)

	Boulder	Cobble	Gravel	Sand	Silt × 10^{-3}	Clay × 10^{-3}
Very large	2–4 m		30–50	1–2		
Large/coarse	1–2 m	100–250	15–30	0.5–1	30–60	2–4
Medium	0.5–1 m		10–15	0.5–0.25	15–30	1–2
Small/Fine	o.25–0.5 m	50–100	5–10	0.125–0.25	8–15	0.5–1
Very fine			2–5	0.06–0.125	4–8	0.25–0.5

Settling velocities u_v at different temperature in clear water $\nu = \nu_m$

Temperature (°C)	d_s (mm)	$\tilde{d}_s = d_s \left[\frac{(\rho_s - \rho_f) g}{\rho_f \nu_m^2} \right]^{1/3}$	u_v (mm/s)
0	2000	34316	5303
	250	4279	1874
	50	858	749
	20	343	599
	10	171	421
	5	85	295
	1	17	106
	0.5	8.6	61
	0.1	1.71	6.08
10	2000	42257	5303
	20	423	599
	10	211	423
	5	105	296
	1	21	109
	0.5	10.5	66.4
	0.1	2.1	8.1
20	2000	50593	5303
	20	505	599
	10	253	423
	5	126	298
	1	25	112
	0.5	12.6	70.3
	0.1	2.53	10.2

Angle of repose in degrees

d_s (mm)	100–2000	10–100	1–10	0.1–1	0.01–0.1
Degrees	42	38–41	32–41	30–32	30

14.3 Graphs

14.3.1 The graphs for the Einstein sediment transport method with a guide for the usage of Einstein's method

All the classical theories are somehow trivial in their application, however not so Einstein's theory shown in chap. 3 (Sect. 3.2.2). and therefore a short scheme is given on how to proceed.

Einstein formulated a relation between the intensity of the sediment transport and the flow intensity and it is this relation. The empirical relation is discussed in Chap. 14. (Sect. 14.3.1). Ψ_{s*} has to be calculated by using Eq. 3.24 for each size fraction. With each of these values Φ_{s*} can be evaluated using the graph mentioned there.

The bedload transport of fraction i_B is then given by Eq. (3.21) using Eq. (3.23)

$$i_B\tilde{q}_{rs} = i_B\Phi_s\rho_s g\sqrt{gd_s^3\frac{\rho_s-\rho_f}{\rho_f}} \tag{14. 1}$$

Now one has to sum up twice, first by integration over the cross section to get i_BQ_s and later over all fractions to get the total bed load Q_s.

The main effort one has to achieve is therefore to calculate Ψ_{s*} using Eq. 3.24 based on Eq. 3.22, and this needs a cascade of calculations. The hiding factor ξ (Eq. 3.27) is again an empirical value and can be found in the graph Chap. 14. (Sect. 14.3.1.2). The value X needed in this plot is a roughness correction given as a multiplication factor as stated in Eq. 3.27 and given graphically in Chap. 14 (Sect. 14.3.1.3). The Y in Eq. 3.34 is a pressure correction given in Eq. 3.27 and plotted in Chap. 14 (Sect. 14.3.1.4). The b values are given in Eq. 3.27 too and for that one needs Δ the apparent roughness of the bed, d_s/X that is known when X is evaluated.

Flow versus sediment transport intensity Eqs 3.33–34

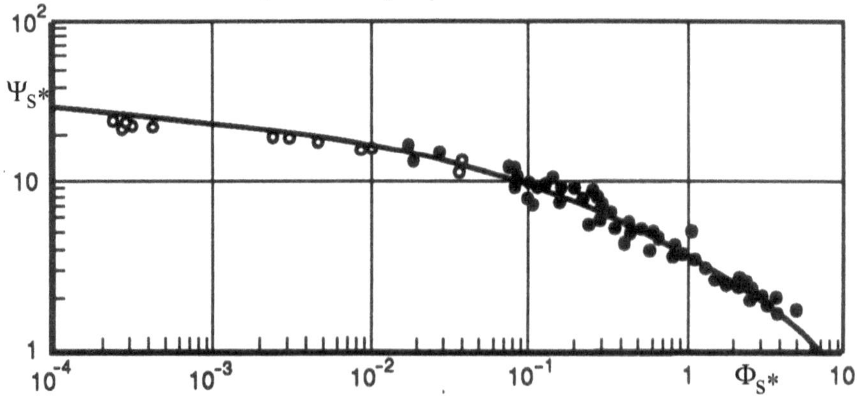

The hiding factor Eq. 3.37

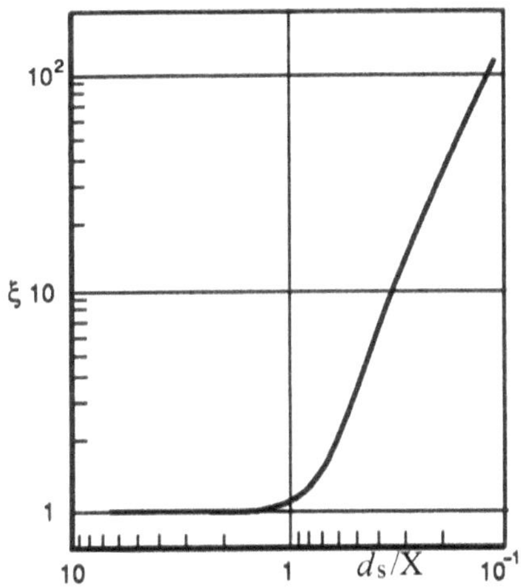

The roughness influence Eq. 3.37

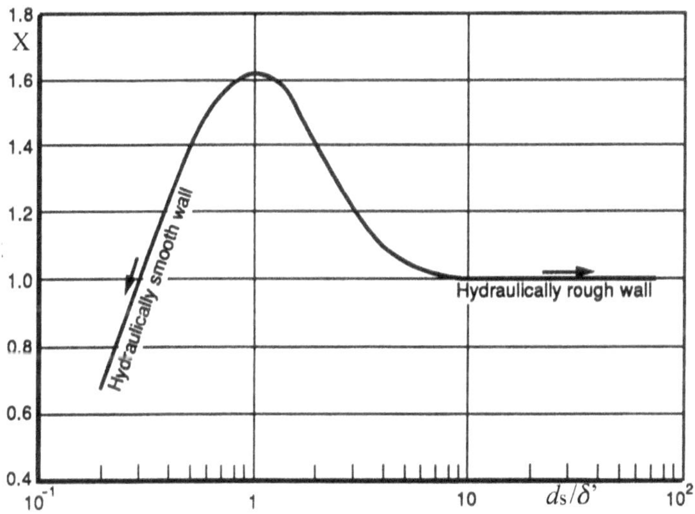

The pressure correction Eq. 3.37

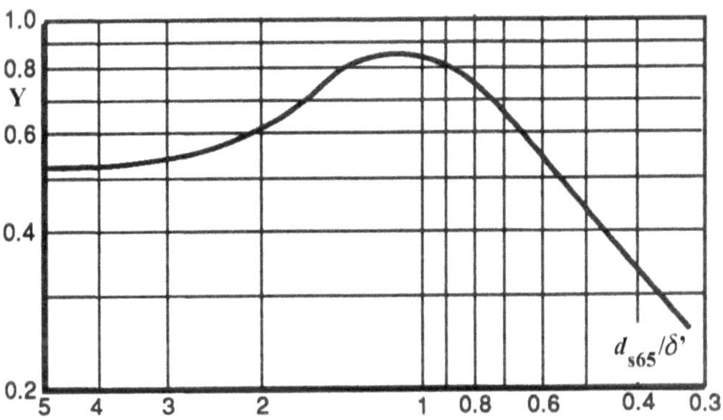

Moody diagram

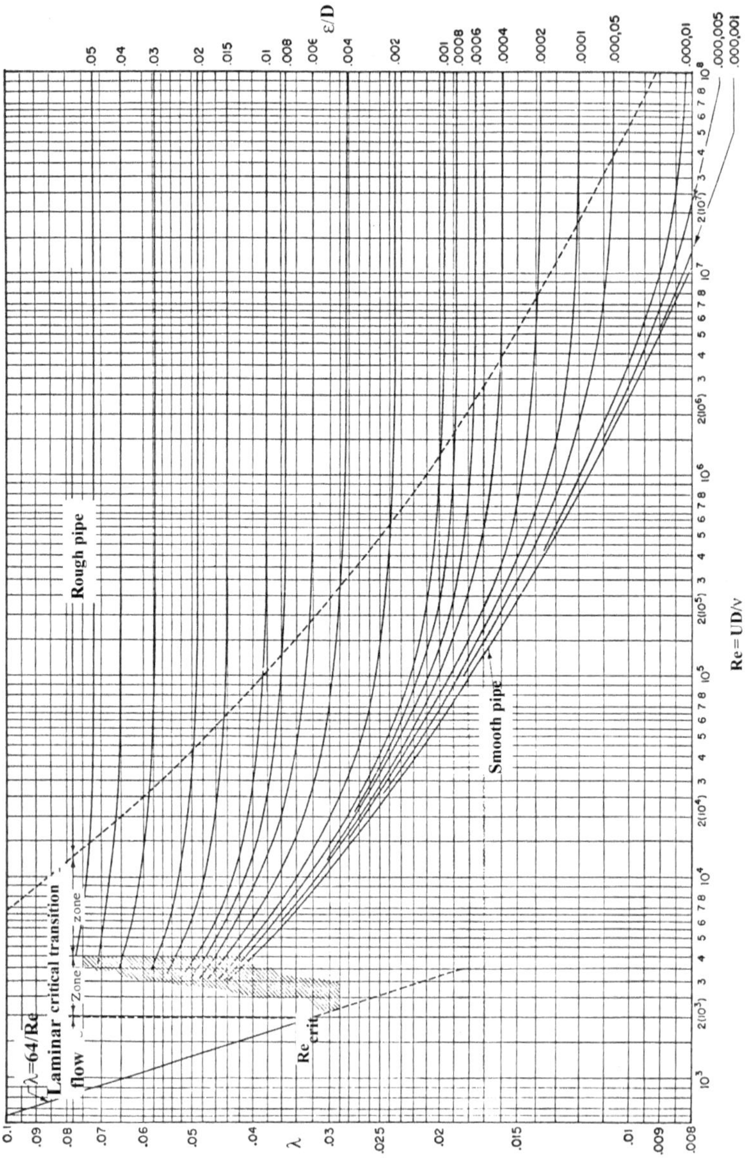

14.4 Symbols

For the representation of complex systems often the numbers of the letters of the alphabet is too small, and therefore one has often to use them for different values. In case of a double meaning normally we use the first indications for any other one the meaning is explained in place. Not all symbols are listed; very special ones only used once are described in the text.

A	area (cross section)	Γ	circulation
A_E	Einstein's probability factor	Φ	angle of repose
A_d	drainage area	Φ_s	intensity of sediment transport
A,B	constants	Λ	wave length (dunes)
Ar	Archimedes number	Λ_i	length scales
B	channel width	Π	Buckingham Π-function
C	Chezy friction coefficient	Θ	dimensionless wall-shear stress
CD	resistance coefficient	Ψ_s	flow intensity
D	diffusion coefficient	Ω	mean vorticity
D	deformation tensor	γ	specific weight
D	diameter	γ	relative intermittency
D_f	fractal dimension	$\underline{\underline{\dot{\gamma}}}$	shear rate
$D*$	sedimentological grain diam.	δ	boundary layer thickness
E	energy spectrum	δ_{ij}	Kronecker delta tensor
F	force	ε	dissipation rate
Fr	Froude number	ε	diffusion vector (turbulent)
F_i	structure function	ε	wall roughness
G	weight	ε	eccentricity
G_r	gradation coefficient	$\dot{\varepsilon}$	rate of extension
Ga	Galilei number	ε_{ijk}	ε-tensor (rotation)
H	water depth	ε_m	eddy viscosity
He	Hedstrom number	κ	v. Karman constant
H_L	Bernoulli sums	κ	circulation
I	identity tensor		
I_e	transport of momentum/event	λ	friction factor
J	Jaccobi determinant	λ	expansion coefficient
K	kurtosis or flatness factor	λ_i	length scales
K	Strickler friction coefficient	ν	kinematic viscosity
K_L	bend loss coefficient	ρ	density
L	length scale	σ	deviation
M	angular momentum	σ	surface tension
$O(..)$	order of magnitude	σ_i	Lyapunov characteristic exp.
N	number of grains	$\underline{\underline{\sigma}}$	shear tensor (σ_{ij})
N	node	μ	dynamical viscosity

P	mean pressure	pH	electrolytic characteristic
P	probability	p_o	porosity
P	stream power	p_0	representative pressure value
Pe	Péclet number	q	flux density
Q	discharge	r_{al}	colloidal radius
Q_i	quadrants	s	skewness
R'	hydraulic radius	s_{ij}	rate of strain tensor (turbulent)
Re	Reynolds number	t	time
Ro	Rouse number	u	local velocity of the fluid
Ri	Richardson number	u_τ	wall shear velocity
R_{ij}	correlation tensor	$\langle u_i' u_j' \rangle$	Reynolds stress tensor
R_x	autocorrelation	x, y, z	the coordinate
S	slope		x: in flow direction
S	saddle		y: in span direction
St	Stokes number		z: in vertical direction
Str	Strouhal number	μ_E	extensional or elongational viscosity
S_{ij}	rate of strain tensor	τ_T	total shear stress
T	temperature	τ_t	turbulent shear stress
T	stress tensor	τ_v	viscous stress
T_B	burst period (T_P)	τ_{ij}	plane shear stress
T_R	Trouton ratio	ξ	hiding factor
T_e	event time	ζ	drag coefficient
U	mean velocity	ω	local vorticity
V	volume	$()_{At}$	attachment
W	weight	$()_N$	Newtonian
X	roughness influence, Einstein	$()_{bl}$	bed load
Y	pressure influence, Einstein	$()_c$	critical value
$a, b, \ldots$	constant	$()_c$	convectional
a, b	axis of an ellipse	$()_d$	based on grain diameter
c_F	form-factor	$()_f$	for the fluid
c	heat capacity	$()_g$	gradiation
c_V	concentration/volume	$()_K$	Kolmogorov
d_H	hydraulic diameter	$()_{kin}$	kinetic energy
$\mathrm{d}V$	surface	$()_{ks}$	rough
d_i	principal value of D	$()_{int}$	integral
d_s	grain size	$()_n$	normal
d_{ij}	deviatoric stress tensor	$()_r$	relative
$d_{s(n)}$	grain size (percentage)	$()_r$	roughness
e	energy density	$()_s$	for the sediment
e	void fraction	$()_{su}$	suspension
e_{ij}	rate of strain tensor	$()_t$	turbulent
f	friction factor	$()_V$	void
g	gravitational acceleration	$()_v$	viscous
h	bed height	$()_w$	at the wall

i, j, k, l	orientation parameters	$()_\delta$	boundary layer
I	fraction index	$\underline{()}$	vector
k	wave number	$\underline{\underline{()}}$	tensor
k_s	equivalent sand-roughness	$(\hat{\cdot})$	dimensionless form
L	scaling parameter	$()'$	fluctuations
N	Grass overlapping factor	$()'$	relative
n_s	number density of the grains	$()^+$	viscous units
p	pressure		

Subject Index

Mechanics

FLUID MECHANICS AND ITS APPLICATIONS

Series Editor: R. Moreau

Aims and Scope of the Series

The purpose of this series is to focus on subjects in which fluid mechanics plays a fundamental role. As well as the more traditional applications of aeronautics, hydraulics, heat and mass transfer etc., books will be published dealing with topics which are currently in a state of rapid development, such as turbulence, suspensions and multiphase fluids, super and hypersonic flows and numerical modelling techniques. It is a widely held view that it is the interdisciplinary subjects that will receive intense scientific attention, bringing them to the forefront of technological advancement. Fluids have the ability to transport matter and its properties as well as transmit force, therefore fluid mechanics is a subject that is particularly open to cross fertilisation with other sciences and disciplines of engineering. The subject of fluid mechanics will be highly relevant in domains such as chemical, metallurgical, biological and ecological engineering. This series is particularly open to such new multidisciplinary domains.

1. M. Lesieur: *Turbulence in Fluids*. 2nd rev. ed., 1990 ISBN 0-7923-0645-7
2. O. Métais and M. Lesieur (eds.): *Turbulence and Coherent Structures*. 1991 ISBN 0-7923-0646-5
3. R. Moreau: *Magnetohydrodynamics*. 1990 ISBN 0-7923-0937-5
4. E. Coustols (ed.): *Turbulence Control by Passive Means*. 1990 ISBN 0-7923-1020-9
5. A.A. Borissov (ed.): *Dynamic Structure of Detonation in Gaseous and Dispersed Media*. 1991 ISBN 0-7923-1340-2
6. K.-S. Choi (ed.): *Recent Developments in Turbulence Management*. 1991 ISBN 0-7923-1477-8
7. E.P. Evans and B. Coulbeck (eds.): *Pipeline Systems*. 1992 ISBN 0-7923-1668-1
8. B. Nau (ed.): *Fluid Sealing*. 1992 ISBN 0-7923-1669-X
9. T.K.S. Murthy (ed.): *Computational Methods in Hypersonic Aerodynamics*. 1992 ISBN 0-7923-1673-8
10. R. King (ed.): *Fluid Mechanics of Mixing*. Modelling, Operations and Experimental Techniques. 1992 ISBN 0-7923-1720-3
11. Z. Han and X. Yin: *Shock Dynamics*. 1993 ISBN 0-7923-1746-7
12. L. Svarovsky and M.T. Thew (eds.): *Hydroclones*. Analysis and Applications. 1992 ISBN 0-7923-1876-5
13. A. Lichtarowicz (ed.): *Jet Cutting Technology*. 1992 ISBN 0-7923-1979-6
14. F.T.M. Nieuwstadt (ed.): *Flow Visualization and Image Analysis*. 1993 ISBN 0-7923-1994-X
15. A.J. Saul (ed.): *Floods and Flood Management*. 1992 ISBN 0-7923-2078-6
16. D.E. Ashpis, T.B. Gatski and R. Hirsh (eds.): *Instabilities and Turbulence in Engineering Flows*. 1993 ISBN 0-7923-2161-8
17. R.S. Azad: *The Atmospheric Boundary Layer for Engineers*. 1993 ISBN 0-7923-2187-1
18. F.T.M. Nieuwstadt (ed.): *Advances in Turbulence IV*. 1993 ISBN 0-7923-2282-7
19. K.K. Prasad (ed.): *Further Developments in Turbulence Management*. 1993 ISBN 0-7923-2291-6
20. Y.A. Tatarchenko: *Shaped Crystal Growth*. 1993 ISBN 0-7923-2419-6
21. J.P. Bonnet and M.N. Glauser (eds.): *Eddy Structure Identification in Free Turbulent Shear Flows*. 1993 ISBN 0-7923-2449-8
22. R.S. Srivastava: *Interaction of Shock Waves*. 1994 ISBN 0-7923-2920-1
23. J.R. Blake, J.M. Boulton-Stone and N.H. Thomas (eds.): *Bubble Dynamics and Interface Phenomena*. 1994 ISBN 0-7923-3008-0

Mechanics

FLUID MECHANICS AND ITS APPLICATIONS

Series Editor: R. Moreau

24. R. Benzi (ed.): *Advances in Turbulence V.* 1995 ISBN 0-7923-3032-3
25. B.I. Rabinovich, V.G. Lebedev and A.I. Mytarev: *Vortex Processes and Solid Body Dynamics.* The Dynamic Problems of Spacecrafts and Magnetic Levitation Systems. 1994 ISBN 0-7923-3092-7
26. P.R. Voke, L. Kleiser and J.-P. Chollet (eds.): *Direct and Large-Eddy Simulation I.* Selected papers from the First ERCOFTAC Workshop on Direct and Large-Eddy Simulation. 1994 ISBN 0-7923-3106-0
27. J.A. Sparenberg: *Hydrodynamic Propulsion and its Optimization.* Analytic Theory. 1995 ISBN 0-7923-3201-6
28. J.F. Dijksman and G.D.C. Kuiken (eds.): *IUTAM Symposium on Numerical Simulation of Non-Isothermal Flow of Viscoelastic Liquids.* Proceedings of an IUTAM Symposium held in Kerkrade, The Netherlands. 1995 ISBN 0-7923-3262-8
29. B.M. Boubnov and G.S. Golitsyn: *Convection in Rotating Fluids.* 1995 ISBN 0-7923-3371-3
30. S.I. Green (ed.): *Fluid Vortices.* 1995 ISBN 0-7923-3376-4
31. S. Morioka and L. van Wijngaarden (eds.): *IUTAM Symposium on Waves in Liquid/Gas and Liquid/Vapour Two-Phase Systems.* 1995 ISBN 0-7923-3424-8
32. A. Gyr and H.-W. Bewersdorff: *Drag Reduction of Turbulent Flows by Additives.* 1995 ISBN 0-7923-3485-X
33. Y.P. Golovachov: *Numerical Simulation of Viscous Shock Layer Flows.* 1995 ISBN 0-7923-3626-7
34. J. Grue, B. Gjevik and J.E. Weber (eds.): *Waves and Nonlinear Processes in Hydrodynamics.* 1996 ISBN 0-7923-4031-0
35. P.W. Duck and P. Hall (eds.): *IUTAM Symposium on Nonlinear Instability and Transition in Three-Dimensional Boundary Layers.* 1996 ISBN 0-7923-4079-5
36. S. Gavrilakis, L. Machiels and P.A. Monkewitz (eds.): *Advances in Turbulence VI.* Proceedings of the 6th European Turbulence Conference. 1996 ISBN 0-7923-4132-5
37. K. Gersten (ed.): *IUTAM Symposium on Asymptotic Methods for Turbulent Shear Flows at High Reynolds Numbers.* Proceedings of the IUTAM Symposium held in Bochum, Germany. 1996 ISBN 0-7923-4138-4
38. J. Verhás: *Thermodynamics and Rheology.* 1997 ISBN 0-7923-4251-8
39. M. Champion and B. Deshaies (eds.): *IUTAM Symposium on Combustion in Supersonic Flows.* Proceedings of the IUTAM Symposium held in Poitiers, France. 1997 ISBN 0-7923-4313-1
40. M. Lesieur: *Turbulence in Fluids.* Third Revised and Enlarged Edition. 1997 ISBN 0-7923-4415-4; Pb: 0-7923-4416-2
41. L. Fulachier, J.L. Lumley and F. Anselmet (eds.): *IUTAM Symposium on Variable Density Low-Speed Turbulent Flows.* Proceedings of the IUTAM Symposium held in Marseille, France. 1997 ISBN 0-7923-4602-5
42. B.K. Shivamoggi: *Nonlinear Dynamics and Chaotic Phenomena.* An Introduction. 1997 ISBN 0-7923-4772-2
43. H. Ramkissoon, *IUTAM Symposium on Lubricated Transport of Viscous Materials.* Proceedings of the IUTAM Symposium held in Tobago, West Indies. 1998 ISBN 0-7923-4897-4
44. E. Krause and K. Gersten, *IUTAM Symposium on Dynamics of Slender Vortices.* Proceedings of the IUTAM Symposium held in Aachen, Germany. 1998 ISBN 0-7923-5041-3
45. A. Biesheuvel and G.J.F. van Heyst (eds.): *In Fascination of Fluid Dynamics.* A Symposium in honour of Leen van Wijngaarden. 1998 ISBN 0-7923-5078-2

Mechanics

FLUID MECHANICS AND ITS APPLICATIONS

Series Editor: R. Moreau

46. U. Frisch (ed.): *Advances in Turbulence VII.* Proceedings of the Seventh European Turbulence Conference, held in Saint-Jean Cap Ferrat, 30 June–3 July 1998. 1998 ISBN 0-7923-5115-0
47. E.F. Toro and J.F. Clarke: *Numerical Methods for Wave Propagation.* Selected Contributions from the Workshop held in Manchester, UK. 1998 ISBN 0-7923-5125-8
48. A. Yoshizawa: *Hydrodynamic and Magnetohydrodynamic Turbulent Flows.* Modelling and Statistical Theory. 1998 ISBN 0-7923-5225-4
49. T.L. Geers (ed.): *IUTAM Symposium on Computational Methods for Unbounded Domains.* 1998 ISBN 0-7923-5266-1
50. Z. Zapryanov and S. Tabakova: *Dynamics of Bubbles, Drops and Rigid Particles.* 1999 ISBN 0-7923-5347-1
51. A. Alemany, Ph. Marty and J.P. Thibault (eds.): *Transfer Phenomena in Magnetohydrodynamic and Electroconducting Flows.* 1999 ISBN 0-7923-5532-6
52. J.N. Sørensen, E.J. Hopfinger and N. Aubry (eds.): *IUTAM Symposium on Simulation and Identification of Organized Structures in Flows.* 1999 ISBN 0-7923-5603-9
53. G.E.A. Meier and P.R. Viswanath (eds.): *IUTAM Symposium on Mechanics of Passive and Active Flow Control.* 1999 ISBN 0-7923-5928-3
54. D. Knight and L. Sakell (eds.): *Recent Advances in DNS and LES.* 1999 ISBN 0-7923-6004-4
55. P. Orlandi: *Fluid Flow Phenomena.* A Numerical Toolkit. 2000 ISBN 0-7923-6095-8
56. M. Stanislas, J. Kompenhans and J. Westerveel (eds.): *Particle Image Velocimetry.* Progress towards Industrial Application. 2000 ISBN 0-7923-6160-1
57. H.-C. Chang (ed.): *IUTAM Symposium on Nonlinear Waves in Multi-Phase Flow.* 2000 ISBN 0-7923-6454-6
58. R.M. Kerr and Y. Kimura (eds.): *IUTAM Symposium on Developments in Geophysical Turbulence* held at the National Center for Atmospheric Research, (Boulder, CO, June 16–19, 1998) 2000 ISBN 0-7923-6673-5
59. T. Kambe, T. Nakano and T. Miyauchi (eds.): *IUTAM Symposium on Geometry and Statistics of Turbulence.* Proceedings of the IUTAM Symposium held at the Shonan International Village Center, Hayama (Kanagawa-ken, Japan November 2–5, 1999). 2001 ISBN 0-7923-6711-1
60. V.V. Aristov: *Direct Methods for Solving the Boltzmann Equation and Study of Nonequilibrium Flows.* 2001 ISBN 0-7923-6831-2
61. P.F. Hodnett (ed.): *IUTAM Symposium on Advances in Mathematical Modelling of Atmosphere and Ocean Dynamics.* Proceedings of the IUTAM Symposium held in Limerick, Ireland, 2–7 July 2000. 2001 ISBN 0-7923-7075-9
62. A.C. King and Y.D. Shikhmurzaev (eds.): *IUTAM Symposium on Free Surface Flows.* Proceedings of the IUTAM Symposium held in Birmingham, United Kingdom, 10–14 July 2000. 2001 ISBN 0-7923-7085-6
63. A. Tsinober: *An Informal Introduction to Turbulence.* 2001 ISBN 1-4020-0110-X; Pb: 1-4020-0166-5
64. R.Kh. Zeytounian: *Asymptotic Modelling of Fluid Flow Phenomena.* 2002 ISBN 1-4020-0432-X
65. R. Friedrich and W. Rodi (eds.): *Advances in LES of Complex Flows.* Prodeedings of the EUROMECH Colloquium 412, held in Munich, Germany, 4-6 October 2000. 2002 ISBN 1-4020-0486-9
66. D. Drikakis and B.J. Geurts (eds.): *Turbulent Flow Computation.* 2002 ISBN 1-4020-0523-7
67. B.O. Enflo and C.M. Hedberg: *Theory of Nonlinear Acoustics in Fluids.* 2002 ISBN 1-4020-0572-5

Mechanics

FLUID MECHANICS AND ITS APPLICATIONS

Series Editor: R. Moreau

68. I.D. Abrahams, P.A. Martin and M.J. Simon (eds.): *IUTAM Symposium on Diffraction and Scattering in Fluid Mechanics and Elasticity*. Proceedings of the IUTAM Symposium held in Manchester, (UK, 16-20 July 2000). 2002 ISBN 1-4020-0590-3
69. P. Chassaing, R.A. Antonia, F. Anselmet, L. Joly and S. Sarkar: *Variable Density Fluid Turbulence.* 2002 ISBN 1-4020-0671-3
70. A. Pollard and S. Candel (eds.): *IUTAM Symposium on Turbulent Mixing and Combustion.* Proceedings of the IUTAM Symposium held in Kingston, Ontario, Canada, June 3-6, 2001. 2002 ISBN 1-4020-0747-7
71. K. Bajer and H.K. Moffatt (eds.): *Tubes, Sheets and Singularities in Fluid Dynamics.* 2002 ISBN 1-4020-0980-1
72. P.W. Carpenter and T.J. Pedley (eds.): *Flow Past Highly Compliant Boundaries and in Collapsible Tubes.* IUTAM Symposium held at the Univerity of Warwick, Coventry, United Kingdom, 26-30 March 2001. 2003 ISBN 1-4020-1161-X
73. H. Sobieczky (ed.): *IUTAM Symposium Transsonicum IV.* Proceedings of the IUTAM Symposium held in Göttingen, Germany, 2-6 September 2002. 2003 ISBN 1-4020-1608-5
74. A.J. Smits (ed.): *IUTAM Symposium on Reynolds Number Scaling in Turbulent Flow.* Proceedings of the IUTAM Symposium held in Princeton, NJ, U.S.A., September 11-13, 2002. 2003 ISBN 1-4020-1775-8
75. H. Benaroya and T. Wei (eds.): *IUTAM Symposium on Integrated Modeling of Fully Coupled Fluid Structure Interactions Using Analysis, Computations and Experiments.* Proceedings of the IUTAM Symposium held in New Jersey, U.S.A., 2-6 June 2003. 2003 ISBN 1-4020-1806-1
76. J.-P. Franc and J.-M. Michel: *Fundamentals of Cavitation.* 2004 ISBN 1-4020-2232-8
77. T. Mullin and R.R. Kerswell (eds.): *IUTAM Symposium on Laminar Turbulent Transition and Finite Amplitude Solutions.* 2005 ISBN 1-4020-4048-2
78. R. Govindarajan (ed.): *Sixth IUTAM Symposium on Laminar-Turbulent Transition.* Proceedings of the Sixth IUTAM Symposium on Laminar-Turbulent Transition, Bangalore, India, 2004. 2006 ISBN 1-4020-3459-8
79. S. Kida (ed.): *IUTAM Symposium on Elementary Vortices and Coherent Structures: Significance in Turbulence Dynamics.* Proceedings of the IUTAM Symposium held at Kyoto International Community House, Kyoto, Japan, 26-28 October, 2004. 2006 ISBN 1-4020-4180-2
80. S.S. Molokov, R. Moreau and H.K. Moffatt (eds.): *Magnetohydrodynamics.* Historical Evolution and Trends. 2006 ISBN 1-4020-4832-7
81. S. Balachandar and A. Prosperetti (eds.): *IUTAM Symposium on Computational Approaches to Multiphase Flow.* Proceedings of an IUTAM Symposium held at Argonne National Laboratory, October 4-7, 2004. 2006 ISBN 1-4020-4976-5
82. A. Gyr and K. Hoyer: *Sediment Transport.* A Geophysical Phenomenon. 2006 ISBN 1-4020-5015-1

FSC
www.fsc.org
MIX
Papier aus verantwortungsvollen Quellen
Paper from responsible sources
FSC® C105338

If you have any concerns about our products,
you can contact us on
ProductSafety@springernature.com

In case Publisher is established outside the EU,
the EU authorized representative is:
Springer Nature Customer Service Center GmbH
Europaplatz 3, 69115 Heidelberg, Germany

Printed by Libri Plureos GmbH
in Hamburg, Germany